Metal-Organic Chemical Vapor Deposition of Electronic Ceramics

MATERIALS RESEARCH SOCIETY SYMPOSIUM PROCEEDINGS VOLUME 335

Metal-Organic Chemical Vapor Deposition of Electronic Ceramics

Symposium held November 29-December 3, 1993, Boston, Massachusetts, U.S.A.

EDITORS:

Seshu B. Desu
Virginia Polytechnic Institute and State University
Blacksburg, Virginia, U.S.A.

David B. Beach
IBM T.J. Watson Research Center
Yorktown Heights, New York, U.S.A.

Bruce W. Wessels
Northwestern University
Evanston, Illinois, U.S.A.

Suleyman Gokoglu
NASA Lewis Research Center
Cleveland, Ohio, U.S.A.

MATERIALS RESEARCH SOCIETY
Pittsburgh, Pennsylvania

Single article reprints from this publication are available through
University Microfilms Inc., 300 North Zeeb Road, Ann Arbor, Michigan 48106

CODEN: MRSPDH

Copyright 1994 by Materials Research Society.
All rights reserved.

This book has been registered with Copyright Clearance Center, Inc. For further
information, please contact the Copyright Clearance Center, Salem, Massachusetts.

Published by:

Materials Research Society
9800 McKnight Road
Pittsburgh, Pennsylvania 15237
Telephone (412) 367-3003
Fax (412) 367-4373

Library of Congress Cataloging in Publication Data

Materials Research Society Meeting (1993 : Boston, Mass.).
Symposium Y.
Metal-organic chemical vapor deposition of electronic ceramics : Symposium Y held on
 Nov. 29 to Dec. 3, 1993, at Boston, Massachusetts, U.S.A. / editors, Seshu B.
 Desu, David B. Beach, Bruce W. Wessels, Suleyman Gokoglu.
   p. cm.—(Materials Research Society symposium ; v. 335)
   "Presented at the Materials Research Society Fall 1993 ... symposium"—CIP pref.
   Includes bibliographical references and index.
   ISBN 1-55899-234-0
   1. Electronic ceramics—Congresses. 2. Vapor-plating—Congresses. 3. Thin
films—Congresses. I. Desu, Seshu B. II. Beach, David B. III. Wessels, Bruce W.
IV. Gokoglu, Suleyman V. Materials Research Society. VI. Title. VII. Series:
Materials Research Society Symposium Proceedings ; v. 335.
TK7871.15.C4M38 1994   94-5807
621.3815'2—dc20   CIP

Manufactured in the United States of America

D
621.38102'8
MET

Contents

PREFACE ix

ACKNOWLEDGMENTS xi

MATERIALS RESEARCH SOCIETY SYMPOSIUM PROCEEDINGS xii

PART I: MOCVD OF NONOXIDE ELECTRONIC CERAMICS

*PLASMA ENHANCED METAL-ORGANIC CHEMICAL VAPOR
DEPOSITION OF GERMANIUM NITRIDE THIN FILMS 3
   David M. Hoffman, Sri Prakash Rangarajan, Satish D. Athavale,
   Demetre J. Economou, Jia-Rui Liu, Zongshuang Zheng, and Wei-Kan Chu

*RECENT ADVANCES IN THE CVD OF METAL NITRIDES AND OXIDES 9
   Roy G. Gordon

LPCVD OF InN ON GaAs(110) USING HN_3 AND TMIn: COMPARISON
WITH Si(100) RESULTS 21
   Y. Bu and M.C. Lin

PART II: FERROELECTRIC MATERIALS

EPITAXIAL GROWTH OF $Ba_{1-x}Sr_xTiO_3$ THIN FILMS ON
$YBa_2Cu_3O_{7-x}$ ELECTRODE BY PE-MOCVD 29
   C.S. Chern, S. Liang, Z.Q. Shi, S. Yoon, A. Safari, P. Lu,
   and B.H. Kear

LOW TEMPERATURE MOCVD OF ORIENTED $PbTiO_3$ FILMS ON Si(100) 35
   H. Chen, B.M. Yen, G.R. Bai, D. Liu, and H.L.M. Chang

PREPARATION OF $Ba_{1-x}Sr_xTiO_3$ THIN FILMS BY METALORGANIC
CHEMICAL VAPOR DEPOSITION AND THEIR PROPERTIES 41
   S.R. Gilbert, B.W. Wessels, D.A. Neumayer, T.J. Marks, J.L. Schindler,
   and C.R. Kannewurf

GROWTH OF $BaTiO_3$ THIN FILMS BY MOCVD 47
   Debra L. Kaiser, Mark D. Vaudin, Greg Gillen, Cheol-Seong Hwang,
   Lawrence H. Robins, and Lawrence D. Rotter

DONOR-DOPED LEAD ZIRCONATE TITANATE ($PbZr_{1-x}Ti_xO_3$) FILMS 53
   Seshu B. Desu, Dilip P. Vijay, and In K. Yoo

GROWTH OF (001)-ORIENTED SBN THIN FILMS BY SOLID
SOURCE MOCVD 59
   Z. Lu, R.S. Feigelson, R.K. Route, R. Hiskes, and S.A. DiCarolis

VAPOR DEPOSITION OF LITHIUM TANTALATE WITH VOLATILE
DOUBLE ALKOXIDE PRECURSORS 65
   Kueir-Weei Chour, Guangde Wang, and Ren Xu

PART III: POSTER SESSION

RAMAN SPECTRA OF MOCVD-GROWN FERROELECTRIC $PbTiO_3$
THIN FILMS 75
   Z.C. Feng, B.S. Kwak, A. Erbil, and L.A. Boatner

*Invited Paper

DEPOSITION OF DIELECTRIC THIN FILMS BY IRRADIATION
OF CONDENSED REACTANT MIXTURES 81
  J.F. Moore, D.R. Strongin, M.W. Ruckman, and M. Strongin

INVESTIGATION OF SECOND HARMONIC GENERATION IN MOCVD
GROWN BARIUM TITANATE THIN FILMS 87
  Bipin Bihari, Xin Li Jiang, Jayant Kumar, Gregory T. Stauf,
  and Peter C. Van Buskirk

LEAD TITANATE THIN FILMS PREPARED BY METALLORGANIC
CHEMICAL VAPOR DEPOSITION (MOCVD) ON SAPPHIRE, Pt, AND
RuO_x SUBSTRATES 93
  Warren C. Hendricks, Seshu B. Desu, and Chien H. Peng

DEPOSITION AND PROPERTIES OF AlN FILMS PREPARED BY
LOW PRESSURE CVD WITH A METALORGANIC PRECURSOR 101
  M.J. Cook, P.K. Wu, N. Patibandla, W.B. Hillig, and J.B. Hudson

ELECTRICAL AND DIELECTRIC CHARACTERISTICS OF $SrTiO_3$ THIN
FILMS GROWN BY PE-MOCVD TECHNIQUE 107
  Z.Q. Shi, C. Chern, S. Liang, Y. Lu, and A. Safari

MOCVD GROWTH OF EPITAXIAL $SrTiO_3$ THIN FILMS ON
$YBa_2Cu_3O_{7-x}/LaAlO_3$ 113
  S. Liang, C. Chern, Z.Q. Shi, Y. Lu, P. Lu, B.H. Kear, and A. Safari

DEPOSITION OF TITANIUM OXIDE FILMS FROM METAL-ORGANIC
PRECURSOR BY ELECTRON CYCLOTRON RESONANCE PLASMA-ASSISTED
CHEMICAL VAPOR DEPOSITION 117
  Atsushi Nagahori and Rishi Raj

HETEROEPITAXIAL $Si-ZrO_2-Si$ BY MOCVD 123
  Anton C. Greenwald, Nader M. Kalkhoran, Fereydoon Namavar,
  Alain E. Kaloyeros, and Ioannis Stathakos

PART IV: MODELING

*THERMOCHEMICAL DATA FOR CVD MODELING FROM AB INITIO
CALCULATIONS 131
  Pauline Ho and Carl F. Melius

*THERMODYNAMIC SIMULATION OF MOCVD $YBa_2Cu_3O_{7-x}$ THIN
FILM DEPOSITION 139
  C. Bernard, F. Weiss, A. Pisch, and R. Madar

LOW TEMPERATURE CVD OF TiN FROM $Ti(NR_2)_4$ AND NH_3: FTIR
STUDIES OF THE GAS-PHASE CHEMICAL REACTIONS 159
  Bruce H. Weiller

THERMODYNAMIC CONSTRAINTS FOR THE IN SITU MOCVD GROWTH
OF SUPERCONDUCTING Tl-Ba-Ca-Cu-O THIN FILMS 165
  William L. Holstein

NUMERICAL ANALYSIS OF AN IMPINGING JET REACTOR FOR THE
CVD AND GAS-PHASE NUCLEATION OF TITANIA 171
  Suleyman A. Gokoglu, G.D. Stewart, J. Collins, and D.E. Rosner

THEORETICAL MODELING OF CHEMICAL VAPOR DEPOSITION OF
SILICON CARBIDE IN A HOT WALL REACTOR 177
  Feng Gao and Ray Y. Lin

*Invited Paper

GROWTH OF EPITAXIAL β-SiC FILMS ON POROUS Si(100) SUBSTRATES
FROM MTS IN A HOT WALL LPCVD REACTOR 183
    Chien C. Chiu, Seshu B. Desu, and Gang Chen

PART V: PRECURSOR DESIGN AND DELIVERY: REACTORS

*SYNTHETIC STRATEGIES FOR MOCVD PRECURSORS FOR HTcS
THIN FILMS 193
    Harry A. Meinema, Klaas Timmer, Hans L. Linden, and Carel I.M.A. Spee

ADVANCES IN PRECURSOR DEVELOPMENT FOR CVD OF BARIUM-
CONTAINING MATERIALS 203
    Brian A. Vaartstra, R.A. Gardiner, D.C. Gordon, R.L. Ostrander,
    and A.L. Rheingold

PRECURSOR DELIVERY FOR THE DEPOSITION OF SUPERCONDUCTING
OXIDES: A COMPARISON BETWEEN SOLID SOURCES AND AEROSOL 209
    O. Thomas, F. Weiss, J. Hudner, R. Haase, C. Dubourdieu,
    E. Mossang, N. Didier, and J.P. Senateur

STRUCTURE AND KINETICS STUDY OF MOCVD LEAD OXIDE (PbO)
FROM LEAD BIS-TETRAMETHYLHEPTADIONATE ($Pb(thd)_2$) 215
    Warren C. Hendricks, Seshu B. Desu, and Ching Yi Tsai

LIQUID DELIVERY OF LOW VAPOR PRESSURE MOCVD PRECURSORS 221
    R.A. Gardiner, P.C. Van Buskirk, and P.S. Kirlin

SYNTHESIS AND DECOMPOSITION OF A NOVEL CARBOXYLATE
PRECURSOR TO INDIUM OXIDE 227
    Aloysius F. Hepp, Maria T. Andras, Stan A. Duraj, Eric B. Clark,
    David G. Hehemann, Daniel A. Scheiman, and Phillip E. Fanwick

LOCAL EQUILIBRIUM PHASE DIAGRAMS FOR SiC DEPOSITION
IN A HOT WALL LPCVD REACTOR 233
    Chien C. Chiu, Seshu B. Desu, Zhi J. Chen, and Ching Yi Tsai

DEVELOPMENT AND IMPLEMENTATION OF LARGE AREA, ECONOMICAL
ROTATING DISK REACTOR TECHNOLOGY FOR METALORGANIC
CHEMICAL VAPOR DEPOSITION 241
    G.S. Tompa, P.A. Zawadzki, M. McKee, E. Wolak, K. Moy, R.A. Stall,
    A. Gurary, and N.E. Schumaker

PART VI: HIGH T_c SUPERCONDUCTORS

*SOME CONSIDERATIONS OF MOCVD FOR THE PREPARATION OF
HIGH T_c THIN FILMS 249
    Michael L. Hitchman, Sarkis H. Shamlian, Douglas D. Gilliland,
    David J. Cole-Hamilton, Simon C. Thompson, Stephen L. Cook,
    and Barbara C. Richards

*PLASMA-ENHANCED MOCVD OF SUPERCONDUCTING OXIDES 261
    Kenji Ebihara, Tomoyuki Fujishima, Masanobu Shiga, and Quanxi Jia

METAL-ORGANIC CHEMICAL VAPOR DEPOSITION OF EPITAXIAL
$Tl_2Ba_2Ca_2Cu_3O_{10-x}$ THIN FILMS 273
    Bruce J. Hinds, Jon L. Schindler, Bin Han, Deborah A. Neumayer,
    Donald C. Degroot, Tobin J. Marks, and Carl R. Kannewurf

*Invited Paper

OPTIMIZATION OF J_c FOR PHOTO-ASSISTED MOCVD PREPARED
YBCO THIN FILMS BY ROBUST DESIGN 279
    P.C. Chou, Q. Zhong, Q.L. Li, A. Ignatiev, C.Y. Wang, E.E. Deal,
    and J.G. Chen

IN SITU HETEROEPITAXIAL $Bi_2Sr_2CaCu_2O_8$ THIN FILMS PREPARED
BY METALORGANIC CHEMICAL VAPOR DEPOSITION 285
    Frank DiMeo Jr., Bruce W. Wessels, Deborah A. Neumayer,
    Tobin J. Marks, Jon L. Schindler, and Carl R. Kannewurf

HIGH QUALITY YBaCuO THIN FILM GROWTH BY LOW-TEMPERATURE
METALORGANIC CHEMICAL VAPOR DEPOSITION USING NITROUS
OXIDE 291
    Hideaki Zama, Jun Saga, Takeo Hattori, and Shunri Oda

PART VII: OPTOELECTRONIC MATERIALS; OXIDE CERAMICS

*ELECTRO-OPTIC MATERIALS BY SOLID SOURCE MOCVD 299
    R. Hiskes, S.A. DiCarolis, J. Fouquet, Z. Lu, R.S. Feigelson,
    R.K. Route, F. Leplingard, and C.M. Foster

*MOCVD OF MAGNETO-OPTICAL CERAMICS 311
    M. Okada, S. Katayama, and T. Kamiya

*MOCVD OF GROUP III CHALCOGENIDE COMPOUND SEMICONDUCTORS 317
    Andrew R. Barron

CHEMICAL VAPOR DEPOSITION OF VANADIUM OXIDE THIN FILMS 329
    Ogie Stewart, Joan Rodriguez, Keith B. Williams, Gene P. Reck,
    Narayan Malani, and James W. Proscia

ATOMIC LAYER CONTROLLED DEPOSITION OF Al_2O_3 FILMS EMPLOYING
TRIMETHYLALUMINUM (TMA) AND H_2O VAPOR 335
    A.C. Dillon, A.W. Ott, S.M. George, and J.D. Way

DEPOSITION OF CERIUM DIOXIDE THIN FILMS ON SILICON
SUBSTRATES BY ATOMIC LAYER EPITAXY 341
    Heini Mölsä and Lauri Niinistö

*GROUP 2 ELEMENT CHEMISTRY AND ITS ROLE IN OMVPE OF
ELECTRONIC CERAMICS 351
    William S. Rees Jr.

AUTHOR INDEX 363

SUBJECT INDEX 365

*Invited Paper

Preface

This symposium showcased progress in the field of metal-organic chemical vapor deposition (MOCVD) of thin-film materials. The symposium covered all aspects of MOCVD of electronic ceramics, including reactor design, ab-initio calculations, precursor chemistry and delivery, materials and processing correlations, and applications. Invited and contributing speakers from industry and academia presented papers covering the latest advances in MOCVD of electronic ceramics.

The symposium provided insight into the current trends emerging from this exciting technology. There was a strong emphasis on oxide and nitride materials including ferroelectrics, high T_c superconductors, and metal nitrides. Recent developments in in-situ characterization of MOCVD were also discussed. In addition, control of stoichiometry, and conformality, and the deposition of metastable phases was discussed at length.

Most of the papers presented at the Metal-Organic Chemical Vapor Deposition (MOCVD) of Electronic Ceramics symposium, held during the 1993 Fall Meeting of the Materials Research Society, November 29 to December 3, Boston, Massachusetts, were included in this volume.

Seshu B. Desu
David B. Beach
Bruce W. Wessels
Suleyman Gokoglu

January 1994

Acknowledgments

The symposium organizers would like to acknowledge the contributing and invited authors for outstanding quality of their presentations and proceedings manuscripts. The invited speakers include:

> D.M. Hoffman
> R.G. Gordon
> A. Erbil
> P. Ho
> C. Bernard
> B.M. Gallois
> H.A. Meinema
> W.S. Rees Jr.
> M.L. Hitchman
> A.E. Kaloyeros
> K. Ebihara
> R. Hiskes
> M. Okada
> A.R. Barron
> P.W. Chiu

In addition, the symposium chairs are indebted to the session chairs for their effort in overseeing the sessions, and guiding subsequent discussions and reviewers for their timely and insightful reviews of the manuscripts.

The symposium organizers wish to express their appreciation to the following organizations which provided financial support, enabling us to present the Metal-Organic Chemical Vapor Deposition of Electronic Materials symposium:

> NASA Lewis Research Center
> IBM Research Division
> Sandia National Laboratories

The symposium organizers wish to extend special thanks to Mr. Justin Gaynor for his help in editing the proceedings. A special thanks is reserved for the Materials Research Society and their staff, as well as the 1993 Fall MRS Meeting Chairs for the development of another outstanding conference.

MATERIALS RESEARCH SOCIETY SYMPOSIUM PROCEEDINGS

Volume 297— Amorphous Silicon Technology—1993, E.A. Schiff, M.J. Thompson, P.G. LeComber, A. Madan, K. Tanaka, 1993, ISBN: 1-55899-193-X

Volume 298— Silicon-Based Optoelectronic Materials, R.T. Collins, M.A. Tischler, G. Abstreiter, M.L. Thewalt, 1993, ISBN: 1-55899-194-8

Volume 299— Infrared Detectors—Materials, Processing, and Devices, A. Appelbaum, L.R. Dawson, 1993, ISBN: 1-55899-195-6

Volume 300— III-V Electronic and Photonic Device Fabrication and Performance, K.S. Jones, S.J. Pearton, H. Kanber, 1993, ISBN: 1-55899-196-4

Volume 301— Rare-Earth Doped Semiconductors, G.S. Pomrenke, P.B. Klein, D.W. Langer, 1993, ISBN: 1-55899-197-2

Volume 302— Semiconductors for Room-Temperature Radiation Detector Applications, R.B. James, P. Siffert, T.E. Schlesinger, L. Franks, 1993, ISBN: 1-55899-198-0

Volume 303— Rapid Thermal and Integrated Processing II, J.C. Gelpey, J.K. Elliott, J.J. Wortman, A. Ajmera, 1993, ISBN: 1-55899-199-9

Volume 304— Polymer/Inorganic Interfaces, R.L. Opila, A.W. Czanderna, F.J. Boerio, 1993, ISBN: 1-55899-200-6

Volume 305— High-Performance Polymers and Polymer Matrix Composites, R.K. Eby, R.C. Evers, D. Wilson, M.A. Meador, 1993, ISBN: 1-55899-201-4

Volume 306— Materials Aspects of X-Ray Lithography, G.K. Celler, J.R. Maldonado, 1993, ISBN: 1-55899-202-2

Volume 307— Applications of Synchrotron Radiation Techniques to Materials Science, D.L. Perry, R. Stockbauer, N. Shinn, K. D'Amico, L. Terminello, 1993, ISBN: 1-55899-203-0

Volume 308— Thin Films—Stresses and Mechanical Properties IV, P.H. Townsend, J. Sanchez, C-Y. Li, T.P. Weihs, 1993, ISBN: 1-55899-204-9

Volume 309— Materials Reliability in Microelectronics III, K. Rodbell, B. Filter, P. Ho, H. Frost, 1993, ISBN: 1-55899-205-7

Volume 310— Ferroelectric Thin Films III, E.R. Myers, B.A. Tuttle, S.B. Desu, P.K. Larsen, 1993, ISBN: 1-55899-206-5

Volume 311— Phase Transformations in Thin Films—Thermodynamics and Kinetics, M. Atzmon, J.M.E. Harper, A.L. Greer, M.R. Libera, 1993, ISBN: 1-55899-207-3

Volume 312— Common Themes and Mechanisms of Epitaxial Growth, P. Fuoss, J. Tsao, D.W. Kisker, A. Zangwill, T.F. Kuech, 1993, ISBN: 1-55899-208-1

Volume 313— Magnetic Ultrathin Films, Multilayers and Surfaces/Magnetic Interfaces— Physics and Characterization (2 Volume Set), C. Chappert, R.F.C. Farrow, B.T. Jonker, R. Clarke, P. Grünberg, K.M. Krishnan, S. Tsunashima/ E.E. Marinero, T. Egami, C. Rau, S.A. Chambers, 1993, ISBN: 1-55899-211-1

Volume 314— Joining and Adhesion of Advanced Inorganic Materials, A.H. Carim, D.S. Schwartz, R.S. Silberglitt, R.E. Loehman, 1993, ISBN: 1-55899-212-X

Volume 315— Surface Chemical Cleaning and Passivation for Semiconductor Processing, G.S. Higashi, E.A. Irene, T. Ohmi, 1993, ISBN: 1-55899-213-8

MATERIALS RESEARCH SOCIETY SYMPOSIUM PROCEEDINGS

Volume 316— Materials Synthesis and Processing Using Ion Beams, R.J. Culbertson, K.S. Jones, O.W. Holland, K. Maex, 1994, ISBN: 1-55899-215-4
Volume 317— Mechanisms of Thin Film Evolution, S.M. Yalisove, C.V. Thompson, D.J. Eaglesham, 1994, ISBN: 1-55899-216-2
Volume 318— Interface Control of Electrical, Chemical, and Mechanical Properties, S.P. Murarka, T. Ohmi, K. Rose, T. Seidel, 1994, ISBN: 1-55899-217-0
Volume 319— Defect-Interface Interactions, E.P. Kvam, A.H. King, M.J. Mills, T.D. Sands, V. Vitek, 1994, ISBN: 1-55899-218-9
Volume 320— Silicides, Germanides, and Their Interfaces, R.W. Fathauer, L. Schowalter, S. Mantl, K.N. Tu, 1994, ISBN: 1-55899-219-7
Volume 321— Crystallization and Related Phenomena in Amorphous Materials, M. Libera, T.E. Haynes, P. Cebe, J. Dickinson, 1994, ISBN: 1-55899-220-0
Volume 322— High-Temperature Silicides and Refractory Alloys, B.P. Bewlay, J.J. Petrovic, C.L. Briant, A.K. Vasudevan, H.A. Lipsitt, 1994, ISBN: 1-55899-221-9
Volume 323— Electronic Packaging Materials Science VII, R. Pollak, P. Børgesen, H. Yamada, K.F. Jensen, 1994, ISBN: 1-55899-222-7
Volume 324— Diagnostic Techniques for Semiconductor Materials Processing, O.J. Glembocki, F.H. Pollak, S.W. Pang, G. Larrabee, G.M. Crean, 1994, ISBN: 1-55899-223-5
Volume 325— Physics and Applications of Defects in Advanced Semiconductors, M.O. Manasreh, M. Lannoo, H.J. von Bardeleben, E.L. Hu, G.S. Pomrenke, D.N. Talwar, 1994, ISBN: 1-55899-224-3
Volume 326— Growth, Processing, and Characterization of Semiconductor Heterostructures, G. Gumbs, S. Luryi, B. Weiss, G.W. Wicks, 1994, ISBN: 1-55899-225-1
Volume 327— Covalent Ceramics II: Non-Oxides, A.R. Barron, G.S. Fischman, M.A. Fury, A.F. Hepp, 1994, ISBN: 1-55899-226-X
Volume 328— Electrical, Optical, and Magnetic Properties of Organic Solid State Materials, A.F. Garito, A. K-Y. Jen, C. Y-C. Lee, L.R. Dalton, 1994, ISBN: 1-55899-227-8
Volume 329— New Materials for Advanced Solid State Lasers, B.H.T. Chai, T.Y. Fan, S.A. Payne, A. Cassanho, T.H. Allik, 1994, ISBN: 1-55899-228-6
Volume 330— Biomolecular Materials By Design, H. Bayley, D. Kaplan, M. Navia, 1994, ISBN: 1-55899-229-4
Volume 331— Biomaterials for Drug and Cell Delivery, A.G. Mikos, R. Murphy, H. Bernstein, N.A. Peppas, 1994, ISBN: 1-55899-230-8
Volume 332— Determining Nanoscale Physical Properties of Materials by Microscopy and Spectroscopy, M. Sarikaya, M. Isaacson, H.K. Wickramasighe, 1994, ISBN: 1-55899-231-6
Volume 333— Scientific Basis for Nuclear Waste Management XVII, A. Barkatt, R. Van Konynenburg, 1994, ISBN: 1-55899-232-4
Volume 334— Gas-Phase and Surface Chemistry in Electronic Materials Processing, T.J. Mountziaris, P.R. Westmoreland, F.T.J. Smith, G.R. Paz-Pujalt, 1994, ISBN: 1-55899-233-2
Volume 335— Metal-Organic Chemical Vapor Deposition of Electronic Ceramics, S.B. Desu, D.B. Beach, B.W. Wessels, S. Gokoglu, 1994, ISBN: 1-55899-234-0

Prior Materials Research Society Symposium Proceedings available by contacting Materials Research Society

PART I

MOCVD of Nonoxide Electronic Ceramics

PLASMA ENHANCED METAL-ORGANIC CHEMICAL VAPOR DEPOSITION OF GERMANIUM NITRIDE THIN FILMS

DAVID M. HOFFMAN,* SRI PRAKASH RANGARAJAN,* SATISH D. ATHAVALE,¶ DEMETRE J. ECONOMOU,¶ JIA-RUI LIU,¥ ZONGSHUANG ZHENG,¥ AND WEI-KAN CHU¥
*Department of Chemistry, University of Houston, Houston, TX 77204
¶Department of Chemical Engineering, University of Houston, Houston, TX 77204
¥Texas Center for Superconductivity, University of Houston, Houston, TX 77204

ABSTRACT

Amorphous germanium nitride thin films are prepared by plasma enhanced chemical vapor deposition from tetrakis(dimethylamido)germanium, $Ge(NMe_2)_4$, and an ammonia plasma at substrate temperatures as low as 190 °C with growth rates >250 Å/min. N/Ge ratios in the films are 1.3 and the hydrogen contents are 13 atom %. The hydrogen is present primarily as N-H. The refractive indexes are close to the bulk value of 2.1, and the band gap, estimated from transmission spectra, is 4.8 eV.

INTRODUCTION

Germanium nitride films have been deposited in thermal chemical vapor deposition (CVD) processes from $GeCl_4$ and ammonia at 400–600 °C, GeH_4 and ammonia at 550 °C, and GeH_4 and hydrazine at 440 °C [1–6]. Lower temperatures of deposition (250–400 °C) have been achieved by using GeH_4/NH_3 or GeH_4/N_2 precursor systems in combination with plasma enhancement [7, 8], but these processes give low growth rates (<40 Å/min) and germane requires special handling.

We and others have recently begun to examine homoleptic dialkylamido complexes, $M(NMe_2)_n$, as precursors to nitride thin films in low temperature plasma enhanced CVD processes [9–11]. The germanium member of this series of compounds, $Ge(NMe_2)_4$, displays properties that make it a potentially viable precursor. It is, for example, a volatile liquid which can be readily prepared in large quantities from inexpensive reagents and purified by distillation. Importantly, it is only moderately air sensitive and therefore much safer to handle than germane. In this paper we report the use of $Ge(NMe_2)_4$ and an ammonia plasma to deposit germanium nitride films at temperatures as low as 190 °C with growth rates greater than 250 Å/min.

EXPERIMENTAL

Depositions were carried out by using a low pressure cold wall tubular reactor. The plasma was excited with an induction coil connected to a 13.56 MHz power supply through an impedance matching network. Ammonia was passed through the plasma, and the plasma-activated ammonia was then mixed with the organometallic precursor downstream of the coil just before the substrate [9]. The substrate was bonded to an aluminum holder, and the temperature was monitored by a thermocouple inserted into a thermowell drilled near the top surface of the holder. The ammonia flow rate was regulated by a mass flow controller. The $Ge(NMe_2)_4$ flow rate was adjusted by carefully opening the precursor flask, which was kept at room temperature. A diffuse plasma glow encompassed the region of the precursor feedthrough as well as the substrate. Before each deposition, base pressures of <2 mtorr were achieved, and the leak rate was confirmed to be less than 5 x 10^{-3} sccm. Typical operating pressures were from 0.09 to 0.27 torr and the plasma power was in the range 15–20 W.

Ge(NMe$_2$)$_4$ was prepared from GeCl$_4$ and LiNMe$_2$ as follows. A dilute solution of GeCl$_4$ (4.35 g, 0.020 mol) in hexane (40 mL) was slowly added via an addition funnel to a stirring slurry of LiNMe$_2$ (4.18 g, 0.082 mol) in hexane (150 mL). The reaction mixture was stirred at room temperature for 1 day, and then the mixture was filtered through a glass frit. The hexane was stripped under vacuum leaving a pale yellow liquid. Distillation (bp 50 °C/10^{-2} torr; reported bp 203 °C/760 torr [12]) gave a colorless liquid (3.96 g, 0.016 mol, yield 80%). A single resonance is observed in the ^{1}H NMR spectrum (2.65 ppm/benzene-d_6). Combustion analysis results were slightly high for nitrogen: Calculated for Ge(NMe$_2$)$_4$: C, 38.6; H 9.7; N 22.5; Found: C 38.5; H 10.1; N 23.3. This compound was previously prepared from GeX$_4$ (X = Br, Cl) and HNMe$_2$ [12].

RESULTS

Film depositions from Ge(NMe$_2$)$_4$ and an ammonia plasma were successfully carried out on silicon and quartz substrates at 190 and 302 °C (Table I). The films are amorphous by X-ray diffraction. All of the films show good adhesion to the substrates as judged by the Scotch tape test. Film growth rates are >250 Å/min despite the low power used.

Table I. Growth Rates and Compositions of Germanium Nitride Films Deposited from an Ammonia Plasma and Ge(NMe$_2$)$_4$ on Silicon.[a]

Temperature (°C)	Growth Rate (Å/min)[b]	N/Ge ratio	H content (atom %)[c]	Refractive Index
190	268	1.29 ± 0.10	13 ± 1.5	2.09
302	312	1.30 ± 0.10	13 ± 1.5	2.24

[a] Flow rates during deposition: NH$_3$, 70 sccm; Ge(NMe$_2$)$_4$, 5–6 sccm.
[b] Not optimized.
[c] To convert from H atom concentrations we assumed a density of 5.24 g/cm^{-3}.

Nitrogen to germanium ratios in the films were determined by backscattering spectrometry (BS) for films deposited on silicon [13]. A spectrum is shown in Figure 1. The films are nearly stoichiometric (Table I) and have uniform depth composition. Carbon contamination is less than 3 atom % and oxygen less than 1 atom %. The hydrogen content, determined by elastic recoil detection (ERD), is approximately 13 atom % (standard: silicon implanted with hydrogen).

In Figure 2 is shown an IR spectrum for the film deposited on silicon at 190 °C. There are two prominent bands at 723 and ≈3300 cm^{-1}. The band at 3300 cm^{-1} was previously assigned to N-H stretching, and the band at 723 cm^{-1} to Ge-N of α-Ge$_3$N$_4$ [14]. In the IR spectrum of the film deposited at 302 °C we also see broad, weak intensity bands at 2108 and 1173 cm^{-1}. The 1173 cm^{-1} band may arise from N-H bending motions (cf. to the N-H bend in SiN$_x$ films, which is reported to occur at ≈1180 cm^{-1}) [15], and the band at 2108 cm^{-1} is in the region where one would expect the Ge-H stretching band [16].

An approximately 2000-Å thick film deposited at 291 °C on quartz shows high transmittance (75–80%) in the visible region. The optical band gap, obtained by plotting (dE)^2 vs. E where E is the photon energy and d is the optical density [17], is approximately 4.8 eV (Figure 3). This is higher than the bandgap of 4.0 eV reported by Hantzpergue and Remy [18]. The refractive indexes of the films, measured by ellipsometry (6328 Å, He-Ne laser), are close to the value of 2.1 reported for bulk Ge$_3$N$_4$ [19].

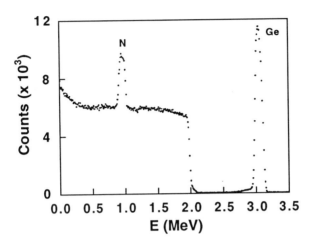

Figure 1. BS spectrum for a germanium nitride film deposited from Ge(NMe$_2$)$_4$ and an NH$_3$ plasma on silicon at 302 °C (3.72-MeV He^{2+} at nitrogen resonance).

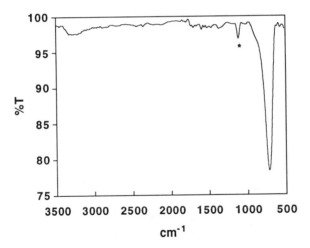

Figure 2. Infrared spectrum for a germanium nitride film deposited from Ge(NMe$_2$)$_4$ and an NH$_3$ plasma on silicon at 190 °C. The starred peak is due to the silicon substrate.

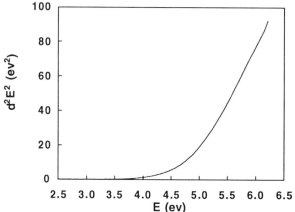

Figure 3. Plot to determine the band gap for a germanium nitride film deposited from Ge(NMe$_2$)$_4$ and an NH$_3$ plasma on quartz at 291 °C.

In a control experiment, Ge(NMe$_2$)$_4$ and NH$_3$ gave no deposition at 287 °C without the plasma. This is consistent with earlier atmospheric pressure thermal CVD experiments [20].

DISCUSSION

It is encouraging that we can deposit germanium nitride films from Ge(NMe$_2$)$_4$ and an NH$_3$ plasma at very low temperatures with little or no carbon contamination. In a simplistic view, the film deposition probably proceeds similarly to the known M(NR$_2$)$_n$/NH$_3$ thermal processes; that is, via M–NH$_2$ and M=NH intermediates [21, 22]. The intermediates in this case would result from reactions between plasma-produced NH$_x$ radicals and the Ge(NMe$_2$)$_4$ precursor.

On the basis of our IR data, it appears that most of the hydrogen is present in the form of N-H. Incomplete decomposition of putative M-NH$_x$ intermediates could account for the presence of hydrogen in this form [21]. The small amount of hydrogen present as Ge-H may result from combination of H atoms produced in the plasma with germanium containing intermediates.

CONCLUSIONS

An ammonia plasma and Ge(NMe$_2$)$_4$ deposits amorphous germanium nitride films at substrate temperatures as low as 190 °C. Growth rates are >250 Å/min. Presumably higher rates will be possible with higher plasma power. The films are nearly stoichiometric with little or no carbon and oxygen contamination and an approximately 13 atom % hydrogen content. The hydrogen is incorporated mostly as N-H. This Ge(NMe$_2$)$_4$ based process offers an attractive alternative to the conventional low temperature plasma enhanced chemical vapor deposition methods involving germane.

ACKNOWLEDGEMENTS

We are grateful to the State of Texas for support of this work through the Texas Center for Superconductivity at the University of Houston. DMH also acknowledges support from the Robert Welch Foundation and the Texas Advanced Research Program. DMH is a 1992–4 Alfred P. Sloan Research Fellow. DJE acknowledges support from the Texas Advanced Technology Program.

REFERENCES

1. H. Nagai, T. Niimi, J. Electrochem. Soc. 115, 671 (1968).
2. T. Yashiro, J. Electrochem. Soc. 119, 780 (1972).
3. A.B. Young, J. J. Rosenberg and I. Szendro, J. Electrochem. Soc. 134, 2867 (1987).
4. K. P. Pande, C. C. Shen, Appl. Phys. A 28, 123 (1982).
5. V. A. Gritsenko, Phys. Status Solid A 28, 387 (1975).
6. Y. Pauleau, J. -C. Remy, C. R. Acad. Sci. Paris 280, 1215 (1975).
7. G. A. Johnson and V. J. Kapoor, J. Appl. Phys. 69, 3616 (1991).
8. D. B. Alford and L. G. Meiners, J. Electrochem. Soc. 134, 979 (1987).
9. D. M. Hoffman, S. P. Rangarajan, S. D. Athavale, S. C. Deshmukh, D. J. Economou, J. R. Liu, Z. S. Zheng, W. K. Chu, submitted for publication.
10. See also: H. Schuh, T. Schlosser, P. Bissinger and H. Schmidbaur, Z. Anorg. Allg. Chem. 619, 1347 (1993).
11. See also: Y. Mikata and T. Moriya, Eur. Pat. Appl. EP 464,515 (1991); JP Appl. 171,156 (1990); Chem. Abstr. 116: 163137c (1992).
12. H. H. Anderson, J. Am. Chem. Soc. 74, 1421 (1952); J. Mack, C. H. Yoder, Inorg. Chem. 8, 278 (1969).
13. W.-K. Chu, J. W. Mayer, M. -A. Nicolet, Backscattering Spectrometry (Academic Press: New York, N. Y., 1978).
14. G. A. Johnson, V. J. Kapoor, J. Appl. Phys. 69, 3616 (1991) and references therein.
15. K. -C. Lin, S. -C. Lee, J. Appl. Phys. 72, 5474 (1992) and references therein.
16. K. Nakamoto Infrared Spectra of Inorganic and Coordination Compounds, 2nd ed.(Wiley: New York, N. Y., 1970).
17. T. H. Yuzuriha, D. W. Hess, Thin Solid Films 140, 199 (1986).
18. J. J. Hantzpergue, J. C. Remy, Thin Solid Films 30, 205 (1975).
19. Q. Hua, J. Rosenberg, J. Ye, E. S. Yang, J. Appl. Phys. 53, 8969 (1982).
20. L. Atagi, D. M. Hoffman, unpublished results.
21. D. M. Hoffman, Polyhedron, in press.
22. R. Fix, R. G. Gordon, D. M. Hoffman, J. Am. Chem. Soc. 112, 7833 (1990).

RECENT ADVANCES IN THE CVD OF METAL NITRIDES AND OXIDES

ROY G. GORDON
Department of Chemistry, Harvard University, Cambridge, MA 02138.

ABSTRACT

Use of metal-organic precursor materials has permitted thermal CVD of many metal nitrides at remarkably low temperatures and without corrosive by-products. Metallic TiN, VN, Nb_3N_4 and Mo_2N_3; semiconducting GaN and Sn_3N_4; and insulating AlN, Zr_3N_4, Hf_3N_4 and Ta_3N_5 can be deposited at temperatures typically in the range 100 to 400 C. Deposits free of carbon are obtained by transamination reactions of metal dialkylamido precursors with a sufficiently large excess of ammonia. The resulting TiN films are good diffusion barriers, and provide low contact resistance between metals and silicon.

Transparent semiconducting oxide films, such as SnO_2, ZnO and TiO_2, are often made by MOCVD for use in solar cells, energy-efficient window coatings and electro-optical displays. These wide band-gap semiconductors can be doped to n-type conductivity by a variety of dopant elements. Fluorine is the dopant which produces materials with the highest electron mobility, conductivity and transparency, by substituting for oxygen. Certain organic fluorine compounds have been found to be very effective fluorine dopants for these CVD reactions, yielding films with very shallow donors having nearly 100 % electrical activity.

Precursors for the CVD of alkaline earth metal oxides, particularly barium, lack the volatility and stability needed for reproducible deposition of superconducting, ferroelectric or magnetic oxides. Use of ammonia or volatile amines as carrier gases greatly enhances the volatility and stability of beta-diketonates of barium, strontium and calcium, providing high and stable transport rates for source temperatures below 100 C, even for non-fluorinated ligands.

CVD OF METAL NITRIDES

Because of the many applications of metal nitrides, a wide variety of CVD methods have been devised for their preparation. The hard refractory metal, titanium nitride, will be considered in this review as an example of an important metal nitride which has received much attention. It is used as a wear-resistant coating on machine tools, as a solar control film on windows, as a corrosion-resistant and decorative coating, and as a diffusion barrier in microelectronics. Successive advances have allowed CVD processes to be used for each of these applications.

CVD of Titanium Nitride from Titanium Tetrachloride

The earliest known CVD route to TiN uses the reaction of titanium tetrachloride with nitrogen and hydrogen at very high temperatures, typically 900 to 1000 C[1].

$$TiCl_4 + 1/2\, N_2 + 2\, H_2 \rightarrow TiN + 4\, HCl$$

This reaction has long been used to coat tungsten carbide machine tools, to improve their wear-resistance. This substrate is sufficiently refractory to resist the high temperatures required, and sufficiently corrosion-resistant to resist the hydrochloric acid byproduct. However, most other potential substrates cannot stand up to these aggressive conditions. Glass substrates would melt, and dopants would diffuse in silicon substrates.

Thermodynamic calculations show that this reaction cannot be spontaneous below temperatures of about 750 C. Thus, it is not possible to find a catalyst which could make this reaction operate at temperatures below 750 C. In order to find lower temperature CVD reactions, different reactants are needed. For example, replacement of the nitrogen and hydrogen by ammonia gives us the following reaction, which thermodynamics allows to be spontaneous above about 310 C:

$$TiCl_4 + 4/3\, NH_3 \rightarrow TiN + 4\, HCl + 1/6\, N_2$$

Unfortunately, use of this CVD reaction fails when the reactants are mixed at ambient temperatures; instead, a rapid acid-base reaction precipitates a salt or adduct, by reactions such as

$$TiCl_4 + 2\, NH_3 \rightarrow TiCl_4 \cdot 2\, NH_3$$

This problem was avoided simply by preheating the reactants to temperatures around 250 to 300 C, at which temperatures this mixture does not immediately form solid byproducts[2]. The reactants stay in a homogeneous vapor phase for at least a brief period during which they flow into a CVD reactor where they deposit a relatively pure TiN film at substrate temperatures above about 600 C. This temperature is low enough to allow commercial deposition of solar control coatings on large areas of window glass, and diffusion barriers on silicon microelectronic devices[3].

Films can be deposited by the reaction of titanium tetrachloride and ammonia at temperatures as low as 400 C, but at the lower temperatures the deposition rate becomes very low and chlorine contamination of the film becomes a problem[4]. Another problem is that cooler parts of the reactor can collect powered salt adduct and/or solid ammonium chloride, and these particulates can contaminate the semiconductor devices being coated.

CVD of TiN from Titanium Dialkylamides

In order to avoid these problems, we sought a non-chlorine-containing precursor which would react rapidly at temperatures below 400 C. Tetrakis(dialkylamido)titanium compounds had been suggested[5] as CVD precursors to TiN, but they produce instead porous, high-resistance films which are predominantly carbon.[6] We found that the CVD reaction of ammonia with the tetrakis(dialkylamido)titanium compounds produced pure TiN films quite free of carbon, at temperatures as low as 200 C.[7]

$$Ti(NR_2)_4 + NH_3 \rightarrow TiN + HNR_2 + \;?$$

Here R represents a methyl or ethyl group. These films are just as effective diffusion barriers between metals and silicon as are the commonly used sputtered TiN films.[8]

The ability to coat uniformly inside holes and trenches is another property ("step coverage') needed for microelectronic applications of TiN. Sputtered TiN has poor step coverage, which is adequate for current microelectronic devices with wide holes, but not for future production of devices with narrow holes. Collimated sputtering can improve the step coverage, at the cost of poor sidewall coverage, reduced deposition rate and increased maintenance. The reaction of tetrakis(dialkylamido)titanium compounds with ammonia produces coatings with better step coverage than sputtering, but insufficient step coverage for very narrow holes.[9] Reaction of remote-plasma-activated hydrogen atoms with tetrakis(dialkylamido)titanium compounds has recently been reported to greatly improve the step coverage, at the cost of introducing 35 atomic % carbon and increasing its electrical resistivity.[10]

Proposed mechanism for CVD TiN from Titanium Dialkylamides

The chemical mechanism of the CVD reaction between tetrakis-(dialkylamido)titanium compounds and ammonia is at least partly understood. The reaction begins very rapidly as soon as the reactants are mixed in the gas phase. By analogy to similar solution-phase reactions[11], it was suggested[12] that these initial gas-phase reactions are transaminations:

$$Ti(NR_2)_4 + NH_3 \rightarrow Ti(NR_2)_3NH_2 + R_2NH$$

$$Ti(NR_2)_3NH_2 + NH_3 \rightarrow Ti(NR_2)_2(NH_2)_2 + R_2NH$$

$$Ti(NR_2)_2(NH_2)_2 + NH_3 \rightarrow Ti(NR_2)(NH_2)_3 + R_2NH$$

$$Ti(NR_2)(NH_2)_3 + NH_3 \rightarrow Ti(NH_2)_4 + R_2NH$$

Flow tube experiments[13] have confirmed that these fast reactions produce dimethylamine, and have measured the rate constants for these fast transamination reactions. This mechanism also predicted that the nitrogen in the film comes from the ammonia, rather than from the the nitrogen initially bound to the titanium. Isotopic tracer experiments[14] have subsequently verified this source of the nitrogen, giving further support to the transamination mechanism.

The resulting titanium amide is not a known compound. It is likely to undergo rapid unimolecular decomposition reactions, eliminating ammonia:

$$Ti(NH_2)_4 \rightarrow HN=Ti(NH_2)_2 + NH_3$$

$$HN=Ti(NH_2)_2 \rightarrow HN=Ti=NH + NH_3$$

These elimination reactions should be somewhat endothermic, because they convert two titanium-nitrogen single bonds into one double bond which should be less than twice as strong. The resulting titanium diimide, HN=Ti=NH, should have a high sticking coefficient on the surface of the growing film, because the titanium atom is only shielded by two ligands. Thus the titanium diimide is a likely growth species, which would lead to a material with composition $(TiN_2H_2)_n$. Loss of some nitrogen and hydrogen from the solid would then lead to the observed composition of nitrogen-rich titanium nitride with some residual hydrogen.[15] The high sticking coefficient of the proposed titanium diimide intermediate would explain the poor step coverage shown by this reaction.

Dimerization reactions are also likely for the titanium diimide:

$$\begin{array}{c} HN=Ti=NH \\ + \\ HN=Ti=NH \end{array} \rightarrow \begin{array}{c} HN-Ti=NH \\ |\quad| \\ HN=Ti-NH \end{array}$$

These would be followed by polymerization reactions:

$$\begin{array}{c} HN-Ti=NH \\ |\quad| \\ HN=Ti-NH \\ + \\ HN=Ti=NH \end{array} \rightarrow \begin{array}{c} HN-Ti=NH \\ |\quad| \\ HN-Ti-NH \\ |\quad| \\ HN=Ti-NH \end{array}$$

$$\begin{array}{c} HN-Ti=NH \\ |\quad| \\ HN-Ti-NH \\ |\quad| \\ HN=Ti-NH \\ + \\ HN=Ti=NH \end{array} \rightarrow \begin{array}{c} HN-Ti=NH \\ |\quad| \\ HN-Ti-NH \\ |\quad| \\ HN-Ti-NH \\ |\quad| \\ HN=Ti-NH \end{array}$$

These polymerization reactions should be exothermic, because they exchange a Ti=N double bond for two single Ti-N bonds. They are likely to be fast with rates that are nearly independent of temperature. Extensive polymerization could lead to the formation of particles. This mechanism suggests that the gas residence time in the CVD reactor should be as short as possible, in order to limit the growth of particles. These gas-phase polymers may also add to the surface of the growing film, giving more material of the same composition, $(TiN_2H_2)_n$. The physical properties of the film deposited from polymers may, however, differ from those of film deposited from monomer, because the sticking coefficients of the polymeric species may differ from those of the monomer. Thus the step coverage and density of films may be reduced by gas-phase polymerization occurring before film deposition.

CVD of Other Metal Nitrides

Similar low-temperature reactions were found to produce other early transition metal nitrides[16], including the metals TiN, VN, Nb_3N_4 and Mo_2N_3 and the insulators Zr_3N_4, Hf_3N_4 and Ta_3N_5, in which the cations retain their highest oxidation state. A few main-group metals also show similar low-energy reaction pathways, which produce the insulator AlN[17], and the semiconductors GaN[18] and Sn_3N_4.[19] Other main-group elements, such as boron and silicon, require considerably higher temperatures to produce nitrides.

TRANSPARENT CONDUCTING OXIDES

The oxides of tin, indium, zinc and titanium are transparent, wide band-gap semiconductors. They may be made conductive by doping with various n-type dopants. The oldest example is antimony-doped tin oxide[20], sprayed films of which were used to defrost bomber windows in the second world war, during which time the process was kept secret. Aluminum-doped zinc oxide[21] is another commonly used example. These metallic dopants are easily introduced during film growth, and have high solid-state solubility. Although these metallic dopants are electrically active, they reduce the mobility of the conduction electrons because they induce strong electronic perturbations to the conduction bands, which are formed primarily from orbitals of the metals. The metal centers can also produce optical absorption bands which reduce their transparency; for example antimony-doped tin oxide shows an absorption in the red region, which makes the films appear blue.

Fluorine-doped tin oxide

Fluorine can also act as an n-type dopant in semiconducting metal oxides. Fluorine doping has the advantage that its electronic perturbation affects mainly the valence band, leaving the electrons to travel rather freely in the conduction band with high mobility. The size of the fluoride ion matches almost exactly that of the oxide ion which it replaces; thus distortions of the lattice by fluorine-doping are negligible. Further, fluorine doping does not produce any localized electronic absorption bands. In fact, it increases the transparency in the blue and ultraviolet region, by the Moss-Burstein effect.

Despite these advantages of fluorine-doping, most commercial transparent conductors were made with metallic dopants until recently. It proved to be difficult to make metal oxide films with substitutional fluorine doping. Some commercial fluorine-doped tin oxide was made by a spray process,[22] but the amount of fluorine that could be incorporated was limited; it was widely believed that the solubility of fluorine in tin oxide was too small to be an optimal dopant. Another problem with the spray process was the difficulty in achieving a uniform and homogeneous coating over large areas. Also, there were difficulties with other impurities carried in from the spray solution or diffused from the hot glass surface, which limited the conductivity and optical clarity of the sprayed films.

The discovery[23] of suitable fluorine dopants for CVD of fluorine-doped tin oxide with high fluorine content removed these limitations of the sprayed material. It allowed production of more conductive material with more uniform properties. The first such process to be used commercially was the reaction

$$(CH_3)_4Sn + O_2 + CF_3Br \rightarrow SnO_2{:}F + \ldots$$

The fluorine dopant, bromotrifluoromethane, is a very inert fluorocarbon which is not normally thought of as a fluorinating agent. However, during the CVD process, it is activated by methyl groups released by the thermal decomposition of the tetramethyltin:[24]

$$(CH_3)_4Sn \rightarrow (CH_3)_3Sn + CH_3$$

$$CH_3 + CF_3Br \rightarrow CF_3 + CH_3Br$$

The trifluoromethyl radicals thus generated can transfer fluorine to the growing film by mechanisms which are not fully characterized. Several atomic per cent of fluorine can be introduced into the film by this reaction. The amount of fluorine in the film is easily varied by changing the initial concentration of bromotrifluoromethane.[25] The highest conductivity is typically obtained by substitution of one to two per cent of the oxygen ions by fluorine ions.[26]

This reaction has been widely used to produce SnO_2:F transparent conductors for amorphous silicon solar cells, flat panel displays and other applications. The CVD is typically carried out at atmospheric pressure in belt furnaces, such as those produced by the Watkins-Johnson Company. The production of bromotrifluoromethane is being stopped because of its ability to deplete the stratospheric ozone. Therefore other fluorocarbons are being substituted for this fluorine dopant in commercial production. In the past few years, CVD of SnO_2:F has been scaled up to widths of more than three meters during on-line production in glass factories. For reasons such as cost, toxicity and flammability, other organo-tin precursors are being used in place of the tetramethyltin in large-scale production. Most of this SnO_2:F-coated glass is being used for its ability to reflect thermal infrared radiation, which improves the thermal insulating ability of windows.

Fluorine-doped Zinc Oxide

Transparent, conductive zinc oxide films been made with many different metallic dopants, including boron, aluminum, indium, and silicon. However, the full potential of zinc oxide as a transparent conductor was not revealed until fluorine-doping was achieved by a CVD process.[27] The ZnO:F material thus produced is the most conductive transparent material known. For example, a ZnO:F film having 5 ohms per square absorbs only 3 per cent of the visible light passing through it. Equally transparent films of SnO_2:F or other materials have resistances at least twice as large.

The CVD process for producing highly conductive fluorine-doped zinc oxide films uses the reaction of diethylzinc, ethanol and hexafluoropropene

$$Et_2Zn + EtOH + CF_2=CFCF_3 \rightarrow ZnO:F + \ldots$$

operating at temperatures of about 350 to 480 C. The fluorine content is easily adjusted in the range up to a few atomic per cent by varying the gas-phase concentration of the hexafluoropropene. The ratio of conductivity to optical absorption is maximized for ZnO:F containing about 0.5 atomic per cent fluorine and deposited at about 470 C. It is remarkable that under these conditions, essentially all the fluorine incorporated in the films is electrically active in releasing conduction electrons. This unusually high doping efficiency helps produce the high mobility, conductivity and transparency of this material. At lower deposition temperatures, fluorine can still be incorporated into the films, but the doping efficiency, mobility, conductivity and transparency are reduced.

The mechanism of this CVD reaction has not been fully determined. However, some aspects of the reaction can be inferred from various observations. When commercial anhydrous grade ethanol is used, ZnO:F can be deposited on most surfaces, although the grain size, amount of texture and diffuse light scattering varied, depending on which brand or batch of alcohol was used. When the alcohol was first dried rigorously by distilling it from magnesium ethoxide, then more film could be grown on a previously deposited zinc oxide film, but no film at all would nucleate on clean glass substrates. These observations suggest that water plays a crucial catalytic role in this reaction.

These observations are consistent with the following mechanism. It is known that at temperatures in the 300 to 500 C range, zinc oxide surfaces catalyze the decomposition of alcohols:

$$CH_3CH_2OH \rightarrow H_2O + CH_2=CH_2$$

Water vapor readily reacts with diethylzinc vapor at temperatures above about 100 C to deposit zinc oxide

$$Et_2Zn + H_2O \rightarrow ZnO + 2EtOH$$

Together, these two reactions constitute a catalytic cycle producing zinc oxide, with water vapor acting as the chain carrier. Thus in order to initiate this catalytic cycle, some water needs to be present. This trace of water can be introduced along with the ethanol, in which case no zinc oxide need be present initially. Alternatively, if a zinc oxide substrate is present initially, the required water is formed *in situ* by the catalytic decomposition of ethanol at the surface of the zinc oxide.

Some applications, such as flat panel displays, require smooth ZnO:F films, whereas others, such as thin-film solar cells, require rough, textured films. By making use of the reaction mechanism, one can influence the growth of the deposited film to produce small grains and a smooth film, or large grains and a textured film. To make a smooth film with small grains, a glass substrate is first coated with thin film of zinc oxide at least a few hundred Angstroms thick, using the reaction of water vapor with diethylzinc at a relatively low temperature, say 200 C. Then a thicker layer of smooth, fine-grained ZnO:F can be grown from the reaction of commercial anhydrous ethanol, diethylzinc and hexafluoropropene at temperatures around 400 C. To make a large-grained, textured film, the initial ZnO deposit formed by the water reaction is kept very thin, so that it consists of isolated nuclei. Then the thick ZnO:F layer is grown using rigorously anhydrous ethanol at temperatures around 480 C.

TRANSPORT OF ALKALINE EARTH METALS

Many important compounds contain the alkaline earth metals calcium, strontium and/or barium. For example, they are essential ingredients in the high T_c superconducting oxides, high-dielectric constant barium-strontium titanates, and many magnetic ferrites.

However, alkaline earth compounds are notably low in their volatility. Thus finding suitable sources for CVD of alkaline-earth-containing materials has been difficult. Barium presents the greatest challenge, since almost all barium compounds are non-volatile at temperatures of use in CVD. The most widely used barium sources are from the family of bis(beta-diketonates)

in which the R groups are chosen to be bulky and/or fluorinated, in order to promote volatility. For example, if the R groups are tert-butyl, the compound is bis(2,2,6,6-tetramethyl-3,5-heptanedionato)barium, which sublimes with decomposition at about 250 C. Fluorinated R groups, such as perfluoropropyl, increase the volatility slightly, but tend to produce fluorine-containing deposits.

These bis(beta-diketonates) are not fully satisfactory CVD sources, since they decompose during sublimation, and thus the amount of material delivered tends to decrease with time. However, their performance can be enormously enhanced by using a volatile amine, such as ammonia or trimethylamine, as a carrier gas.[28] The amine greatly increases the transport rate while at the same time suppressing the decomposition reaction. Useful and stable transport rates can be obtained in this way at bis(2,2,6,6-tetramethyl-3,5-heptanedionato)barium source temperatures below 100 C. By using this amine transport method and air as an oxidant, reproducible CVD of pure barium oxide was achieved. No nitrogen or carbon was found in the deposited barium oxide. Similar, but smaller, enhancements were found for the transport of strontium and calcium compounds.

Precaution should of course be taken to stay outside the regions of flammability for CVD gas mixtures containing both oxygen and amines. For example, ammonia has a flammable range of concentrations from about 15 to 28 mole per cent in air, and 13 to 79 mole per cent in oxygen.

This amine transport method works better at higher amine gas pressures. If, for example, the CVD reactor is operating at atmospheric pressure, the vapor supply lines normally operate at slightly over one atmosphere of pressure, and the amine-assisted transport operates very effectively. In the case of low-pressure CVD, it is usual to operate the vapor supply lines also at low pressures, under which conditions amine-assisted transport is not as effective. Therefore, the supply lines to low pressure CVD reactors should also be operated at a pressure near one atmosphere of amine carrier gas, and a high-temperature leak valve should reduce the pressure just before the gas mixture enters the CVD chamber.

---

[1] W. Schintlmeister, O. Pacher and K. Pfaffinger, J. Electrochem. Soc. **123**, 924 (1976); W. Schintlmeister, O. Pacher, T. Krall, W. Wallgram and T. Raine, Powder Metallurgy International **13**, 71 (1981).

[2] Roy G. Gordon, U.S. Patent No. 4,535,000 (13 August, 1985); S. R. Kurtz and R. G. Gordon, Thin Solid Films **140**, 277 (1986).

[3] Arthur Sherman, in Chemical Vapor Deposition of Refractory Metals and Ceramics, edited by T. M. Besman and B. M. Ballois, (Mater. Res. Soc. Proc., Pittsburgh, PA, 1989) p. 323; Arthur Sherman, J. Electrochem. Soc. **137**, 1892 (1990).

[4] Rama I. Hegde, Robert W. Fiordalice and Philip J. Tobin, Applied Phys. Lett. **62**, 2326 (1993).

[5] K. Sugiyama, S. Pac, Y. Takahashi, S. Motojima, J. Electrochem.Soc. **122**, 1545 (1975).

[6] R. M. Fix, R. G. Gordon, D. M. Hoffman, Chem. of Materials **2**, 235 (1990).

[7] R. M. Fix, R. G. Gordon, D. M. Hoffman, Mater. Res. Soc Symp. Proc. **168**, 357 (1990); R. M. Fix, R. G. Gordon, D. M. Hoffman, Chem. of Materials **3**, 1138 (1991).

[8] J. N. Musher and R. G. Gordon, J. Electronic Materials **20**, 1105 (1991).

[9] G. S. Sandhu, T. T. Doan, Mater. Res. Soc. Conf. Proc. ULSI-VII, Pittsburgh, PA, 1992) pp. 323-328; T. S. Cale, G. B. Raupp, J. T. Hillman, M. J. Rice, Mater. Res. Soc. Conf. Proc. ULSI-VIII, Pittsburgh, PA, 1993) pp. 195-202.

[10] A. Intemann, H. Koerner, F. Koch, J. Electrochem. Soc. **140**, 3215 (1993).

[11] D. C. Bradley and E. G. Torrible, Can. J. Chem. 41, 134 (1963).

[12] R. M. Fix, R. G. Gordon, D. M. Hoffman, J. Am. Chem. Soc. 112, 7833 (1990).

[13] B. H. Weiller and B. V. Partido, Chem. Mater. (submitted, 1993).

[14] J. A. Prybyla, C.-M. Chiang, L. H. Dubois, J. Electrochem. Soc. (in press, 1993).

[15] R. M. Fix, R. G. Gordon, D. M. Hoffman, Chem. of Materials **3**, 1138 (1991).

[16] R. M. Fix, R. G. Gordon, D. M. Hoffman, Chem. of Materials **5**, 614(1993).

[17] U. Riaz, R. G. Gordon, D. M. Hoffman, J. Materials Res. **7**, 1679 (1992).

[18] U. Riaz, R. G. Gordon, D. M. Hoffman, Mater. Res. Soc. Symp. Proc. **242**, 445 (1992).

[19] U. Riaz, R. G. Gordon, D. M. Hoffman, Chem. Materials **4**, 68 (1992).

[20] J. M. Mochel, U.S. Patent 2,564,706 (21 August, 1951).

[21] J. Hu and R. G. Gordon, J. Appl. Phys. 71, 880 (1992).

[22] W. O. Lytle and A. F. Junge, U.S. Patent 2,566,346 (4 September, 1951).

[23] R. G. Gordon, U.S. Patent 4,146,657 (27 March, 1979).

[24] C. J. Giunta, A. G. Zawadzki, R. G. Gordon, J. Phys. Chem. **96**, 5364 (1991).

[25] C. Borman and R. G. Gordon, J. Electrochem. Soc. **136**, 3820 (1989).

[26] J. Proscia and R. G. Gordon, Thin Solid Films **214**, 175 (1992).

[27] Jianhua Hu and Roy G. Gordon, Solar Cells **30**, 437 (1991).

[28] J. M. Buriak, L. K. Cheatham, J. J. Graham, R. G. Gordon and A. R. Barron, Mat. Res. Soc. Symp. **204**, 545 (1991).

LPCVD OF InN ON GaAs(110) USING HN_3 AND TMIn: COMPARISON WITH Si(100) RESULTS

Y. BU AND M.C. LIN
Department of Chemistry, Emory University, Atlanta, GA 30322

ABSTRACT

Low-pressure chemical vapor deposition (LPCVD) of InN and laser-assisted LPCVD on GaAs(110) and Si(100) using HN_3 and trimethyl indium (TMIn) has been studied with XPS, UPS and SEM. Without 308-nm excimer laser irradiation, InN film was built on the GaAs but not on Si surface under the present low-pressure conditions. When the photon beam was introduced, InN films with In:N atomic ratio of 1.0±0.1 and a thickness of more than 20 Å (the limit of the electron escaping depth for the In_{3d} X-ray photoelectrons) were formed on Si(100) surface. In both cases, the formation of surface nitrides at the initial film growth processes was clearly indicated in the XPS spectra. The He(II) UP spectra taken from InN films on GaAs and Si are nearly identical and agree well with the result of a pseudo-potential calculation for the InN valence band. The corresponding SEM pictures showed smooth InN films on GaAs(110), while grains with diameter of ~100 nm were observed for InN on Si(100).

INTRODUCTION

The growth of III-V nitride semiconductor materials has been the subject of intense interest in the past decade. Because of their wide direct band-gaps, these materials have a potential for use in electronic and optoelectronic devices such as visible-light semiconductor lasers, ultraviolet light detectors and laser diodes. However, few articles have been published regarding the growth of InN crystals or thin films, probably because InN has a low dissociation temperature and thus low-temperature growth is required [1,2].

Recently it has been shown that the CVD of InN and InGaN on sapphire substrates can be achieved by using TMIn, TEGa and NH_3 under extremely high V/III ratios ($>10^4$) [3]. Other methods, such as reactive rf-sputtering and ion plating have been used in an effort to produce InN films [4-7]. Using low energy In^+ and N_2^+ ion beams, Bello et al. [8] were able to grow InN on a Si(100) surface; however, the resulting InN showed an N:In atomic ratio of <0.4. In this paper, we report the results of LPCVD of InN on GaAs(110) and Si(100) surfaces using HN_3 and TMIn, aided by 308-nm photon excitation in the latter case.

EXPERIMENTAL

The present experiment was carried out in a custom-designed ultra-high vacuum (UHV) system (Leybold, Inc.) which is composed of two compartments, one for the surface analysis and the other for film deposition as described elsewhere [9,10]. HN_3 and TMIn samples were prepared in the same manner as described in references [11] and [12].

GaAs(110) single crystals from Laser Diode, Inc. were cut into 0.5 cm x 1.5 cm samples. They were cleaned with 5% hydrogen fluoride (HF) solution and then annealed at 1500 K *in vacuo* bombarded with 2.5 keV Ar^+ ion beam followed by annealing at 770 K. TMIn and HN_3 samples were introduced into the system through two separate 1/8" stainless steel tubes, whose ends were ~1/2" above the surface. After a designated deposition time, the sample was moved into the surface analysis chamber for X-ray photoelectron spectroscopy (XPS) and ultraviolet photoelectron spectroscopy (UPS) measurements and then moved back to the deposition chamber for continuing film deposition. When the deposition was finished, the samples were taken out of the vacuum chamber and stored in N_2 atmosphere before scanning electron microscopy (SEM) measurements.

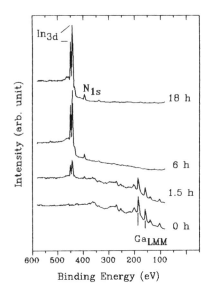
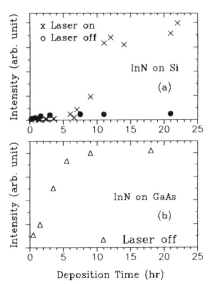

Fig.1 XP spectra for InN on GaAs at the indicated deposition times.

Fig.2 The dependence of $In_{3d-5/2}$ XPS signal intensity on InN deposition time.

RESULTS

Figure 1 shows XP spectra taken from a clean GaAs(110) and after InN deposition for 1.5, 6 and 18 h. The deposition was carried out at $P_{TMIn} = 2 \times 10^{-7}$, $P_{HN3} = 1 \times 10^{-6}$ torr and $T_s = 700$ K. For the clean surface, XPS showed major peaks at 204, 140, 104, 189 and 162 eV due to $As_{3d-1/2}$, $As_{3p-3/2}$, Ga_{3p} photoelectrons and Ga_{LMM} and Ga_{LMM} Auger electrons, respectively. These peak intensities decreased continuously as the InN deposition proceeded. Meanwhile, new peaks at 452, 444.4 and 396.6 eV for $In_{3d-3/2}$, $In_{3d-5/2}$, and N_{1S} photoelectrons increased and dominated the spectra after a deposition of ≥6 h.

Figure 2 compares the $In_{3d-5/2}$ XPS signal intensity increase with the length of the deposition time for InN on Si(100) (a) and GaAs(110) (b) surfaces at 700 K. On Si(100), $In_{3d-5/2}$ XPS signal quickly reached and remained at a value corresponding to about 1 monolayer TMIn adsorbed on the surface (solid circles). No InN film formed on Si(100) without laser irradiation within 24 h. However, when a 308-nm photon beam was used, InN film built up on the surface after an induction time of about 9 h as indicated by the increase of the $In_{3d-5/2}$ signal. It then reached a saturation value because of the XPS detecting limit rather than a termination of the film growth [10]. On the other hand, InN film could be formed on GaAs(110) surface with a shorter induction time and without laser assistance.

Figure 3 details the $Ga_{2p-3/2}$ (a), $As_{2p-3/2}$ (b), $In_{3d-5/2}$ (c) and N_{1S} (d) XP spectra by scanning over smaller energy ranges after InN deposition of 0, 1.5, 6 and 18 h. For the clean surface, $Ga_{2p-3/2}$ and $As_{2p-3/2}$ XPS showed peaks at 1116.6 and 1322.0 eV, respectively. These peaks could be fitted well with a single mixed Gaussian-Lorentz distribution. After deposition for 1.5 h, these peak intensities attenuated by ~60% due to the suppressing effect of the InN overlayers. In addition, a new peak at 1324.3 eV appeared in the $As_{2p-3/2}$ XP spectrum. This broad peak can be further deconvoluted into two components at 1323.8 and 1324.6 eV attributable to AsN and AsN_2, respectively [13].

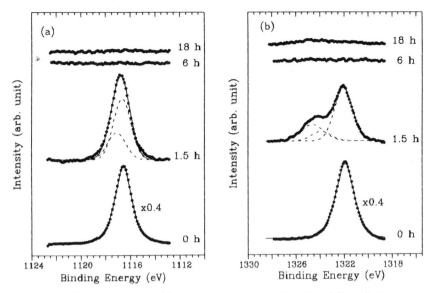

Fig. 3 (a) Ga$_{2p-3/2}$ and (b) As$_{2p-3/2}$ XP spectra taken from InN on GaAs at the indicated deposition times with Mg and A

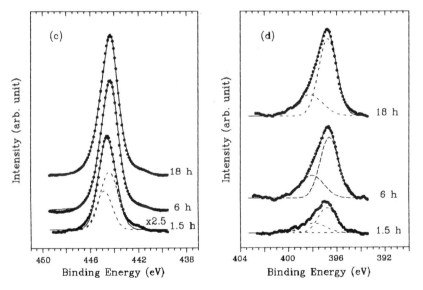

Fig. 3 (c) In$_{3d-5/2}$ and (d) N$_{1s}$ XP spectra taken from InN on GaAs at the indicated deposition times using Mg X-ray source.

The $Ga_{2p-3/2}$, XPS peak became broader, shifted to 1116.9 and could be deconvoluted into two components at 1116.6 and 1117.1 eV due to substrate and nitride Ga atoms, respectively [13, 14]. Meanwhile, the $In_{3d-5/2}$, and N_{1S} photoelectrons presented peaks at 444.6 and 399.9 eV, respectively. The $In_{3d-5/2}$ peak could be deconvoluted into two peaks at 444.4 and 444.7 eV, which are attributable to InN and InN with N atom bonded to the surface atoms, respectively [8, 10]. The N_{1S} XP spectrum could also be fitted with two Gaussian distributions peaking at 396.8 and 398.1 eV. The former peak should have contributions from N atoms bonded to Ga (396.6) and In (396.6) or the combination of them. The 398.1 eV peak is due to the N atoms bonding to As and NH species likely in the form of Ga-NH-As. Both N_{1S} and $In_{3d-5/2}$ peak intensities increased with increasing of the InN deposition time. At 6 h, the 444.4 eV peak dominated $In_{3d-5/2}$ XP spectrum because of the further formation of InN on the surface. The accompanying N_{1S}, XP spectrum also showed a major peak at 396.6 eV due to the InN species, while the 398.1 eV component shifted to 398.2 eV. The latter peak was attributed to the NH species on InN surface, since the substrate XP signals vanished already under the same experimental conditions (see Figs. 1, 3a and 3b). Further deposition caused no obvious XP spectra changes due to the XPS detecting limit.

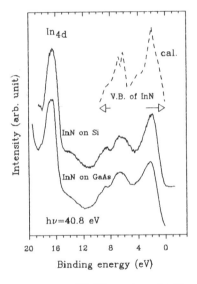

Fig. 4 He(II) UP spectra of InN on GaAs and Si (dashed line is a pseudo-potential calculation)

Fig. 5 SEM of InN on GaAs(top). and Si(bottom) substrates.

Figure 4 shows He(II) UP spectra for InN film on GaAs(110) after 18 h of deposition and on Si(100) after 24 h deposition with 308-nm laser irradiation. For comparison, a pseudo-potential calculation for InN valence band by Foley and Tansley is also plotted in the figure as the dashed line. The 16.6 eV peak due to In_{4d} photoelectrons was not given in the calculated curve.

Figure 5 compares SEM pictures taken from InN films on GaAs (a) and Si (b). The samples used here are the same as those for Fig. 4. In contrast to the case of InN on Si(100), where islands of about 100 nm in diameter were observed, InN film on GaAs(110) showed a much smoother surface with no grains being observed at a magnification of 100,000. The particle in the picture, likely a speck of dust, was chosen for the sake of focus.

DISCUSSION

XPS

The growth of InN on GaAs(110) at 700 K has been monitored using XPS technique and the formation of surface nitrides at the initial deposition process was clearly indicated in the XP spectra as shown in Fig. 3. After a deposition time of 1.5 h, $Ga_{2p-3/2}$ XPS intensity decreased by ~60% and the peak could be deconvoluted into two components at 1116.6 and 1117.1 eV due to Ga atoms in substrate and those bonded to N atoms, respectively. The 1117.1 eV peak is about 0.7 eV lower than the value of 1117.8 observed for GaN [13,14], likely because of the fact that the N atom is bonded to an In atom simultaneously. Similar changes were noted in the $As_{2p-3/2}$ XP spectrum. In addition to the substrate As peak at 1322.0 eV, two peaks at 1323.8 and 1324.6 eV appeared in the spectrum. These peaks have also slightly lower binding energies than those of 1324.0 and 1325.4 eV observed for As nitrides without In atoms [13]; this is consistent with the argument that the N atom is bonded to an In atom as well. Further supporting evidence is that the $In_{3d-5/2}$ peak could be deconvoluted into two components at 444.4 and 444.7 eV. The former peak is due to In in InN, while the latter is due to the InN in which the N atom is back-bonded to surface As and/or Ga atoms. The corresponding N_{1S} XPS result is less conclusive, because the binding energies for GaN, InN and AsN are too close to be separated in the present study. Nevertheless, the peak at 396.8 eV, which is higher than the value of 396.6 for In and Ga nitrides, could be partially attributed to the N atoms bonded to In atoms also.

Further deposition of InN caused the increase of InN signals, i.e., the 396.6 and 444.4 eV peaks for N_{1S} and $In_{3d-5/2}$, respectively, and the decrease of all other XPS signals. Meanwhile, the In:N atomic ratio also increased because of the reduced contribution from other nitrides (GaN and AsN for example) to the N_{1S} XP spectrum. After about 6 h of deposition, the XPS signals from the substrate disappeared and N_{1S} as well as $In_{3d-5/2}$ signals reached their saturating values due to the limited electron escaping length for these photoelectrons (~20 Å). In addition, the In:N atomic ratio is now about 0.9, if the photo-emission cross-section for N_{1S} and $In_{3d-5/2}$ is taken into account [15]. This value becomes 1.1, if we exclude the 398.2 eV component for the NH species, which was also observed in the deposition process of InN on Si(100), according to the N_{1S} XP spectrum. The In:N atomic ratio of our deposited film therefore averaged to be 1.0±0.1, indicating a stoichiometric or nearly stoichiometric growth of InN film under the indicated experimental conditions. Further growth of InN cannot be characterized by the XPS method.

Another interesting and important observation is the absence of an obvious C XPS signal, which suggests that C contamination is not as serious as that implied by the thermal decomposition study of TMIn on Si substrates. We found that partial CH_x species remained on the surface at temperatures higher than the In atom desorption temperature [12]. Similar observations were reported for TMAl and TMGa on Si substrates [16,17]. Clearly, the reaction mechanisms are different in the thermal decomposition of TMIn on Si from that in the InN film deposition process. In the latter case, the formation of the InN bond may facilitate the release of CH_3 radicals.

UPS

The InN films were also characterized with the UPS technique. Representative spectra for InN on both GaAs(110) and Si(100) are shown in Fig. 4. The two spectra are nearly identical, indicating that the substrate contribution to the spectra is negligible. This observation is consistent with those observed from XPS measurements, where no substrate signals could be detected either. The He(II) UP spectra also compare favorably with the result of a pseudo-potential calculation for InN valence band [18] (dashed line in Fig. 4). A peak at ~2.1 eV appeared in all spectra, however, the doublet at 6.5 and 7.1 eV was not resolved in our UP spectra. Instead, a single broad peak centered at ~6.6 eV was observed and the shoulder at ~8.5 eV was more pronounced. Accordingly, the overall valence band widths of our UP spectra are greater than that gleaned from the calculation. The most intense peak at ~16.6 eV is originated from In_{4d} electrons, which was not presented in the theoretical calculation but was observed in the UP spectrum for TMIn adsorbed on Si substrates [12].

CONCLUSION

The deposition of InN thin films on GaAs(110) using HN_3 and TMIn has been studied with XPS, UPS and SEM. In contrast to InN on Si(100), InN films could be formed on GaAs(110) at 700 K without 308-nm laser irradiation, which is necessary for the growth of InN film on Si(100). Under the present experimental conditions, $In_{3d-5/2}$ XPS signal reached a saturation value after ~6 h of deposition due to the XPS detecting limit rather than the termination of film growth. Meanwhile, the In:N atomic ratio increased to 1.0±0.1 and remained unchanged with further film deposition, indicating a stoichiometric or nearly stoichiometric growth of InN on GaAs(110). Similar results were found for the deposition of InN on Si(100) with 308-nm laser assistance. The He(II) UP spectra taken from InN films on GaAs(110) and Si(100) are nearly identical and agree reasonably well with a pseudo-potential calculation for the InN valence band. The corresponding SEM analysis showed smooth InN films on GaAs(110), however, grains with diameter of ~100 nm were observed for InN on Si(100) at a magnification of ≥20,000 X.

The fact that InN film can be deposited on GaAs(110) using TMIn and HN_3 beams, but not on Si(100) without laser irradiation may be due to the following factors which inhibit the growth of InN on Si(100): 1. NH species was found to be stable up to 800 K on Si substrate; 2. The different chemical properties of Si and InN; and 3. The lattice mismatching for Si(100)-2x1 and InN as listed below:

Si(100)-2x1 a_1=2.3Å, a_2=5.4Å   GaAs(110)   a=3.997Å   InN   a=3.548Å
              b=3.8Å   b=5.66Å   c=5.76Å
Si(100)   a=b=3.8Å

ACKNOWLEDGMENTS

The authors gratefully acknowledge the support of this work by the Office of Naval Research. We also thank Mr. L. Ma for his preliminary study on the deposition of InN on GaAs substrates.

REFERENCES

1. J.B. MacChesney, P.M. Bridenbaugh and P.B. O'Connor; Mater. Res. Bull., **5** 783 (1970).
2. A. Wakahara, T. Tsuchiva and A. Yoshida; Vacuum **41** 1071 (1990).
3. T. Matsuoka, T. Hanaka, T. Sasaki and A. Katsui; Inst. Phys. Conf. Ser. No. 106: Chapter 3, 141 (1989); Optoelectronics-Devices and Technologies **V5**, 53 (1990).
4. T.L. Tansley and C.P. Foley; J. Appl. Phys. **59** 3241 (1986).
5. B.R. Natarajan, A.H. Eltonkhy, J.E. Green and T.L. Barr; Thin Solid Films 69 201 (1980).
6. H.J. Hovel and J.J. Cuomo; Appl. Phys. Lett. **20** 71 (1972).
7. J.W. Trainor and K. Rose; J. Electron. Mater. **3** 821 (1974).
8. I. Bello, W.M. Lau, R.P.W. Lawson and K.K. Foo; J. Vac. Sci. Technol. **A10** 1642 (1992).
9. Y. Bu, D.W. Shinn and M.C. Lin; Surf. Sci. **276** 184 (1992).
10. Y. Bu, L. Ma and M.C. Lin; J. Vac. Sci. Technol. **A11**, 2931 (1993).
11. J.C.S. Chu, Y. Bu and M.C. Lin; Surf. Sci. **284** 281 (1993).
12. Y. Bu, J.C.S. Chu and M.C. Lin; Mater. Lett. **14** 207 (1992).
13. Y. Bu and M.C. Lin; in preparation.
14. X.-Y. Zhu, M. Wolf, T. Huett, and J.M. White, J. Chem. Phys. **97**, 5856 (1992)
15. H. Berthou and C.K. Jorgensen; Analytical Chem. **47**, 482 (1975).
16. T.R. Gow, R. Lin, L.A. Cadwell, F. Lee, A.L. Backman and R.I. Masel; Chem. Mater. **1** 406 (1989).
17. F. Lee, A.L. Backman, R. Lin, T.R. Gow and R.I. Masel; Surf. Sci. **216** (1989) 173.
18. C.P. Foley and T.L. Tansley; Phys. Rev. **B33** 1430 (1986).

PART II

Ferroelectric Materials

EPITAXIAL GROWTH of $Ba_{1-x}Sr_xTiO_3$ THIN FILMS ON $YBa_2Cu_3O_{7-x}$ ELECTRODE BY PE-MOCVD

C.S. CHERN*, S. LIANG*, Z.Q. SHI* S. Yoon**, A. Safari** P. Lu***, and B.H. Kear***
*EMCORE Corporation, 35 Elizabeth Ave, Somerset, NJ 08873
**Dept. of Ceramic Engineering, Rutgers University, Piscataway, NJ 08854
***Dept. of Materials Science & Engineering, Rutgers University., Piscataway, NJ 08854

ABSTRACT

Plasma-enhanced metalorganic chemical vapor deposition (PE-MOCVD) has been successfully employed for the deposition of (100) oriented barium strontium titanate (BST) thin films on a variety of substrate and electrode materials. The incorporation of O_2 plasma, which was used as oxidation reactant, has helped to reduce the required temperature for deposition of high-quality STO and BST thin films. This low temperature processing may make it possible to integrate BST on Si and GaAs. BST films with low leakage current densities of about 10^{-7} A/cm^2 at 2-volt (about 10^5 V/cm) operation were obtained from PE-MOCVD processing. Moreover, the BST results of capacitance-temperature (C-T) measurements show that most of the PE-MOCVD BST films have Curie temperatures of about 30-35°C and a peak dielectric constant of 600-800 at zero bias voltage. The sharp transition in the C-T data indicates that the BST films may have a high induced pyroelectric coefficient at room temperature, which is highly desirable for uncooled IR imaging arrays. The x-ray diffraction and Rutherford backscattering spectrometry results show that the BST film composition reproducibility was well controlled at around $Ba_{0.75}Sr_{0.25}TiO_3$ with a 4% variation. Device quality BST thin films with the thickness of 1000-2000 Å were produced. These results indicate that PE-MOCVD has high potential to be further developed and promoted as a production deposition technique providing high permittivity dielectric thin films for microelectronics and IR sensor industries.

I. INTRODUCTION

Thin film perovskite oxide materials have great potential for many device applications. High-quality epitaxial $YBa_2Cu_3O_{7-x}$ (YBCO) superconducting thin films are considered as a material applicable in IR bolometers, high-frequency antennas and receivers, multichip modules, and interconnects for integrated circuits. Single crystalline $BaTiO_3$ (BTO) has the highest electro-optical coefficient [1], which is crucial for fabrication of devices such as light modulators and optical switches, of all the other materials. $SrTiO_3$ (STO) which is paraelectric at room temperature and have dielectric constant as high as 320, is suitable for fabrication of the storage capacitor for high-density memory device and high resolution read-out circuits. $Ba_{1-x}Sr_xTiO_3$ (BST), which is a solid solution of BTO and STO and has a dielectric constants as high as 10,000 in bulk [2], is a candidate materials for high-density capacitors integrated on dynamic random access memories (DRAM's) and uncooled IR sensing and imaging devices [3]. Sr concentration in the BST materials determines the Curie temperature, at which the dielectric constant exhibit a peak maximum. Therefore, devices using BST as a high permittivity dielectric can be operated at variable ambient temperatures by precisely controlling the Sr concentration. Uniaxially oriented crystalline BST with appropriate Sr concentration will be an excellent candidate materials for this class of application. Moreover, low driving power and integration compatibility with other circuits are always desirable for potential applications. It is thus clear that epitaxial or highly oriented crystalline thin films of perovskite oxides possess a variety of potential uses in sensor, computer, microelectronics, and telecommunication device industries. However, lack of a mature large-scale fabrication technique for high-quality crystalline perovskite thin films hinders the advancement and development of the device technologies using this types of materials. Large area fabrication is a key issue for advancement and development of the 'perovskite technology'. Metalorganic chemical vapor deposition (MOCVD) and plasma-enhanced metalorganic chemical vapor deposition (PE-MOCVD) has been employed to deposit high quality perovskite YBCO thin films on large area substrates [4-6]. This paper reported that PE-MOCVD has been successfully employed to deposit device quality BST thin films.

II. EXPERIMENTAL

We have successfully deposited $Ba_{1-x}Sr_xTiO_3$ (BST) thin films in an EMCORE oxide PE-MOCVD system. The system, which incorporates a 300 Watt microwave power supply, can perform both thermal, and PE-MOCVD deposition processes, as shown in Figure 1. β-diketonate complex of $Ba(dpm)_2$, $Sr(dpm)_2$, and titanium isopropoxide (TIP) were used as barium, strontium, and titanium sources for the PE-MOCVD process. The evaporation temperature for $Ba(dpm)_2$, $Sr(dpm)_2$, and TIP were maintained at 225 °C, 220°C and 23.5 °C, respectively. Due to the high vapor pressure of TIP at 23.5°C, the TIP bubbler pressure was kept at 450 Torr to provide stabilized TIP transport flux and to match that of $Ba(dpm)_2$ and that of $Sr(dpm)_2$. Argon was used as the process carrier gas at a flow rate of 100 sccm for $Ba(dpm)_2$, 100 sccm for $Sr(dpm)_2$, and 70 sccm for TIP, respectively. Ultra high purity (99.995%) oxygen was used for the growth with a flow rate of 300 sccm. Reactor chamber pressure during the deposition was 10 Torr with a nominal oxygen partial pressure of about 5 Torr. A microwave oxygen plasma (2.45GHz, 300 Watt) was used in order to lower the substrate temperature required for epitaxial growth of $Ba_{1-x}Sr_xTiO_3$ thin films. Epitaxial $Ba_{1-x}Sr_xTiO_3$ films were achieved at a substrate temperature of 680 °C with a growth rate of approximately 1000 Å/hr in this study.

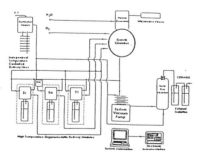

Fig. 1 PE-MOCVD system schematic

III. RESULTS

III.1 Electrical Properties of BST Films

Barium strontium titanate (BST) films were epitaxially grown on YBCO/LaAlO$_3$ and Pt/MgO substrate using PE-MOCVD. The substrate temperature was controlled at 650-700°C and the deposition process was described as in section II. The film thickness was determined by a Tencor alpha-step profiler and cross-section SEM. To measure the electrical and dielectric properties of the thin films, a Au or Pt pad with a diameter of 0.6 mm was sputtered on the top of the BST film to form a metal-insulator-metal (MIM) capacitor.

The current-voltage (I-V) characteristics of the capacitors were measured at room temperature using a Keithley model 614 electrometer and a 230 programmable voltage source. The dc leakage current density of the thin film was determined from I-V curve at a voltage of +2.0V. The capacitance-voltage (C-V) and capacitance-frequency (C-f) characteristics of the capacitor were obtained using an HP 4194A Impedance /Gain-Phase Analyzer in the frequency range of 10 Hz to 10 MHz. The relative permittivity, ε_r, of the film was calculated from the C-V curve and by knowing the film thickness as well as the capacitor area. The capacitance-temperature (C-T) and dissipation factor-temperature (tanδ-T) measurements at frequencies of 1, 10 and 100 KHz were taken by using HP 4194A Impedance /Gain-Phase Analyzer in the temperature range of -80 to 100°C. The Curie temperature of the BST films could be determined from the C-T measurement, as shown in Fig. 2. The frequency dependence of the relative permittivity and dissipation factor of the BST films were measured at room temperature from 10 Hz to 10 MHz. Figure 3 shows the current-voltage (I-V) characteristics of the BST film with a thickness of 1500Å. The dc leakage current density is about 1×10^{-8} A/cm^2 at 5×10^4 V/cm and lower than 1×10^{-7} A/cm^2 at 10^5 V/cm. This indicates that the films have a very good insulating property and can meet the requirement for device applications. The frequency dependence of the relative permittivity (ε_r) exhibits a monotonic decrease with increasing frequency. The ε_r value varies from 450 to 400 with increasing frequency from 100 Hz to 100 KHz. The tanδ, however, shows a minimum value of 4.8% at a frequency of around 10 KHz. At frequency higher than 10 KHz, tanδ increased with increasing frequency. This phenomena can be explained by invoking a loss mechanism from the contact electrode or by the low electrical conductivity of YBCO electrode at room temperature. The results of I-V and C-f measurements confirm that the PE-MOCVD produced BST films may be made into thin film capacitors with high capacitance. Generally, low dielectric loss, low leakage current, and insignificant frequency dependence on capacitance are highly desirable for most device applications. A new electrode system or an improved growth process can further advance the fabrication of BST MIM capacitors.

To fabricate IR detector elements from BST, the temperature dependence of the variation of capacitance is a very important characteristic for this particular application. The most interesting result of the BST film is its temperature dependent ε as shown in Figure 2. Capacitance-temperature (C-T) measurements were taken at different frequencies of 10^3, 10^4, and 10^5 Hz with zero bias voltage. The temperature range for this C-T test is from -50 to 100 °C. As the temperature increases from -50 °C to 20°C, the dielectric constant of the film increases slowly from 360 to 470; as temperature increases from 25 °C to 35 °C, the dielectric constant increases drastically from 470 to 590. Very sharp transitions were observed at same temperature for the other two different frequencies. This indicates that the deposited film has a ferroelectric behavior and has a phase transition near room temperature. At transition temperature, relative permittivity of the BST film exhibits a 100% increase. The results of C-T measurements provide the evidence that the desired Ba-Sr-Ti-O composition was achieved.

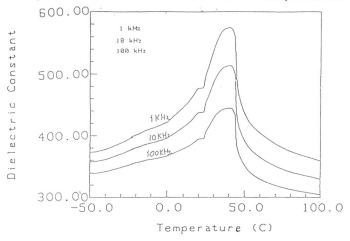

Fig. 2: Temperature dependence of the capacitance (C-T measurement) of a BST film deposited on YBCO/LaAlO$_3$. The Curie temperature of the film is about 30°C.

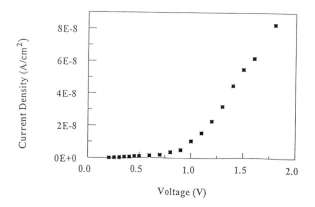

Fig. 3 I(J)-V characteristic of a BST film deposited by PE-MOCVD

III.2 Composition and Structural Characterizations:

The crystal structure of the films was characterized by X-ray diffraction (XRD) using a Simens-500 diffractometer with Cu Kα radiation. The surface morphology of the films was investigated by scanning electron microscopy (SEM). The composition of the films was determined by EDS and RBS. In order to confirm the formation of BST phase with desired texture and composition, XRD θ–2θ scan was used to determine the crystal phase, preferred orientation, and lattice constant. From the XRD θ–2θ pattern in Fig. 4, it was confirmed that the BST films were (001) oriented with lattice constants of 3.975-3.98Å. From the x-ray data and lattice constant calculation, Sr substitutes for 20-25% of the Ba in the BaTiO$_3$ lattice to form BaTiO$_3$-SrTiO$_3$ solid solution. Fig. 5 displays the XRD ϕ-scan patterns of BST (202), YBCO (108), and LaAlO$_3$ (202) reflections. All of these ϕ-scan patterns exhibit 90° modulation patterns, i.e., peaks occurring every 90°, indicating that the YBCO and BST films do not have in-plane misorientations. These results indicate that BST films are epitaxially grown on the YBCO electrode (template) layer.

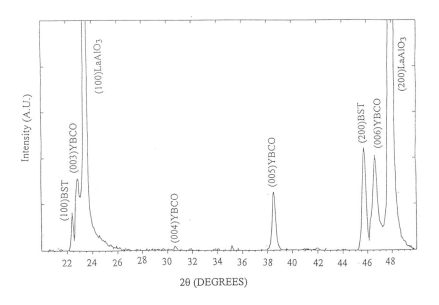

Fig. 4 XRD θ–2θ scan pattern of a BST film deposited on YBCO/LaAlO$_3$ by PE-MOCVD.

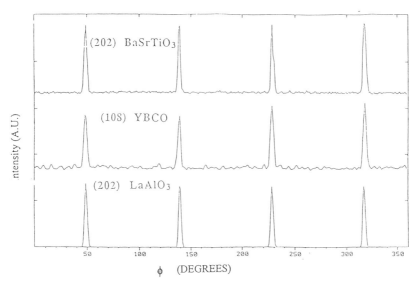

Fig. 5 XRD φ-scan patterns of (202) $LaAlO_3$, (108) YBCO, and (202) BST.

Summary

PE-MOCVD has been successfully used for deposition of epitaxial $Ba_{1-x}Sr_xTiO_3$ (BST) thin films with Sr concentration, x, of 0.2-0.25. All of such BST films exhibit sharp Curie transitions at a temperature of 30-35°C, which indicate that the epitaxial BST thin films may be suitable for uncooled IR focal plane arrays. The BST lattice constant, which is calculated from XRD data, is 3.975-3.98 Å and consisted with the Sr concentration in the BST films. Electrode induced stress effect on Curie transition of the BST films will be further investigated in the future.

References

1. P. Güenter, Ferroelectrics, **24**, 35 (1980).
2. E.N. Bunting, G.R. Shelton, and A.S. Creamer, J. Res. Nat. Bur. Std. **38**, 337 (1947).
3. R. Watton, Ferroelectrics, **92**, 87 (1989).
4. H. Busch, A. Fink, and A. Müller, J. Appl. Phys. **70**, 2449 (1991).
5. R. Hiskes, S.A. Dicarolis, R.D. Jacowitz, Z. Lu, R.S. Feigelson, R.K. Route, and J.L. Young, Int'l. Conf. Crystal Growth 10 (San Diego, CA, August 1992).
6. C.S. Chern, J.S. Martens, Y.Q. Li, B.M. Gallois, P. Lu, and B.H. Kear, Supercon. Sci. Technol. **6**, 460 (1993).

LOW TEMPERATURE MOCVD OF ORIENTED PbTiO$_3$ FILMS ON Si(100)

H. CHEN*, B.M. YEN*, G.R. BAI*, D. LIU*, AND H.L.M. CHANG**
* Department of Material Science and Engineering, University of Illinois at Urbana-Champaign, Urbana, IL 61801 USA
** Material Science Division, Argonne National Laboratory, Argonne, IL 60439 USA

ABSTRACT

Highly oriented PbTiO$_3$ (PT) thin films have been successfully grown on Si(100) using MOCVD technique at as low as 450°C. Titanium isopropoxide, Ti(C$_3$H$_7$O)$_4$, tetraethyl lead, Pb(C$_2$H$_5$)$_4$, and pure oxygen were chosen as precursor materials in this work. The resulting film chemistry and texture were found to be strongly dependent on Pb/Ti source flow ratio and growth temperature.

INTRODUCTION

Lead titanate (PbTiO$_3$) is a perovskite type ferroelectric material having attractive ferroelectric, piezoelectric, and pyroelectric properties. Large pyroelectric coefficient and electromechanical coupling factor have made this material very useful for such applications as pyroelectric sensor, high frequency ultrasonic device, infrared detector, piezoelectric transducer, and nonvolatile memory[1-3]. Thin film type of this material has attracted great attentions recently because of the low operation voltage, large in area, and its ability to integrate with semiconductor VLSI technology. Various techniques, including rf sputtering[4], chemical vapour deposition (CVD)[5], sol-gel process[6], laser ablation[7], metal organic decomposition (MOD)[8], and metal organic chemical vapour deposition (MOCVD)[9-10], etc., have been attempted to grow PbTiO$_3$ films on different substrates. The MOCVD method have been proved to have high growth rate and good uniformity, thus has higher potential for industrial use.

Several studies of highly oriented PbTiO$_3$ films have been reported[1-2, 6-10]. It was found that oriented films exhibit improved piezoelectric and pyroelectric properties, resulting better performances for device applications. In this study, the PT films have been grown by MOCVD at temperatures ranged from 425°C-700°C on Si(100) substrate. Highly oriented PbTiO$_3$ polycrystalline films have been successfully obtained under certain Pb/Ti source flow ratio and substrate temperature. The phases and texture of the grown PT films, as well as microstructure and chemistry are described.

EXPERIMENTAL

The film deposition was carried out in a low pressure, horizontal, cold wall reactor with a resistive substrate heater. Titanium isopropoxide, Ti(C$_3$H$_7$O)$_4$, and tetraethyl lead, Pb(C$_2$H$_5$)$_4$,

Table I. Summary of growth condition, phases, texture, and chemistry of Pb-Ti films

Specimen ID	Temperature (°C)	Ti Source (sccm)	Pb Sourc (sccm)	Phases and Structure	Atomic Ratio (Pb:Ti:O)
0215a	600	50	50	$PbTiO_3$(MO)	1.03:1:2.61
0216b	500	50	50	$PbTiO_3$(MO)	--
0223a	475	50	50	$PbTi_3O_7$(A)	--
0224	475	40	50	$PbTiO_3$(SO)+$PbTi_3O_7$(A)	0.77:1:2.60
0216a	450	50	50	$PbTi_3O_7$(A)	--
0224b	450	40	50	$PbTi_3O_7$(A)	--
0303	450	30	50	$PbTi_3O_7$(A)	0.75:1:2.77
0304	450	20	50	$PbTiO_3$(SO)	1.02:1:2.82
0223b	425	50	50	Amorphous	--
0308	425	20	50	Amorphous	0.98:1:2.14

A: Amorphous, MO: Multi-oriented, SO: Single-oriented.

were used as the metal-organic precursors. The mixture of the organometallic precursor vapor was introduced into the reactor via high purity nitrogen carrier gas. The flow rate of the carrier gas and the metal-organic source temperature for each of the precursor chambers were controlled individually to adjust the film composition and the growth rate. Pure oxygen was used as the oxidant and introduced into the reactor in a separate line. The precursor delivery lines, as well as the inlet flange, were heated and maintained at temperatures higher than the highest source evaporation temperature to prevent condensation of the vapor phase. Polished Si(100) wafers were used as substrates. Detailed growth conditions for the preparation of a series of Pb-Ti films on silicon substrate are summarized in Table I.

The compositions of MOCVD films were determined by means of energy dispersive analysis of x-ray (EDAX) and Auger electron spectroscopy (AES). The depth profiles of the grown films were also analyzed using AES. The film structure and crystallinity were characterized by x-ray diffraction (XRD) and, in some cases, by selected area electron diffraction (SAED) as well. The microstructure of grown films were examined using transmission electron microscopy (TEM), scanning electron microscopy (SEM), and scanning transmission electron microscopy (STEM).

RESULTS AND DISCUSSION
Structure and Chemistry of Pb-Ti Films

X-ray diffraction scans were carried out on each specimen to determine the orientation change of the films with the growth temperature. Results are tabulated in Table I. It was found that the structure and phases of grown Pb-Ti films are highly dependent on growth temperature and Pb/Ti precursor gas flow ratio. With proper growth condition, a stoichiometric $PbTiO_3$ film with highly preferred orientation can be formed. Grain boundaries are clearly visible for polycrystalline films using SEM technique. In contrast, the amorphous films exhibit featureless surface morphology.

Several samples were selected for AES analysis. Besides the existence of Pb, Ti, and O, a significant amount of carbon(C) was found on all inspected sample surfaces. Small concentration of sulfur(S) also existed on the surfaces of samples 0215a and 0308; and Cl was found on both 0304 and 0308 samples. However, all these impurities disappeared after one minute Ar ion sputtering. The depth profile of sample 0303 is shown in Fig.1. The atomic ratios of Pb, Ti, and O obtained from these profiles after 15min sputtering, where the film composition is stable, are listed in Table I. For samples 0215a and 0304, a higher Pb/Ti ratio (1.03) was found. This result is consistent with the XRD data, where relatively strong peak intensity indicates the presence of a significant percentage of oriented $PbTiO_3$ phase. In contrast, the XRD data of sample 0224a showed a relatively lower (001) peak intensity along with an amorphous bump around 28°/2θ, in the mean time, a lower Pb/Ti ratio (0.77) was measured. Therefore, this film is suspected to be composed of both $PbTiO_3$ and $PbTi_3O_7$ phase. A lower Pb/Ti ratio (0.75) was measured in sample 0303, which corresponds to mostly $PbTi_3O_7$ phase. Finally, for sample 0308 a higher Pb/Ti ratio (0.98) was observed. However, XRD didn't show the existence of any crystalline phase. No reaction seemed to have occured between Pb and Ti at this low deposition temperature (425°C).

Effect of Growth Temperature on the Texture and Microstructure of Pb-Ti Films

In order to investigate how the growth temperature affects the orientation of the deposited films, several $PbTiO_3$ films with almost the same Pb/Ti ratio close to 1:1 were prepared at different temperatures. The PT film prepared at 450°C showed an extraordinarily high fiber texture with [100] direction perpendicular to the substrate surface. The degree of texture in the growth direction of this film was examined by an ω scan using x-ray diffraction (Fig.2). The full width at half maximum (FWHM) of the rocking curve is about 4°.

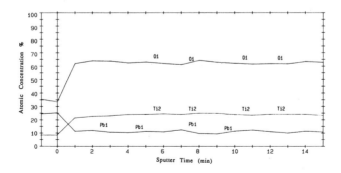

Figure 1. AES depth profile of sample 0303, showing a stable composition after 1min Ar sputtering.

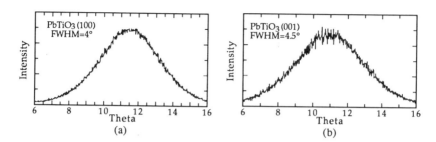

Figure 2. Rocking Curves of sample 0303, corresponding to (a) PbTiO$_3$(100) and (b) PbTiO$_3$ (001), respectively.

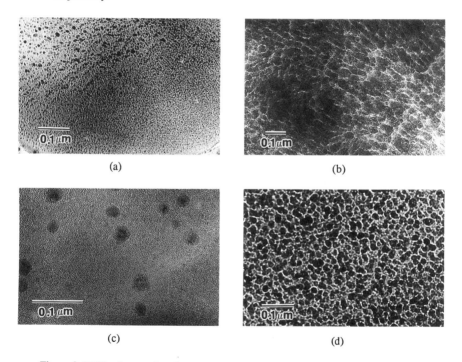

Figure 3. TEM micrographs of samples (a) 0215a, (b) 0224a, (c) 0303, and (d) 0304.

Fig.3 gives the TEM planar-view images of four selected samples (0215a, 0224a, 0303, and 0304). The general morphologies of three polycrystalline films (0215a, 0224a, and 0304) are similar. There are many fine crystalline particles (20-50nm) clustered within each grain, resulting in only ring powder diffraction patterns as observed by SAED. Comparing the images of these films, it can be found that grain size and the particle size decrease as substrate temperature decreases.

The reason why the PbTiO3 films, deposited on silicon substrate at low temperatures (≤475°C), showed a single preferred orientation in contrast to those fabricated at high temperatures (≥500°C) may be explained as follows. Because of the oxidization enviroment in the MOCVD reactor, a native silicon dioxide layer on silicon substrate surface was inevitable. Because the native silicon dioxide layer is amorphous, the nucleation centers for PT grain growth are expected to be random, irrespective of the growth temperature. However, the growth rate is anisotropic. At high temperatures above the Curie point ($T_c \approx 490°C$), the adatoms have sufficient energy to migrate and diffuse on the substrate surface so that the nucleation centers with various orientations can all grow. Eventually, the resulted film is polycrystalline with multiple preferred orientations. On the other hand, at lower growth temperatures, the adatoms are less mobile, thereby resulting in suppression of growth in some directions. We have previously observed that for the phases with a tetragonal structure, such as TiO_2, PbO, and $Y_1Ba_2Cu_3O_{7-x}$, the [100] direction has a more rapid growth rate than other directions. The present finding, i.e., a [100] oriented PT film on Si grown below T_c in the tetragonal phase, is consistent with the earlier results.

Effect of Pb/Ti Source Flow Ratio on the Structure and Phases of Pb-Ti Films

For the growth temperatures above 500°C, polycrystalline PT films were fabricated with Pb/Ti(gas flow)=1. In the intermediate temperature range (450°C-475°C), equal precursor gas ratio was unable to form the stoichiometric PbTiO3; an amorphous phase seemed to be dominant. The existence of excess Ti atom in these films, found by AES, implies that a higher Pb/Ti source flow ratio was necessary. At 475°C, a Pb/Ti ratio of 1.25 was required for the PbTiO3 formation while as high as 2.5 was needed at 450°C. No PbTiO3 phase has been produced below 450°C.

The effect of Pb/Ti flow ratio on the Pb-Ti films can be further understood by comparing x-ray diffraction scans taken from samples 0224b, 0303, and 0304 (Fig.4), for which all samples were grown at 450°C but with different Pb/Ti flow ratio. A gradual structural evolution is seen. The diffraction bump near 28°/2θ was observed in both 0224b and 0303 samples. This bump envelops several $PbTi_3O_7$ peaks, indicating that TiO_2 was able to react with PbO at 450°C. The higher bump seen in sample 0303 suggests that a stronger reaction occurred. Therefore, a further increase of Pb/Ti flow ratio resulted in the formation of stoichiometric PT films.

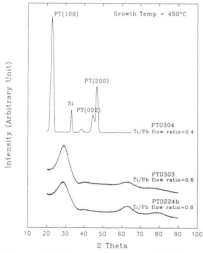

Figure 4. XRD traces of samples 0224b, 0303, and 0304, showing a gradual structural evolution.

This is evident for sample 0304 in Fig.4.

TEM micrograph, as shown in Fig.3(c), reveals no evidence of crystallinity for sample 0303, which is consistent with XRD result. However, there are some isolated dark spots (10-40nm) observed throughout the sample. STEM microprobe examination indicated a higher Pb concentration in dark spots as compared with that in the amorphous matrix, suggesting these spots are probably the nucleation sites for the crystallization of $PbTiO_3$ phase.

The amorphous phase was found, as stated above, when the film is richer in Ti at low growth temperatures ($\leq 475°C$). However, this is not the case for high temperature deposition. At high temperatures, both $PbTiO_3$ and TiO_2, instead of $PbTi_3O_7$, appeared, which suggests that $PbTi_3O_7$ is perhaps a metastable phase. It might decompose into a mixture of $PbTiO_3$ and TiO_2 at higher growth temperatures.

CONCLUSIONS

The final phases and crystallinity of MOCVD grown Pb-Ti films on Si(100) were found to be highly dependent on the Pb/Ti source flow ratio and deposition temperature. A increasing Pb/Ti flow ratio was required to obtain stoichiometric PT films as growth temperature decreased. With proper Pb/Ti ratio, stoichiometric $PbTiO_3$ films can be grown at as low as 450°C with O_2 as oxidant. The highly oriented PT films grown at low growth temperatures below the Curie point can be attributed to the anisotropic growth rates in the tetragonal phase.

ACKNOWLEDGEMENTS

This work is supported by the U.S. Department of Energy. The use of facilities at Material Science Laboratory, University of Illinois at Urbana-Champaign and Argonne National Laboratory are gratefully acknowledged.

REFERENCES

1. K.Iijima, Y.Tomita, R.Takayama, and I.Ueda, J. Appl. Phys. **60** [1] 361 (1986)
2. H.Takeuchi and K.Kushida, IEEE 7th Int. Symp. on Appl. of Ferroelectrics, 115 (1990)
3. M.Okuyama and Y.Hamakawa, Ferroelectrics, **63**, 243 (1985)
4. T.Ogawa, A.Senda, and T.Kasanami, Jpn. J. Appl. Phys. Suppl. 28-2, 11 (1989)
5. K.Y. Hsieh, L.L.H.King, S.Rou, and A.I.Kingon, Mat. & Manu. Proc. **7**[1] 101 (1992)
6. C.Chen and D.F.Ryder Jr., J. Am. Ceram. Soc. **72** [8] 1495 (1989)
7. S.G.Ghonge, E.Goo, and R.Ramesh, Appl. Phys. Lett. **62** [15] 1742 (1993)
8. B.W.Vest and J.Xu, Proc. of 6th IEEE Int. Symp. on Appl. of Ferroelectrics, 374 (1986)
9. G.R.Bai, H.L.M.Chang, H.K.Kim, C.M.Foster, and D.J. Lam, Appl. Phys. Lett. **61** [4] 408 (1992)
10. M.Okada, S.Takai, M.Amemiya, and K.Tominaga, Jpn. J. Appl. Phys. **28** [6] 1030 (1989)

PREPARATION OF $Ba_{1-x}Sr_xTiO_3$ THIN FILMS BY METALORGANIC CHEMICAL VAPOR DEPOSITION AND THEIR PROPERTIES

S.R. GILBERT*, B.W. WESSELS*, D.A. NEUMAYER**, T.J. MARKS**,
J.L. SCHINDLER***, AND C.R. KANNEWURF***
Northwestern University, Evanston, Illinois 60208
* Department of Materials Science and Engineering
** Department of Chemistry
*** Department of Electrical Engineering and Computer Science

ABSTRACT

$Ba_{1-x}Sr_xTiO_3$ thin films were deposited over the entire solid solution range by low pressure metal-organic chemical vapor deposition. The metal-organic precursors employed were titanium tetraisopropoxide and barium and strontium(hexafluoroacetylacetonate)$_2$•tetraglyme. The substrates used were $LaAlO_3$ and (100) p-type Si. $Ba_{1-x}Sr_xTiO_3$ films deposited on $LaAlO_3$ were epitaxial, while the films deposited on Si showed no texture. Auger spectroscopy indicated that single phase $Ba_{1-x}Sr_xTiO_3$ films did not contain detectable levels of fluorine contamination. The dielectric constant was found to depend upon the solid solution composition x, and values as large as 220 measured at a frequency of 1 MHz were obtained. The resistivities of the as-deposited films ranged from 10^3 to 10^8 Ω-cm. Temperature dependent resistivity measurements indicated the films were slightly oxygen deficient.

INTRODUCTION

The intense research devoted to perovskite oxide thin films in recent years has been motivated by their potential application in a wide variety of electrical and optical devices. Of special interest are the perovskite compounds $BaTiO_3$ and $SrTiO_3$, which form a continuous solid solution [1]. $BaTiO_3$ possesses a Curie temperature T_c of 393K and is ferroelectric at room temperature [2]. $SrTiO_3$, on the other hand, is paraelectric at room temperature and exhibits a T_c of about 33K [2]. The Curie temperature of the $Ba_{1-x}Sr_xTiO_3$ solid solution varies linearly with composition. Because many of the material properties depend strongly upon T_c, they can be tailored across a broad range. The dielectric constant in bulk ceramics, for example, exhibits an extremely large maximum at T_c, with dielectric constants as high as 22000 reported [3]. The solid solution composition corresponding to the room temperature phase boundary between the ferroelectric and paraelectric phases in bulk specimens is approximately x=0.30 [1]. For devices operating at or near 300K, it is this composition that holds the most technological importance. For many of these applications, thin films are required.

To date $Ba_{1-x}Sr_xTiO_3$ solid solution thin films have been prepared almost exclusively by RF sputtering [4,5] and pulsed laser deposition [6-9] techniques. A technique that shows considerable promise for the synthesis of high quality $BaTiO_3$ and $SrTiO_3$ thin films is metal-organic chemical vapor deposition (MOCVD) [10-15]. MOCVD is an attractive growth technique because of its high deposition rates, ease of compositional control, and ability to deposit high quality films with low defect densities. Recently we reported the deposition of $Ba_{1-x}Sr_xTiO_3$ polycrystalline and epitaxial thin films by MOCVD [16]. In the present study, the crystalline structure and the parameters influencing the phases stabilized and composition of the $Ba_{1-x}Sr_xTiO_3$ thin films were investigated. The electrical properties of the $Ba_{1-x}Sr_xTiO_3$ thin films were also characterized.

EXPERIMENTAL PROCEDURE

The $Ba_{1-x}Sr_xTiO_3$ thin films were deposited in a low pressure, two zone, horizontal quartz reactor. Details of the growth system have been described elsewhere [10]. The metal-organic precursors employed were liquid titanium tetraisopropoxide (TPT) and solid barium and strontium (hfa)$_2$•tetraglyme (hfa = hexafluoroacetylacetonate). The barium and strontium sources were placed in separate quartz boats, and heated by independently controlled resistive heating elements. A stainless steel bubbler heated by a recirculating constant temperature bath was used to contain the titanium source. The metal-organic sources were transported to the reaction zone by an argon carrier gas. Oxygen bubbled through deionized water served as the reactant gas. The total reactor pressure was controlled by the total set flow rate, and monitored using a Baratron gauge placed immediately downstream of the quartz reactor. The substrates were placed on a SiC-coated graphite susceptor located in the reaction zone, and were heated using a water cooled infrared lamp. The growth temperature was monitored using a chromel-alumel thermocouple placed inside a sealed quartz tube, upon which the susceptor was positioned. The substrates used were (100) p-type Si and (100) LaAlO$_3$. Deposition was carried out on both thermally oxidized and bare Si substrates. The thickness of the as-deposited films was on the order of 0.2 to 1.5µm. Typical growth conditions employed for this study are given in Table I.

Table I Typical growth conditions for the preparation of Ba$_{1-x}$Sr$_x$TiO$_3$ thin films.

Growth Temperature (°C)	800
Substrates	(100) p-type Si and (100) LaAlO$_3$
Total Pressure (Torr)	3.8 - 4.0
Temperature of Ba(hfa)$_2$•tetraglyme (°C)	110 - 120
Temperature of Sr(hfa)$_2$•tetraglyme (°C)	99 - 115
Temperature of TPT Bubbler (°C)	39-44
Temperature of Water Bubbler (°C)	17 - 18
Total Flow Rate (sccm)	120
Oxygen Flow Rate (sccm)	47 - 50

The phases stabilized and degree of texturing were examined with a Rigaku powder diffractometer using Cu Kα radiation. A double crystal diffractometer was employed to study the relative crystalline quality of epitaxial films from the full width at half maximum (FWHM) of the (200) rocking curves using monochromatic Cu Kα_1 radiation. In addition, the degree of in-plane epitaxy was examined using an x-ray diffractometer equipped with a four circle goniometer. The film thickness was measured with a Tencor Alphastep profilometer on an etched portion of the film. Both the (Ba+Sr):Ti ratio and the composition x of the films deposited on silicon were analyzed using energy dispersive x-ray spectroscopy (EDX). The solid solution composition x of the films deposited on LaAlO$_3$ was also calculated from the measured lattice parameter. Auger electron spectroscopy (AES) was used to detect the presence of lighter elements such as fluorine incorporated into the films from the metal-organic precursors during growth.

Metal-insulator-semiconductor (MIS) Ba$_{1-x}$Sr$_x$TiO$_3$/Si and Ba$_{1-x}$Sr$_x$TiO$_3$/SiO$_2$/Si diodes were fabricated for capacitance-voltage measurements. An evaporated Au dot served as the contact to the Ba$_{1-x}$Sr$_x$TiO$_3$ thin film, while evaporated Al served as the ohmic contact to the backside of the Si substrate. The dc electrical resistivity of the films deposited on LaAlO$_3$ was measured using a four probe technique.

RESULTS AND DISCUSSION

<u>Deposition Studies</u>

The phases stabilized in the as-deposited films depended on several parameters, including the (Ba+Sr):Ti reactant flow ratio, the oxygen partial pressure, and the deposition temperature. Figure 1 shows the x-ray diffraction patterns of films grown at different (Ba+Sr):Ti ratios at 800°C and an oxygen partial pressure of 1.6 Torr. For films grown under Ti-deficient conditions, the Ba$_{1-x}$Sr$_x$TiO$_3$ and (BaSr)F$_2$ phases were stabilized. Under near stoichiometric conditions, single phase Ba$_{1-x}$Sr$_x$TiO$_3$ was stabilized. For films grown under Ti-rich conditions, Ba$_{1-x}$Sr$_x$TiO$_3$, (BaSr)$_4$Ti$_{13}$O$_{30}$, and (BaSr)$_2$Ti$_4$O$_9$ were stabilized. Figure 2 shows the range of phases stabilized near 800°C as a function of the (Ba+Sr):Ti reactant flow ratio. The single phase films deposited on SiO$_2$/Si and bare Si substrates showed no texture, as is consistent with results obtained previously for BaTiO$_3$ and Ba$_{1-x}$Sr$_x$TiO$_3$ thin films deposited on (100) Si substrates by MOCVD [15,16], RF sputtering [4], and pulsed laser deposition [6,7]. The growth of textured films directly on the Si substrates is most likely hindered by film-substrate interactions and the presence of an amorphous SiO$_2$ layer [6].

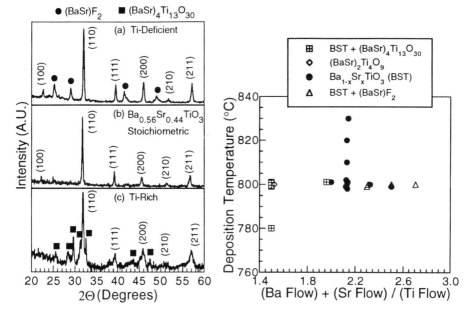

Figure 1. X-ray diffraction patterns of a) Ti-deficient, b) stoichiometric, and c) Ti-rich $Ba_{1-x}Sr_xTiO_3$ thin films deposited on (100) Si substrates at a growth temperature of 800°C and an oxygen partial pressure of 1.6 Torr.

Figure 2. The range of phases stabilized as a function of the (Ba+Sr):Ti reactant flow ratio near a growth temperature of 800°C and an oxygen partial pressure of 1.6 Torr.

The solid solution composition x of the single phase films could be controlled by adjusting the Ba and Sr source temperatures. Films with x varying between 0.0 and 1.0 could be deposited at temperatures of 800°C. For a constant Ba source temperature, the composition x increased nearly linearly with the Sr source temperature.

Single phase $Ba_{1-x}Sr_xTiO_3$ films deposited on $LaAlO_3$ using the growth conditions given in Table I were (h00) oriented with the a-axis perpendicular to the substrate, as indicated by the Θ–2Θ x-ray diffraction pattern in Figure 3. In order to determine if the films deposited on $LaAlO_3$ were epitaxial, an in-plane x-ray phi scan pattern of the film and substrate {220} planes was recorded. The results are shown in Figure 4 for a film with x=0.26. Only the four {220} reflections expected for an (h00) oriented film were present in the phi scan pattern, indicating that the film possessed the requisite four fold symmetry. In addition, the film and substrate {220} reflections were perfectly aligned. Both of these observations indicate that the $Ba_{1-x}Sr_xTiO_3$ films deposited on $LaAlO_3$ were epitaxial, with the thin film orientation following that of the single crystal substrate. In order to determine the degree of alignment perpendicular to the substrate, the x-ray rocking curves of the (200) $Ba_{1-x}Sr_xTiO_3$ reflections were measured using a double crystal diffractometer. The FWHM values obtained for the (200) reflections ranged from 0.47° to 1.90°, as compared to a value of approximately 0.10° for the $LaAlO_3$ substrate.

The extent of fluorine incorporation in the $Ba_{1-x}Sr_xTiO_3$ films was determined by Auger analysis. Prior to the measurement, the films were Ar ion sputtered for 20 minutes to remove any surface contamination. The Auger analysis indicated that fluorine was not present at detectable levels in single phase films. Multiphase films, on the other hand, were found to contain fluorine.

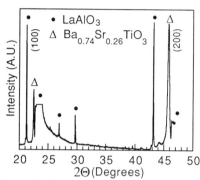
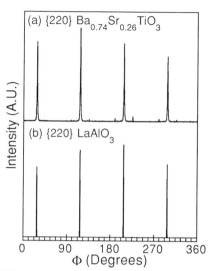

Figure 3. The x-ray diffraction pattern of a $Ba_{0.74}Sr_{0.26}TiO_3$ film deposited on (100) $LaAlO_3$ at 800°C with an oxygen partial pressure of 1.6 Torr.

Figure 4. The in-plane x-ray phi scan patterns of the {220} reflections of a) the $Ba_{0.74}Sr_{0.26}TiO_3$ film and b) the $LaAlO_3$ substrate.

The presence of fluorine is undesirable since it can substitute for oxygen, and act as an unintentional donor dopant.

Electrical Characterization

Electrical resistivity measurements of the $Ba_{1-x}Sr_xTiO_3$ films deposited on $LaAlO_3$ were performed using a four probe technique. The resistivities of the films deposited using a low purity Sr source ranged from 10^3 to 10^8 Ω-cm (Table II), as compared to a value of approximately 10^9 to 10^{10} Ω-cm for bulk, undoped $BaTiO_3$ and $SrTiO_3$. The lower resistivities of the films may arise from trace impurities introduced by the metal-organic precursors, such as aliovalent cation dopants or fluorine substitution at the oxygen sites. Previous work involving the deposition of $BaTiO_3$ thin films using the $Ba(hfa)_2$•tetraglyme precursors has indicated a strong dependence of the resistivity on the chemical purity of the metal-organic precursor, rather than on the presence of fluorine [16]. When the chemical purity of the metal-organic precursors was improved, the resistivity increased to 1×10^7 Ω-cm, which is still lower than that observed for the bulk.

Table II The room temperature electrical resistivity of $Ba_{1-x}Sr_xTiO_3$ films deposited on $LaAlO_3$.

Sample Number	Composition X	Resistivity (Ω-cm)
15	0.27	2.0×10^8
20	0.00	8.6×10^5
22	0.34	3.3×10^6
27	0.17	2.3×10^3
34	0.04	2.4×10^5
36	0.26	6.8×10^4
39	0.20	1.9×10^4

In order to determine the nature of the defect present, the resistivity was measured as a function of temperature for several films. The resistivity depended exponentially on the inverse of temperature. From this temperature dependence, the ionization energy of the dominant thermally activated donor was determined. The donor ionization energy for compositions between x=0 to x=0.20 ranged from 0.07 to 0.14eV. This is comparable to the activation energies previously reported for both bulk and thin film $BaTiO_3$. In that case the defect was attributed to the oxygen vacancy [17]. Apparently the as-deposited films are slightly oxygen deficient. Thus future improvements in the thin film resistivity are expected for growth at higher oxygen partial pressures.

The electrical properties of Au/$Ba_{1-x}Sr_xTiO_3$/SiO_2/p-Si MIS diodes were studied. In order to determine the quality of each MIS diode fabricated, capacitance-voltage curves were routinely measured. The dielectric constants of the $Ba_{1-x}Sr_xTiO_3$ films were determined from the capacitance-voltage curves measured at a frequency of 1MHz. The dielectric constant was found to depend upon the solid solution composition x, with maximum values occurring within the region of x=0.35 to x=0.45, in agreement with work performed on bulk ceramics [18]. To date, the largest dielectric constant measured was 220 for a composition of x=0.44.

The role of carrier injection in the electrical properties of films deposited on Si was investigated using capacitance-voltage hysteresis curves. The hysteresis curve measured for an MIS diode fabricated from a $Ba_{0.81}Sr_{0.19}TiO_3$ film is shown in Figure 5. This film was grown on an SiO_2 buffer layer approximately 400 to 500Å thick. From this curve, it appears that a slight clockwise hysteresis exists. The effect was reproducible and was observed for all films deposited on a thick layer of SiO_2. The clockwise hysteresis is due to a shift in the flatband voltage of the MIS diode with applied bias. A clockwise loop may be induced by a number of phenomena, including a reversible polarization of the ferroelectric thin film, the presence of mobile ions within the $Ba_{1-x}Sr_xTiO_3$ film or SiO_2 layer, or carrier injection between the film and electrode interface [19,20]. In contrast, Figure 6 shows the slight counterclockwise hysteresis measured for a paraelectric film that was observed only for those films grown on substrates with little or no SiO_2 buffer layer. This effect was also reproducible. Because the hysteresis exhibited a counterclockwise sense, the effect was attributed to carrier injection between the film and Si substrate interface. Preliminary results, therefore, indicate that the presence of a sufficiently thick SiO_2 buffer layer prevents unwanted carrier injection between the film and substrate [4,21]. The question, however, of whether or not the clockwise hysteresis observed for ferroelectric films grown on a thick SiO_2 buffer layer was caused by a ferroelectric polarization requires further investigation.

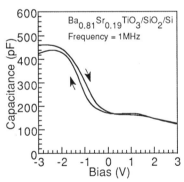
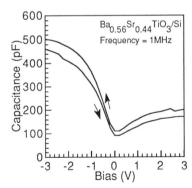

Figure 5. Clockwise C-V hysteresis observed for films deposited on a buffer layer of SiO_2.

Figure 6. Counterclockwise C-V hysteresis observed for films deposited directly on Si.

CONCLUSIONS

In summary, single phase $Ba_{1-x}Sr_xTiO_3$ thin films were synthesized across the entire solid solution range. The phases stabilized depended upon the (Ba+Sr):Ti reactant flow ratios. Thin films deposited under stoichiometric conditions on $LaAlO_3$ were epitaxial, while films grown on silicon substrates showed no texture. Electrical resistivity measurements indicated that the as-deposited films are semi-insulating. The dielectric constant was found to depend upon the composition x, and values as large as 220 were measured at a frequency of 1MHz. Electrical characterization indicated that carrier injection between the $Ba_{1-x}Sr_xTiO_3$ thin film and the Si substrate may be eliminated by oxidizing the Si before growth.

ACKNOWLEDGMENTS

This work was funded by the DOE under Grant No. DE-FG02-85-ER45209 and the Materials Research Center (DMR 91-20521). The MRC Materials Preparation and Crystal Growth Facility was extensively used throughout this study.

REFERENCES

1. J.A. Basmajian and R.C. DeVries, J. Am. Ceram. Soc., **40**, 373 (1957).
2. Y. Xu, Ferroelectric Materials and Their Applications, (North Holland, New York, 1991), pp. 40, 120.
3. N.J. Ali and S.J. Milne, J. Am. Ceram. Soc., **76**, 2321 (1993).
4. K.V. Belenov, E.A. Goloborod'ko, O.E. Zavadovskii,Yu. A. Kontsevoi, V.M. Mukhortov, and Yu.S. Tikhodeev, Sov. Phys. Tech. Phys., **29**, 1037 (1984).
5. Z. Surowiak, Y.S. Nikitin, S.V. Biryukov, I.I. Golovko, V.M. Mukhortov, and V.P. Dudkevich, Thin Solid Films, **208**, 76 (1992).
6. S.R. Summerfelt, in Heteroepitaxy of Dissimilar Materials, edited by R.F.C. Farrow, J.P. Harbison, P.S. Peercy, and A. Zangwill (Mat. Res. Soc. Proc. **221**, Pittsburgh, PA, 1991) pp. 29-34.
7. D. Roy and S.B. Krupanidhi, Appl. Phys. Lett., **62**, 1056 (1993).
8. K.R. Carroll, J.M. Pond, D.B. Chrisey, J.S. Horwitz, R.E. Leuchtner, and K.S. Grabowski, Appl. Phys. Lett., **62**, 1845 (1993); **63**, 1291(E) (1993).
9. V. Mehrotra, S. Kaplan, A.J. Sievers, and E.P. Giannelis, J. Mater. Res., **8**, 1209 (1993).
10. W.A. Feil, B.W. Wessels, L.M. Tonge, and T.J. Marks, J. Appl. Phys., **67**, 3858 (1990).
11. B.S. Kwak, K. Zhang, E.P. Boyd, A. Erbil, and B.J. Wilkens, J. Appl. Phys., **69**, 767 (1991).
12. L.A. Wills, B.W. Wessels, D.S. Richeson, and T.J. Marks, Appl. Phys. Lett., **60**, 41 (1992).
13. P.C. VanBuskirk, R. Gardiner, P.S. Kirlin, and S. Nutt, J. Mater. Res., **7**, 542 (1992).
14. C.S. Chern, J. Zhao, L. Luo, P. Lu, Y.Q. Li, P. Norris, B. Kear, F. Cosandey, C.J. Maggiore, B. Gallois, and B.J. Wilkens, Appl. Phys. Lett., **60**, 1144 (1992).
15. T.W. Kim, M. Jung, Y.S. Koon, W.N. Kang, H.S. Shin, S.S. Yom, J.Y. Lee, Sol. St. Comm., **86**, 565 (1993).
16. B.W. Wessels, L.A. Wills, H.A. Lu, S.R. Gilbert, D.A. Neumayer, and T.J. Marks, in CVD XII, edited by K. Jensen (Electrochemical Society N.J., 1993).
17. L.A. Wills and B.W. Wessels, in Ferroelectrics III, edited by E. Meyer (Mat. Res. Soc. Proc. **310**, Pittsburgh, PA, 1993).
18. U. Syamaprasad, R.K. Galgali, and B.C. Mohanty, Mat. Lett., **7**, 197 (1988).
19. S.Y. Wu, IEEE Trans. Elect. Dev., **21**, 499 (1974).
20. S.M. Sze, Physics of Semiconductor Devices, 2nd ed. (John Wiley and Sons, New York, 1981), p. 390.
21. D.R. Lampe, D.A. Adams, M. Austin, M. Polinsky, J. Dzimianski, S. Sinharoy, H. Buhay, P. Brabant, Y.M. Liu, Ferroelectrics, **133**, 61 (1992).

GROWTH OF BaTiO$_3$ THIN FILMS BY MOCVD

DEBRA L. KAISER*, MARK D. VAUDIN*, GREG GILLEN**, CHEOL-SEONG HWANG*, LAWRENCE H. ROBINS* AND LAWRENCE D. ROTTER*
*Ceramics Division, National Institute of Standards and Technology, Gaithersburg, MD 20899
**Surface and Microanalysis Science Division, National Institute of Standards and Technology, Gaithersburg, MD 20899

ABSTRACT

Polycrystalline thin films of BaTiO$_3$ were deposited on fused quartz substrates at 600°C by metalorganic chemical vapor deposition (MOCVD). The films were characterized by x-ray powder diffraction (XRD), transmission electron microscopy (TEM), secondary ion mass spectroscopy (SIMS) and Raman spectroscopy. Films prepared in the early stages of this study that had appeared to contain only crystalline BaTiO$_3$ by XRD were found to have nonuniform composition and microstructure through the film thickness by SIMS and TEM. The MOCVD system was then modified by installing a process gas bypass apparatus and an elevated pressure bubbler for the titanium isopropoxide precursor. A 1.2 μm thick BaTiO$_3$ film prepared in the modified system demonstrated much improved compositional and microstructural uniformity through the thickness of the film. This film had a columnar microstructure with grain widths of 0.1-0.2 μm and exhibited tetragonality as detected by Raman spectroscopy.

INTRODUCTION

The ferroelectric oxide BaTiO$_3$ has excellent electro-optic properties [1] that make it suitable for many photonic devices, including electro-optic modulators and switches, holographic imaging devices, phase conjugate mirrors and second harmonic generators. Commercialization of these devices for future communications, image processing and optical computing systems requires that high-quality, single crystal films be fabricated reproducibly by a practical processing technique such as MOCVD. Growth of polycrystalline BaTiO$_3$ thin films by MOCVD was first reported by Kwak et. al. [2]. More recently, several groups have achieved epitaxial growth of BaTiO$_3$ on NdGaO$_3$ [3,4], LaAlO$_3$ [4,5] and MgO substrates [6]. Data on the optical properties of the films [6,7] indicate that further work is required to produce films with properties comparable to single crystal BaTiO$_3$.

In this paper, we report on the growth and microstructural characterization of BaTiO$_3$ thin films on fused quartz substrates, with particular emphasis on improving compositional and microstructural uniformity through the film thickness.

EXPERIMENTAL DETAILS

Films were deposited in a research-scale MOCVD system described elsewhere [8]. Briefly, the system consists of three precursor volatilization chambers (each housing a bubbler for one liquid metalorganic) and gas mixing, reaction and exhaust chambers. Prior to each deposition, 4.0 g of titanium isopropoxide (TIP) and 2.5 g of bis(2,2,6,6-tetramethyl-3,5-

heptanedionato) barium hydrate [Ba(thd)$_2$] were loaded into their respective bubblers in an argon atmosphere drybox. During a deposition, each chamber in the system was heated to the temperature given in Table I. (The mixing, reaction and exhaust chambers and the delivery lines were maintained at a temperature 10°C higher than the temperature of the Ba(thd)$_2$ volatilization chamber to prevent condensation of Ba(thd)$_2$.) Argon carrier gas was percolated through the TIP and Ba(thd)$_2$ bubblers. The two gas streams containing the metalorganic vapors were mixed with pure oxygen in the mixing chamber and the resulting process gas flowed into a horizontal, cold wall reaction chamber containing a substrate supported on a SiC-coated graphite susceptor. The susceptor was heated by a radio-frequency induction generator. In the vicinity of the hot substrate, the metalorganic complexes pyrolyze and react with oxygen to form a film on the substrate.

During this study, the MOCVD system was modified to improve compositional and microstructural uniformity through the film thickness. The first modification was an elevated pressure TIP bubbler to stabilize the TIP concentration in the process gas. Increasing the pressure in this bubbler suppresses the vaporization of TIP, so that the argon flowrate through the bubbler can be increased to provide a more stable flux of TIP in the process gas [4]. The pressure in the bubbler is controlled by means of a metering valve located downstream from the bubbler and upstream from the mechanical pump, and the pressure is read with a transducer positioned upstream from the bubbler. By adjusting the metering valve, pressures can be fixed within the range of 6.7 - 106.6 kPa (50 - 800 torr).

The second modification was a process gas bypass apparatus to allow the gas composition to stabilize before deposition. The bypass apparatus consists of stainless steel delivery lines and a quartz U-tube filled with stainless steel wool. The delivery lines are heated to 240°C to prevent condensation of Ba(thd)$_2$ and the U-tube is heated to 600°C. The metalorganics react and deposit in the U-tube and the exhaust gas passes through a cold trap before venting to a fume hood. At the start of a growth run, the process gas stream is diverted through the bypass apparatus for 1 h, after which time the stream is directed into the reactor to begin film deposition.

A series of deposition experiments was performed to determine a set of growth conditions for obtaining BaTiO$_3$ [8]. Table I gives sets of conditions for two films A and B, which were deposited in the MOCVD system before and after modification, respectively. These films were deposited at 600°C on fused quartz substrates (microscope slides), which were selected as inexpensive substrates for the microstructural studies reported here. The growth conditions for films A and B were identical except for the pressure in the TIP bubbler and the carrier gas flowrate through this bubbler.

The microstructure of the films was characterized by several techniques. Crystalline phases were identified by conventional (θ-2θ) x-ray powder diffraction (XRD) techniques. Compositional uniformity and microstructural homogeneity through the film thickness were determined by secondary ion mass spectrometry (SIMS) depth profiling and transmission electron microscopy (TEM) respectively. The presence of tetragonal BaTiO$_3$ was detected by Raman spectroscopy [8]. Energy dispersive x-ray spectroscopy measurements revealed that the films contained ~ 2 mole % Sr which is a known impurity in the Ba(thd)$_2$ precursor.

RESULTS AND DISCUSSION

Microstructural characterization of film A will be discussed first. The XRD pattern shown in Fig. 1 indicates that film A contains only polycrystalline BaTiO$_3$ with a slight <111>

Table I
Growth Conditions

Variable	Film A	Film B
temperatures, °C		
TIP bubbler	25	25
Ba(thd)$_2$ bubbler	230	230
all other chambers	240	240
substrate	600	600
reactor pressure, kPa	6.7	6.7
TIP bubbler pressure, kPa	6.7	86.6
flowrates, sccm		
Ar through TIP	12.5	70
Ar through Ba(thd)$_2$	150	150
O$_2$	300	300
deposition rate, μmh^{-1}	0.27	0.35

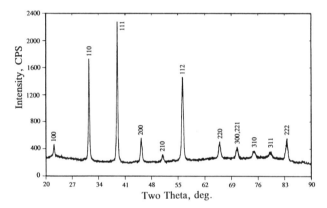

Fig. 1. Conventional θ-2θ XRD pattern for film A. BaTiO$_3$ peaks are indexed (using cubic indices).

texture. Compositional homogeneity through the thickness of this film was then examined by SIMS depth profiling. Profiles for Ba, Ti, Sr and Si taken from the surface of the film (depth = 0) to the substrate (depth > 1.45 μm) are shown in Fig. 2. These profiles are qualitative, i.e., the relative ion counts for the different elements do not indicate their actual concentrations. Peaks in the Ba, Sr and Ti profiles at the surface of the film are likely due to transient charging effects at the start of the SIMS measurement. In the film, the Ba and Sr ion counts remain nearly constant but the Ti ion count varies considerably with depth. These data indicate that the TIP concentration in the process gas was not stable during the deposition run. There is also a transition region at the film/substrate interface in which the ion counts of the four elements change sharply. This transition layer probably resulted from both interdiffusion and variations in the gas composition during the initial stages of the deposition. The shoulder in the Si profile near the film/substrate interface may indicate Si diffusion into the BaTiO$_3$ film.

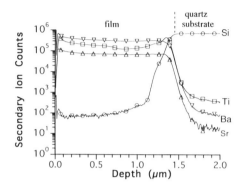

Fig. 2. SIMS depth profile data on film A for Si (○), Ti (□), Ba (▽) and Sr (△).

Fig. 3. Cross-sectional TEM micrograph of film A showing the substrate (S), transition layer (T) and BaTiO$_3$ film (F).

Cross-sectional TEM studies confirmed the existence of a transition layer (Fig. 3) composed of crystalline phases other than BaTiO$_3$. The film was composed of columnar, dendritic grains. Films prepared under other growth conditions [8] in the same MOCVD system were also nonuniform through the thickness.

Film B was deposited in the MOCVD system after the system had been modified to improve the film uniformity; the growth conditions are given in Table I. The XRD pattern for this film (Fig. 4) indicates that the film contains only BaTiO$_3$ and has a preferred <100>

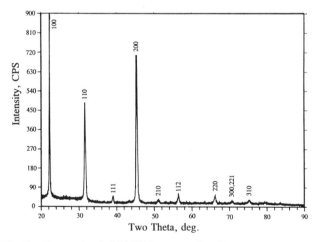

Fig. 4. Conventional θ-2θ XRD pattern for film B. BaTiO$_3$ peaks are indexed (using cubic indices).

texture. The SIMS depth profiles for Ba, Sr, Ti and Si are shown in Fig. 5. As discussed above, the high ion counts for Ba, Sr and Ti at the surface of the film (depth = 0) are an artifact of the SIMS measurement. As observed in film A, there is a shoulder in the Si profile near the film/substrate interface (at depth ≈ 1.2 μm). The nearly flat and parallel Ti, Ba and Sr profiles from the surface to the film/substrate interface indicate good uniformity through the entire film. In addition, the Ba, Sr and Ti ion counts drop much more rapidly at the film/substrate interface in film B than in film A (Fig. 2). Thus, the two modifications to the MOCVD system have greatly improved compositional uniformity through the film thickness.

Microstructural homogeneity of film B was examined by cross-sectional TEM. The film is composed of columnar grains (Fig. 6) with widths in the range of 0.1-0.2 μm. This film has no transition layer and has a more uniform grain size and morphology than film A, indicating that the two modifications to the MOCVD system have also improved microstructural uniformity. Lattice imaging studies revealed the presence of a very thin (< 5 nm) interdiffusion layer at the film/substrate interface.

The films were also studied by Raman spectroscopy. The spectra given in Fig. 7 show two peaks associated with both cubic and tetragonal $BaTiO_3$ as well as one or two peaks that occur only in the tetragonal phase [9]. Peaks associated with non-$BaTiO_3$ phases are also present: the broad peak at 800 cm^{-1} in film B, which may arise from amorphous material, and the narrow line at 1060 cm^{-1} in film A. Thus, the films contain tetragonal $BaTiO_3$, but the cubic phase may also be present.

CONCLUSIONS

A research-scale MOCVD system was used to deposit polycrystalline thin films of $BaTiO_3$ on fused quartz substrates at 600°C at growth rates of 0.25-0.35 μmh^{-1}. The micro-

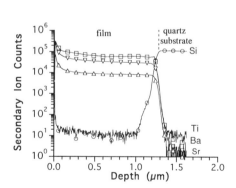

Fig. 5. SIMS depth profile data on film B for Si (○), Ti (□), Ba (▽) and Sr (△).

Fig. 6. Cross-sectional TEM micrograph of film B showing the substrate (S) and $BaTiO_3$ film (F). The white streaks are not pores but are likely amorphous material.

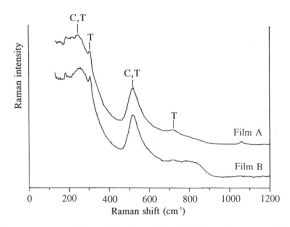

Fig. 7. Raman spectra from films A and B. The baselines are offset for clarity. Observed peaks characteristic of tetragonal BaTiO$_3$ (T) and both the cubic (C) and tetragonal phases are labeled.

structure of the resulting films was characterized by XRD, SIMS, TEM and Raman spectroscopy. To obtain films with good compositional and microstructural uniformity through the thickness, the process gas was diverted through a bypass apparatus at the start of a deposition and an elevated pressure bubbler was used to stabilize the TIP concentration in the gas. A 1.2 μm thick film prepared in this manner was BaTiO$_3$ by XRD and was uniform through the thickness as determined by SIMS and TEM. This film had a columnar microstructure composed of grains 0.1 to 0.2 μm in width with a preferred <100> texture and contained tetragonal BaTiO$_3$ as determined by Raman spectroscopy.

References

1. A.M. Glass, Science **235**, 1003 (1987).
2. B.S. Kwak, K. Zhang, E.P. Boyd, A. Erbil and B.J. Wilkens, J. Appl. Phys. **69**, 767 (1991).
3. P.C. Van Buskirk, R. Gardiner, P.S. Kirlin, and S. Nutt, J. Mater. Res. 7, 542 (1992).
4. C.S. Chern, J. Zhao, L. Luo, P. Lu, Y.Q. Li, P. Norris, B. Kear, F. Cosandey, C.J. Maggiore, B. Gallois, and B.J. Wilkens, Appl. Phys. Lett. **60**, 1144 (1992).
5. L.A. Wills, B.W. Wessels, D.S. Richeson, and T.J. Marks, Appl. Phys. Lett. **60**, 41 (1992).
6. P.C. Van Buskirk, G.T. Stauf, R. Gardiner, P.S. Kirlin, B. Bihari, J. Kumar, and G. Gallatin, Mat. Res. Soc. Proc. **310** (1993).
7. H.A. Lu, L.A. Wills, B.W. Wessels, W.P. Lin, T.G. Zhang, G.K. Wong, D.A. Neumayer, and T.J. Marks, Appl. Phys. Lett. **62**, 1314 (1993).
8. D.L. Kaiser, M.D. Vaudin, G. Gillen, C.-S. Hwang, L.H. Robins, and L.D. Rotter, to be published in J. Crystal Growth, 1994.
9. J.A. Sanjurjo, R.S. Katiyar, and S.P.S. Porto, Phys. Rev. B 22 (1980) 2396.

DONOR–DOPED LEAD ZIRCONATE TITANATE (PbZr$_{1-x}$Ti$_x$O$_3$) FILMS

Seshu B. Desu, Dilip P. Vijay, and In K. Yoo
Department of Materials Science and Engineering
Virginia Polytechnic Institute and State University
Blacksburg, Virginia 24061.

ABSTRACT

Properties of undoped– and doped–ferroelectric PbZr$_{1-x}$Ti$_x$O$_3$ films, with both Pt or RuO$_2$ electrodes, were compared, with a view to understand the reasons for degradation of ferroelectric films. Donor–doped (e.g., Nb^{5+} at Ti^{4+} site) PbZr$_{1-x}$Ti$_x$O$_3$ films, for which the PbO loss has been compensated, showed higher resistance to fatigue and low leakage currents compared to those of undoped films. The fatigue of ferroelectric films, a decrease in switchable polarization with increasing number of polarization reversals, has been attributed to the migration of oxygen vacancies and their entrapment at various interfaces (e.g., electrode/ferroelectric interface) that are present in the film.

INTRODUCTION

Degradation properties such as fatigue, aging, and leakage current have been the major problems affecting lifetime of ferroelectric films. For ferroelectric oxide films, Desu and Yoo [1–4] have noted that oxygen vacancies are a common source for these degradation properties. Their quantitative models [1–4] showed that both the migration of oxygen vacancies and their entrapment at electrode/ferroelectric interface (and/or at grainboundaries and domain boundaries) are important contributing factors. These models [1–4] suggested two possible solutions to overcome fatigue: (1) reducing the tendency for entrapment (i.e., changing the nature of electrode/ferroelectric interface), and (2) minimizing the defect density. Conducting oxide electrodes (e.g., RuO$_2$), which minimize the oxygen vacancy entrapment, have been used to minimize the fatigue problems of PbZr$_{1-x}$Ti$_x$O$_3$ (PZT) films [5]. In this paper we deal with the Improvements in degradation properties by minimizing the defect density.

BACKGROUND

A composition of PbZr$_{0.5}$Ti$_{0.5}$O$_3$ is chosen for this study because it is close to the morphotropic phase transition and thus is highly sensitive to the external influences. For defect density minimization one need to understand the sources of defects, especially oxygen vacancies.

Investigation of bulk materials have shown that the major defects are lead and oxygen vacancies in PZT [6]. It is also argued that the presence of lead and oxygen vacancies is inevitable since PbO volatilizes from PZT during calcination and sintering; this process can be represented by a quasi–chemical reaction (in Kroeger–Vink notation) as follows:

$$Pb_{Pb} + O_O \longleftrightarrow V_O^{\cdot\cdot} + V_{Pb}'' + PbO\uparrow \qquad (1).$$

where Pb_{Pb} and O_O represent respectively occupied Pb and O sites in the lattice, $V_O^{\cdot\cdot}$ and V_{Pb}'' represent vacancies of oxygen and lead, respectively. Furthermore, it has also been shown that PZT solid solutions show p–type conductivity [6].

Studies of bulk PZT materials also determined that 3+ ions which substitute for $(Zr, Ti)^{4+}$ (acceptors) create oxygen vacancies [7–9]; this process can be represented as

$$A_2O_3 \longrightarrow 2A'_{Ti, Zr} + 3O_O + V_O^{\cdot\cdot} \qquad (2)$$

where A_2O_3 is a trivalent oxide, and $A'_{Ti, Zr}$ represents a 3+ ion at $(Ti, Zr)^{4+}$ site. If the processing conditions are limited to high oxygen activities (i.e., $p_{O_2} > 10^{-4}$ atm), reduction can be eliminated as a source of oxygen vacancies [10]. Although such studies have not been carried out for thin films, it may be assumed that loss of PbO and presence of acceptors are the two major sources of oxygen vacancies for thin films also.

Since the vacancy concentrations resulting due to PbO volatilization can be very high, techniques for controlling PbO activities during film formation are very important for minimizing defect concentrations. One common technique is to provide excess PbO during film formation to compensate the loss. However, precise control is often very difficult. In addition to compensating the PbO loss, another technique that can be used to lower the oxygen vacancy concentration is the introduction of donor dopants (e.g., Nb^{5+} at Zr or Ti site). It is well established for the bulk PZT materials that donor dopants not only reduce oxygen vacancies but also compensate for acceptors. The consumption of oxygen vacancy can be written as

$$Nb_2O_5 + V_O^{\cdot\cdot} \longrightarrow 2Nb^{\cdot}_{Ti, Zr} + 5O_O \qquad (3)$$

where $Nb^{\cdot}_{Ti, Zr}$ represents a Nb^{5+} ion at $(Ti, Zr)^{4+}$ site in PZT. The compensation acceptors by donors can be written as

$$A_2O + Nb_2O_5 \longrightarrow 2A'_{Pb} + 2Nb^{\cdot}_{Ti, Zr} + 6O_O \qquad (4)$$

where A'_{Pb} is a +1 ion at Pb^{2+} site (acceptor).

Therefore, for effective lowering of the oxygen vacancy concentration, both donor addition and minimization of PbO loss are necessary. Depending on the acceptor concentration and the nature of heat treatment, there will be a critical donor concentration above which effects of donors can be seen. Without the knowledge of all the defect reaction constants, the critical concentration is difficult to calculate. However, an experimental study may yield valuable information. Since the undoped PZT materials show p–type conductivity, donor doping will also reduce the conductivity (or decrease the leakage current of PZT materials.

RESULTS AND DISCUSSIONS

Figure 1 depicts the fatigue and leakage current behavior of undoped PZT films with both Pt and RuO_2 electrodes. Fatigue of Pt electroded PZT films can be seen from Figure 1a as the drastic decrease in the normalized polarization at around 10^7 switching cycles. Yoo and Desu [3] have quantitatively modeled this fatigue behavior and derived

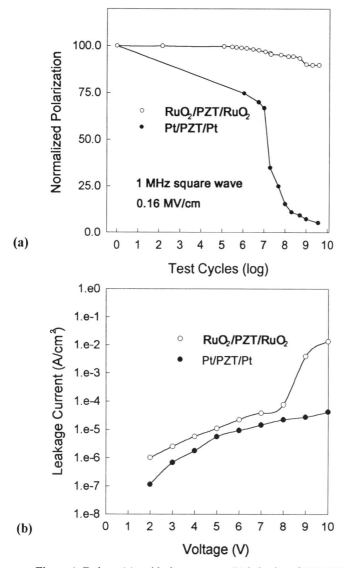

Figure 1: Fatigue **(a)** and leakage current **(b)** behavior of PZT (300 nm thick) capacitors with both Pt and RuO_2 electrodes.

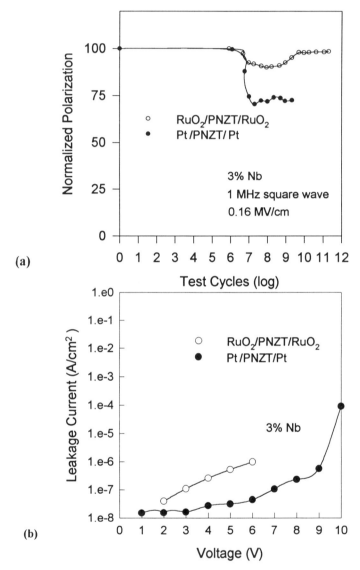

Figure 2: Fatigue **(a)** and leakage current **(b)** behavior of PNZT (300 nm thick) capacitors with both Pt and RuO_2 electrodes.

an equation:

$$P = P_0 (An + 1)^m \tag{5}$$

where P_0 is the initial polarization, P is the polarization after n switching cycles, A is the piling constant and m is the decay constant. They have also shown that oxygen vacancy movement and their entrapment at the electrode/ferroelectric interface are the primary reasons for the fatigue [1–4]. As long as fatigue occurs on account of defect entrapment at the interface, it can be improved by controlling the interface. On of the possible ways to improve the interface is the use of conductive oxides (e.g., RuO_2) as electrode materials because they can consume the oxygen vacancies by increasing their oxygen non–stoichiometry and thus, prevent the space charge development at the interface.

As shown in Figure 1a, no fatigue was found up to $\simeq 10^{10}$ switching cycles for PZT capacitors with RuO_2 electrodes. Although RuO_2 electroded PZT films do not fatigue, they show higher leakage currents when compared to Pt electroded PZT films at similar electrical fields (Figure 1b). These results agree with earlier accounts [1–4 & 11] which indicate that (1) Schottky emission is responsible for leakage current in PZT films and (2) RuO_2 electrodes show lower effective Richardson constant values. As indicated above, the primary carriers responsible for electrical conductivity in PZT are holes.

Since oxygen vacancies contribute to PZT fatigue and holes contribute to leakage current of PZT, it is to be expected that both fatigue and leakage current can be minimized by donor doping of PZT. Figure 2 depicts the fatigue and leakage current behavior of Nb–doped PZT films (PNZT) for both Pt and RuO_2 electrodes. It can be clearly seen that the leakage currents of PNZT films are lower than those of PZT films for both Pt and RuO_2 electrodes (compare Figure 2b and Figure 1b). Furthermore, it can also be noticed that the fatigue behavior of Pt electroded PNZT films is much better than that of Pt electroded PZT films. In the case of RuO_2 electroded PNZT films, although they do not show fatigue over all, a slight decrease in polarization at around 10^7 switching cycles and recovery at around 10^{10} cycles was observed. At this point, we do not have any explanation for this behavior. However, further experiments are underway to understand this fatigue behavior of RuO_2 electroded PNZT films.

SUMMARY

The role of oxygen vacancies in the fatigue of PZT thin films has been discussed. Based on a quantitative fatigue mechanism, it was shown that fatigue can be minimized by (1) by decreasing oxygen vacancy concentration with donor (e.g., Nb) doping and (2) by the use of conductive oxide electrodes (e.g., RuO_2), which can consume oxygen vacancies at the electrode and ferroelectric interface and thus eliminate space charge development. It was also shown that the donor–doping decreases the leakage currents of PZT films. However, for donors to be effective the PbO loss from PZT films should be carefully controlled.

REFERENCES

1) I.K. Yoo and S.B. Desu, Mat. Sci. & Eng., B13, 319 (1992).
2) I.K. Yoo and S.B. Desu, Phys. Stat. Sol., a133, 565 (1992).
3) I.K. Yoo and S.B. Desu, J. Int. Mat. Sys., 4, 490, (1993).
4) S.B. Desu and I.K. Yoo, J. Electrochem. Soc., 140, L133, (1993).
5) D.P. Vijay and S.B. Desu, J. Electrochem. Soc., 140, 2640 (1993).
6) V.V. Prisedsky, V.I. Shishkovsky and V.V. Klimov, Ferroelectrics, 17, 465 (1978).
7) M. Takahashi, Jap. J. Appl. Phys., 9, 1236, (1970).

8) G.H. Dai, P.W. Lu, X.Y. Huang, Q.S. Liu, and W.R. Xue, J. Mat. Sci: Mat. in Elect., 2, 164, (1991).
9) J.J. Dih and R.M. Fulrath, J. Am. Ceram. Soc., 61, 448, (1978).
10) C.K. Kwok and S.B. Desu, Mat. Res. Soc. Symp. Proc., 243, 393, (1992).
11) I. K. Yoo, S. B. Desu and J. Xing, Mat. Res. Soc. Symp. Proc., 310, 165 (1993).

ACKNOWLEDGEMENTS

The assistance of Mr. W. Hendricks and Mr. A. Khan for the deposition of PZT and PNZT films, and Mr. W. Pan for the deposition of RuO_2 electrodes is gratefully acknowledged. This work was partially funded by a grant from ARPA through ONR.

GROWTH OF (001)-ORIENTED SBN THIN FILMS BY SOLID SOURCE MOCVD

Z. LU,[1] R. S. FEIGELSON,[1] R. K. ROUTE,[2] R. HISKES[3] and S. A. DICAROLIS[3]
[1]Materials Science and Engineering Department, Stanford University, CA 94305-2205
[2]Center for Materials Research, Stanford University, CA 94305-4045
[3]Hewlett-Packard Corporation, 3500 Deer Creek Road, Palo Alto, CA 94303

ABSTRACT

By the solid source MOCVD technique, we have deposited 2000 - 3000 Å thick single phase $Sr_xBa_{1-x}Nb_2O_6$ (SBN) films on (100) MgO substrates using tetramethylheptanedionate (thd) sources. X-ray diffraction (XRD) 2θ scans indicated that these films were completely (001) oriented. XRD φ scans, however, showed the films contained four in-plane grain orientations whose volume fractions could be controlled by altering the Sr/(Sr+Ba) and Nb/(Sr+Ba) ratios in the source powders. The in-plane volume fractions did not change with the deposition rate or the cooling rate. Films with composition $Sr_{0.58}Ba_{0.42}Nb_{1.94}O_6$ had mainly two in-plane orientations. Optical waveguiding behavior was demonstrated in these films. Refractive indices were found to be n_o= 2.20 and n_e = 2.13, as compared to n_o = 2.31 and n_e = 2.27 for bulk SBN60.

INTRODUCTION

The tetragonal tungsten bronze $Sr_xBa_{1-x}Nb_2O_6$ (0.25 < x < 0.75), known as SBN, has the point group 4mm and lattice constants a=b=12.480-12.430 Å, c= 3.983-3.913 Å for x=0.25-0.75[1,2]. There are five formula units, 10 distorted NbO_6 octahedra and 10 interstitial sites filled by only five alkaline earth cations in each unit cell. A disordered arrangement of Sr and Ba atoms found in the A2 sites, and disordered oxygen positions in the Sr and Ba planes make the structure of this material very complicated.[2] Bulk SBN is an excellent ferroelectric and photorefractive material which has been widely studied for holographic recording and optical processing.[3,4] High optical quality thin films of SBN are useful in high density electro-optic switching for parallel optical processing applications. There have been few reports on the growth of high quality SBN thin films,[5,6] due partly to the complicated chemistry and structure of this material, and partly to the difficulty in finding a suitable non-reacting, lattice-matched substrate.

In [7], we reported on our initial success in growing strongly (001)-oriented SBN films on r-Al_2O_3 substrates using (100) CeO_2 buffer layers, and a novel solid source MOCVD method that allowed careful control over layer composition. Subsequent experiments revealed that using

(100) MgO buffer layers in place of CeO$_2$ gave better (001) film alignment. This finding led us to evaluate single crystal MgO as a substrate. The results form the basis of this report.

EXPERIMENTAL PROCEDURES

The solid source MOCVD reactor used for these growth experiments, which incorporates a "U" shaped quartz reactor tube, has been described previously.[8] In these experiments, accurately weighed mixtures of tetramethylheptanedionate (thd) sources, Sr(thd)$_2$, Ba(thd)$_2$ and Nb(thd)$_4$, were vaporized at 250-300 °C and carried on a 300 SCCM Ar gas stream to the deposition chamber. Oxyen at 1.0 SLPM was merged with it before it reached the substrate. The entire reactor vessel was maintained at a temperature of 200-300 °C to prevent condensation of the source vapor, and growth was carried out at a pressure of 4.0 torr. Substrates were held on the underside of a radiatively heated inconel susceptor, the temperature of which was monitored by an embedded thermocouple. In these experiments, substrate surface temperatures were held constant at 710 °C. Film thicknesses measured with a profilometer[9] were typically 2000 - 3000 Å for a standard growth period of 80 minutes.

Film compositions were determined by electron microprobe (EPMA) analysis using a single crystal of known composition (Sr$_{0.75}$Ba$_{0.25}$Nb$_2$O$_6$) as a standard. Phase content and orientation were determined using four circle X-ray diffractometry (XRD). Surface microtopography was analyzed by atomic force microscopy (AFM). Microstructures were studied by transmission electron microscopy (TEM), and optical refractive indices were measured by the single prism coupling technique using a rutile prism.

RESULTS AND DISCUSSION

Film composition and X-ray diffraction characterization

By systematically varying the input source composition, thin films with the compositions listed in Table 1 were grown. With the Nb/(Sr+Ba) ratio in the source material fixed at 1.5, the Sr/(Sr+Ba) ratio in the films was found to be higher than that in the source compositions (samples No. 2-5), which indicates that the incorporation of Ba from the Ba(thd)$_2$ solid source is less efficient than that of Sr from the Sr(thd)$_2$ solid source. This is consistent with our previous experience in the growth of YBCO films where it was found that Ba(thd)$_2$ partially decomposed at vaporizer temperatures in the 250-300 °C range. With the Sr/(Sr+Ba) ratio fixed at 0.35, the Nb/(Sr+Ba) ratio in the films was mostly higher than that in the source material (samples No. 4, 6-8), indicating that the incorporation of Nb from the Nb(thd)$_4$ solid source is more efficient than the combination of Sr and Ba. Table I also shows that the Nb/(Sr+Ba) ratio decreases with increasing Sr/(Sr+Ba) and fixed Nb/(Sr+Ba) ratios in the source, and that the Sr/(Sr+Ba) ratio increases with increasing Nb/(Sr+Ba) and fixed Sr/(Sr+Ba) ratios in the source. Differential scanning calorimetry (DSC) of the individual source components and mixtures has suggested that the mixtures undergo chemical reaction during the sublimation process.

Table I SBN film composition adjustment and XRD characterization results

Sample	Source Composition Sr : Ba : Nb	Film Composition Sr : Ba : Nb	Phases (add'l peaks)	In-plane Orientations
1	0.75 : 0.25 : 2.0	0.70 : 0.30 : 2.20	two (31.9°)	N/S
2	0.50 : 0.50 : 1.5	0.72 : 0.28 : 2.19	two (31.9°)	N/S
3	0.40 : 0.60 : 1.5	0.59 : 0.41 : 1.85	one	four
4	0.35 : 0.65 : 1.5	0.58 : 0.42 : 1.94	two (46.8°)	two
5	0.30 : 0.70 : 1.5	0.51 : 0.49 : 1.97	two (46.8°)	four
6	0.35 : 0.65 : 1.7	0.57 : 0.43 : 2.19	two (46.8°)	four
7	0.35 : 0.65 : 1.4	0.51 : 0.49 : 1.85	one	four
8	0.35 : 0.65 : 1.3	0.49 : 0.51 : 1.20	one	two

N/S: no symmetry.

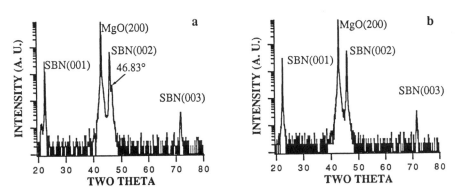

Figure 1 XRD θ-2θ scans of films grown from sources with Sr:Ba:Nb equal to (a) 0.35:0.63:1.5 and (b) 0.40:0.60:1.5.

Table I also lists results from the corresponding XRD θ-2θ scans. Films with Nb/(Sr+Ba) ratios higher than 2.19 usually had a strong peak at about 31.9°, while films with Nb/(Sr+Ba) ratios less than 2.19 but higher than 1.94, usually had a peak at 46.83°, as shown in

Fig 1a. Nb/(Sr+Ba) ratios less than 1.85 but higher than 1.2 usually gave single phase, single orientation films, as shown in Fig 1b. The fact that single phase SBN was found in films with Nb/(Sr+Ba) ratios as low as 1.20 (for Sr/(Sr+Ba) in the range of 0.49-0.58), contradicts the established phase equilibria obtained from bulk crystals which indicates that Nb/(Sr+Ba) cannot be lower than about 1.51.[10]

An XRD ω scan of the SBN (001) peak for sample No.3 showed a FWHM value about 0.65°, indicating good out-of-plane alignment in the film.

XRD ϕ scans of the SBN (221) peak and MgO (202) peak for sample No. 3 shows four-fold symmetry around the film plane normal as one would expect from a (001)-oriented epitaxial film with tetragonal symmetry, Fig 2. However, there are four sets of four peaks each, implying four in-plane crystallite orientations. The MgO (202) peak is located in a position 14° from the two nearer and stronger peaks and 26° from the two farther and weaker peaks. This indicates two pairs of SBN crystallites that are mirror symmetric to the MgO (100) plane. The MgO [100] orientation is therefore ±14° away from the SBN [110] in one set of crystallites, and ±26° away from the SBN [110] in the other set. Attempts to eliminate these many fold in-plane orientations did not succeed. However, it was found that the volume fractions could be controlled to some degree by varying the starting source composition as shown in Figures 3a-c. Films grown from a source composition ratio of 0.35:0.65:1.5 had only two in-plane orientations (±14°), as shown in Fig 3a, although results of the 2θ scans shown previously in Fig 1a indicated that these films contained an additional phase. Films grown from a composition ratio of 0.30:0.70:1.5 had four in-plane orientations with approximately equal volume fractions residing in the ±14° and the ±26° grain orientations (Fig 3b). Films grown from a composition ratio of 0.35:0.65:1.3 had a much stronger outer set of peaks, and therefore a major volume fraction residing in the ±26° orientations (Fig 3c). The number of in-plane orientations was independent of growth rate and cooling rate within the range of conditions studied (25-40 Å/min and ≥100 °C/hr, respectively). No clear relationship between the number of in-plane orientations and film composition was found.

<u>TEM results</u>

To better understand the microstructure and the second phase in sample No.4, a TEM planar study was carried out. The results, in Fig 4, show two grain orientations (labelled "A" and "B") 28° apart, consistent with the XRD results in Fig 3a. There is also a second phase labelled "C" in this film. EDAX microanalysis showed this phase to contain Nb, Ba and a trace of Mg. Selected area electron diffraction analysis of "C" showed its diffraction pattern to be similar to that of $Ba_3Nb_{10}O_{28}$. However, the JCPDS-XRD diffraction peak of $Ba_3Nb_{10}O_{28}$ (002) is at 45.701°, which is about 1° less than the unknown peak found in Fig 1a. This discrepancy may be due to Mg doping of the $Ba_3Nb_{10}O_{28}$ phase, with the Mg being introduced from the substrate through an interface reaction.

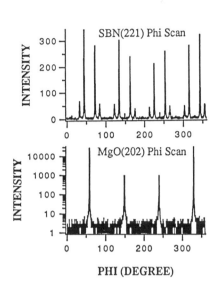

Figure 2 XRD φ scans of sample NO.3.

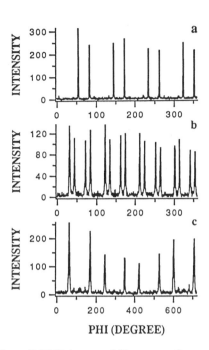

Figure 3 XRD φ scans of films grown from sources with Sr:Ba:Nb about (a) 0.35:0.65:1.5; (b) 0.30:0.70:1.5; (c) 0.35:0.65:1.3.

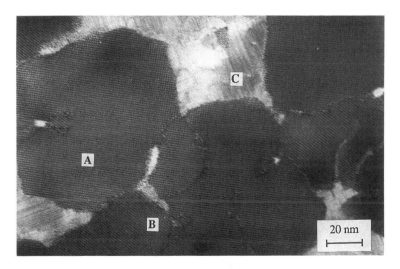

Figure 4 TEM planar view of sample No. 4. Gains labelled "A" and "B" are 28° apart. A second phase is revealed in the area labelled "C".

Optical measurements

All SBN film surfaces were smooth under optical microscopic evaluation. A high density of grains with average size about 0.1 μm was observed by atomic force microscopy (AFM), and rms surface roughness values of 80 Å were typical. Optical waveguiding behavior was verified in these samples with a single prism coupling technique using a 632.8 nm He-Ne laser beam. The guided light was scattered broadly, which we attributed to the high angle grain boundaries and the intragranular defects shown in Fig 4. The scattering prevented an accurate energy loss determination. Refractive indices were found to be $n_o = 2.20$, and $n_e = 2.12$, which are relatively low compared to bulk value for SBN60 of $n_o = 2.31$ and $n_e = 2.27$.

CONCLUSION

Single phase, (001) oriented SBN films were grown on (100) MgO substrates over a range of source composition. Film in-plane orientations could be controlled by varying source compositions. While optical waveguiding was observed for the first time in heteroepitaxial SBN films, additional improvements in film crystalline quality are required for device applications.

ACKNOWLEDGMENTS

This work was supported in part by ONR/ARPA through the Center for Nonlinear Optical Materials at Stanford University, in part by the NSF/MRSEC Program through the Center for Materials Rsearch at Stanford University, and in part by the Hewlett-Packard Corporation. The authors thank T. Kono, A. Marshall and J. Li of Stanford university for helpful discussions in TEM studies, and F. Leplingard of Xerox PARC center for optical measurements.

REFERENCES

[1] K. Megumi, N. Nagatsuma, Y. Kashiwada and Y. Furuhata, J. Materials Science 11, 1583 (1976).
[2] P. B. Jamieson, S. C. Abrahams and J. L. Bernstein, J. Ceramic Physics 48, 5048 (1968).
[3] H. Smith, Holographic Recording Materials (Springer-Verlag, New York, 1977).
[4] R. Fisher, Optical Phase Conjugation (Academic, New York, 1983).
[5] R. R. Neurgaonkar and E. T. Wu, Mat. Res. Bull. 22, 1095 (1987).
[6] R. R. Neurgaonkar, I. S. Santha and J. R. Oliver, Mat. Res. Bull. 26, 983 (1991).
[7] Z. Lu, R. S. Feigelson, R. K. Route, S. A. DiCarolis, R. Hiskes and R. D. Jacowitz, J. Crystal Growth 128, 788 (1992).
[8] R. Hiskes, S. A. DiCarolis, R. D. Jacowitz, Z. Lu, R. S. Feigelson, R. K. Route and J. L. Young, J. Crystal Growth 128, 781 (1992).
[9] Tencor Instruments, Alpha Step 100.
[10] J. R. Carruthers and M. Grasso, J. Electrochem. Soc. 117, 1427 (1970).

VAPOR DEPOSITION OF LITHIUM TANTALATE WITH VOLATILE DOUBLE ALKOXIDE PRECURSORS

KUEIR-WEEI CHOUR, GUANGDE WANG AND REN XU
University of Utah, Department of Materials Science and Engineering, Salt Lake City, Utah 84112

ABSTRACT

Stoichiometric vapor deposition of $LiTaO_3$ is demonstrated by thermal evaporation of volatile $LiTa(OBu^n)_6$, controlled vapor phase partial-hydrolysis, and subsequent polycondensation reactions on a heated substrate surface. Fully dense, amorphous $LiTaO_3$ film of up to 34 μm thick can be obtained at high deposition rate of 23 μm per hour. Single phase $LiTaO_3$ films are obtained by annealing as deposited films. The vapor phase species responsible for the stoichiometric vapor deposition appeared to be $Li_2Ta_2(OBu^n)_{12}$ from variable temperature mass spectrometry results.

INTRODUCTION

MOCVD of multicomponent oxide involves the use of separate metalorganic compounds for each component as precursors. In ordinary operations, the precursor compounds are independently transformed to the vapor phase (by evaporation or other means) in a controlled manner to achieve the desired molar ratio between vapor phase species. Whether a given vapor phase composition results in the desired film composition is determined only after a compositional analysis of the final film is performed. While this popular operation allows for convenient variation of compositions away from stoichiometry, it inevitably requires a trial-and-error practice when precisely stoichiometric film is desired. One possible alternative is to directly use a single metal organic compound, carrying all metal elements in the appropriate ratio, as the precursor. If such precursors do exist in stable form, volatile enough for evaporation, and provided that the subsequent reactions do not disturb the metal-to-metal ratio carried by such molecules through out the deposition process, stoichiometry control for multicomponent oxide film deposition may be significantly simplified. Thus, it is of great interest to investigate possibilities of vapor deposition using stable and volatile molecular precursors which carry metal elements in exactly the same ratio as that of a stoichiometric multicomponent oxide.

Double metal alkoxides are possible candidates for such molecular precursors. It has been established that a large group of such compounds exist in molecular form stable enough to be easily isolated through low pressure sublimation. Table I is a list of a few such compounds, taken after Bradley, Mehrotra and Gaur[1]. Also listed in the table are ferroelectric oxides with stoichiometry corresponding to the given double alkoxide. These double alkoxides, although possessing the desired metal-to-metal ratio in their molecular form, and are reported to be volatile enough for at least low pressure vapor depositions, their thermal stability may not be sufficient for typical deposition process. For example, double alkoxide $M'M''(OR)_{2n}$ may decompose prematurely to produce individual metal alkoxides:

$$M'M''(OR)_{2n} \rightarrow M'OR + M''(OR)_{2n-1}.$$

Since $M'OR$ and $M''(OR)_{2n-1}$ are most likely different in their volatility, such premature thermal decomposition will lead to non-stoichiometric deposition. For all known double alkoxides, such decomposition occur at temperatures much lower than the pyrolytic temperatures of $M'OR$ or $M''(OR)_{2n-1}$. In other words, the customary pyrolysis reactions leading to the final film deposition cannot be used for these precursors without causing variations of M' to M'' ratio prior to film deposition.

In order to avoid such complication, a different reaction scheme leading to the film deposition must be used. In this paper, we report the use of hydrolysis-polycondensation reactions in place of pyrolysis reactions for the stoichiometric vapor deposition of $LiTaO_3$ from volatile $LiTa(OBu^n)_6$ as the precursor.

Table I. List of possible double alkoxide precursors.

Double Alkoxide	V.P. (°C/mmHg)	Ceramic/Application
$LiTa(OEt)_6$	230/0.2	
$LiTa(OPr^i)_6$	160-80/0.1	
$LiTa(OBu^t)_6$	110-20/0.1	$LiTaO_3$/pyroelectricity
$LiNb(OPr^i)_6$	<140/0.2	
$LiNb(OBu^t)_6$	110-20/0.1	$LiNbO_3$/E-O
$KNb(OPr^i)_6$	<200/0.8	
$KNb(OBu^t)_6$	<200/0.8	$KNbO_3$/E-O, NLO

V.P.: Vapor Pressure (°C/mmHg)
E-O: Electro-Optics
NLO: Non-Linear Optics

DEPOSITION APPARATUS AND FILM DEPOSITION

A simple low-pressure deposition apparatus is constructed to perform the following functions: (1) thermal vaporization of precursor compounds; (2) transportation of vaporized double alkoxide by carrier gas to the deposition chamber; (3) controlled generation and delivery of water vapor to the deposition chamber; (4) vapor-phase partial-hydrolysis reactions occur in the vicinity of heated substrate; (5) polycondensation of partially hydrolyzed double alkoxide onto substrate surface; and (6) continued hydrolysis-polycondensation reactions on the heated substrate surface. Figure 1 shows schematically the simple deposition apparatus constructed for this study. The advantage of the hydrolysis-polycondensation reaction scheme is that the hydrolysis reaction is essentially an SN2 reaction. The bridging alkoxy group which links M' to M'' is not readily accessible to H_2O molecules due to steric hindrance, thereby preserving the integrity of M' to M'' linkage through out the deposition process. In a general form, the hydrolysis-polycondensation reaction, similar to those in the well known Sol-Gel processing, can be expressed by the following reaction equations:

$$M'M''(OR)_{2n} + 2mH_2O \rightarrow M'M''(OH)_{2m}(OR)_{2(n-m)} + 2mHOR$$

$$M'M''(OH)_{2m}(OR)_{2(n-m)} \xrightarrow{on-substrate} M'M''O_m(OR)_{2(n-m)} + mH_2O$$

$$M'M''O_m(OR)_{2(n-m)} + (n-m)H_2O \xrightarrow{on-substrate} M'M''O_n + 2(n-m)HOR.$$

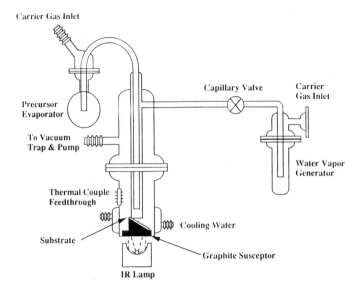

Figure 1. Schematic deposition apparatus

Precursor LiTa(OBut^n)$_6$ (where OBut^n stands for n-butoxy group), a light yellowish solid is prepared with lithium metal granules, 99.5% and Ta(OEt)$_5$, 99.999% from Johnson Matthey Co., and n-butanol, 99.9% from J. T. Baker Co. Lithium metal reacts with n-butanol to form dilute solution of LiOBut^n in n-butanol. Ta(OBut^n)$_5$ is prepared by alcoholysis reaction of Ta(OEt)$_5$ with n-butanol by 24 hours of reflux. The resulting solution is mixed with LiOBut^n in n-butanol, refluxed for 24 hours and then heated to distill out the resulting ethanol and excessive solvent n-butanol. At above 150°C, LiTa(OBut^n)$_6$ is a light yellow liquid which solidifies when cooled to room temperature. The precursor is stored in glove box.

Table II. Typical deposition conditions.

Initial Pressure	0.1-0.4 mbar
Substrate Temperature	450-750°C
Precursor Temperature	195±5°C
Precursor Vapor Pressure (including carrier gas)	3.1-3.3 mbar
Water Vapor Pressure (including carrier gas)	1.3-2.7 mbar

Film deposition is generally performed at the overall pressure of a few milli-bar. Precursor is heated to 195± 5°C. The substrate for deposition, in this study sapphire wafers, are heated with programmable IR line heater via a graphite susceptor to temperatures between 450°C to 750°C. As deposited films are generally of optical quality, fully dense, and amorphous in nature. Samples are removed from the deposition chamber and annealed at various temperatures to achieve full crystallization. Table II summarizes typical deposition conditions used in this study.

RESULTS AND DISCUSSION

One remarkable feature of this method is its high deposition rate. Figure 2(a) shows the SEM cross-sectional view of a film grown on sapphire under the above conditions for

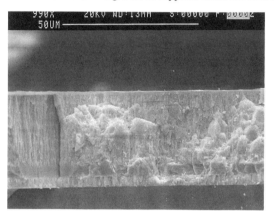

Figure 2(a). SEM picture of LiTaO$_3$ film on sapphire substrate.

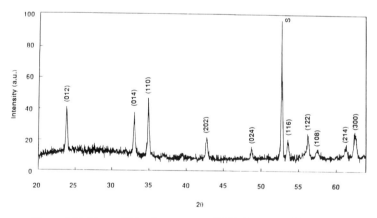

Figure 2(b). X-ray diffraction pattern of LiTaO$_3$ on sapphire substrate.

1.5 hours, annealed at 750°C for 5 hours. Film has measured thickness of 34 μm. The growth rate corresponds to 23 μm per hour which is also the highest observed in this study. Deposition rate as low as 1.8 μm/hour was observed, and even lower rate can be achieved by adjusting the flow rates of precursor and water. Figure 2(b) is the X-ray diffraction pattern of the film. All indexed peaks are from the ilmenite structure of $LiTaO_3$. The peak marked with "S" is that of sapphire (024). It appeared from the X-ray pattern that the film is pure $LiTaO_3$ within the sensitivity of the method.

The films appear to be equally dense before and after annealing, as one would expect from this deposition method. SEM pictures of one film cross-section show essentially same thicknesses before and after annealing. Refractive indices of films measured before and after annealing also confirmed the above observation. Results indicated negligible differences.

Nominal evaporation rate of individual alkoxides, $LiOBut^n$ and $Ta(OBut^n)_5$, and the double alkoxide $LiTa(OBut^n)_6$, as functions of temperature, are studied without carrier gas and water vapor supply from room temperature up to that used for typical depositions. Based on the assumption that the mechanical pump used to evacuate the deposition apparatus has a constant air displacement rate of 160 liter per minute, under typical deposition pressure of a few mbar, gas behavior can be assumed to be nearly ideal. In a dynamic gas flow system, constant volumetric displacement of evacuating pump implies that any flow of gaseous species (in the scale of moles-per-unit time) into the deposition chamber contributes linearly to the overall measurable chamber pressure:

$$P_{tot} = \sum_i P_i = (\frac{RT}{dV/dt})\sum_i (dn_i/dt).$$

where dn_i/dt are the molar rates for (1) apparatus leakage; (2) carrier gas; (3) precursor evaporation. Therefore, simple measure of the overall chamber pressure gives a nominal evaporation rate of precursors at the heater temperature. Figure 3 shows the nominal evaporation rate of $LiOBut^n$, $Ta(OBut^n)_5$, and $LiTa(OBut^n)_6$ as functions of the heater

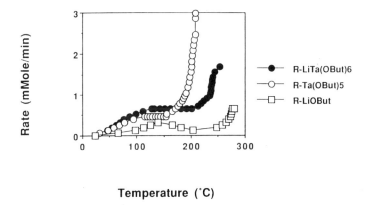

Figure 3. Nominal evaporation rate of individual and double alkoxides.

temperature. Ta(OBut^n)$_5$ appears more volatile than LiOBut^n through out the temperature range. Based on this it can be argued that a mechanical mixture of 1:1 molar ratio of LiOBut^n and Ta(OBut^n)$_5$ will not likely give rise to an equal-molar evaporation of these compounds into the vapor phase. Consequently such mechanical mixture will not result in stoichiometric deposition. It is also interesting to see that LiTa(OBut^n)$_6$ show a relatively constant evaporation rate through a wide evaporator temperature range, from 125°C to 210°C. This perhaps indicates that LiTa(OBut^n)$_6$ is thermally stable at temperature below 210°C, and it has fairly uniform evaporation rate when the evaporator is in this temperature range. It should be noted that the evaporator temperature in this figure is not the true temperature of the precursor, but rather the temperature of the heating mantle. Although a measure of true precursor temperature should be more informative, the above result does provide enough information in this study for the purpose of choosing appropriate evaporator temperatures.

Finally, attempt to identify the exact molecular species responsible for carrying Li and Ta with 1:1 molar ratio from the precursor evaporator to the deposition chamber may prove worthwhile. We do not have elaborate equipment for an in situ study. Instead, we have conducted a mass spectrometry study of LiTa(OBut^n)$_6$ heated from room temperature to 355°C at a constant heating rate. Figure 4 is the mass spectra at 146°C. The mass spectra strongly suggests that the species responsible is the dimmeric form of LiTa(OBut^n)$_6$, the volatile Li$_2$Ta$_2$(OBut^n)$_{12}$, although definite confirmation requires more sophisticated mass spectrometers capable of in situ analysis. Such investigation is under way in our laboratory.

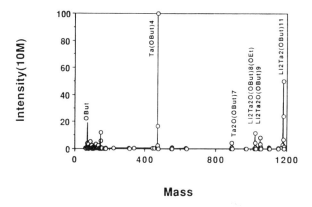

Figure 4. Mass spectrum showing fragmentation patterns of Li$_2$Ta$_2$(OBut^n)$_{12}$.

CONCLUSIONS

In this study, we have demonstrated the possibility of a stoichiometric vapor deposition of LiTaO$_3$ by using a volatile molecular precursor LiTa(OBut^n)$_6$, or perhaps more appropriately, Li$_2$Ta$_2$(OBut^n)$_{12}$. The deposition is conducted by an alternative reaction scheme, namely vapor phase hydrolysis-polycondensation reactions. Such deposition did give rise to high quality stoichiometric LiTaO$_3$ films at deposition rates as high as 23 μm per hour. As deposited films are amorphous, and crystallized upon annealing. The as deposited films appeared to be fully dense, with identical thicknesses and refractive indices before and after annealing. The method should be applicable to all double alkoxides with sufficient thermal stability.

ACKNOWLEDGMENT

This work is partially supported by the Faculty Research Grant of the University of Utah.

REFERENCES

[1] D.C. Bradley, R.C. Mehrotra, and D.P. Gaur, <u>Metal Alkoxides</u>, (Academic Press, London, 1978), pp. 299-334.

PART III

Poster Session

RAMAN SPECTRA OF MOCVD-GROWN FERROELECTRIC PbTiO$_3$ THIN FILMS

Z. C. FENG,[*] B. S. KWAK,[**,+] A. ERBIL,[**] and L. A. BOATNER,[+]
[*]Department of Physics, National University of Singapore, S0511, Singapore;
[**]School of Physics, Georgia Institute of Technology, Atlanta, GA 30332;
[+]Solid State Division, Oak Ridge National Laboratory, Oak Ridge, TN 37831

ABSTRACT

Lead titanate (PbTiO$_3$) has been grown on a variety of substrates by using the metalorganic chemical vapor deposition (MOCVD) technique. The substrates employed included Si, GaAs, MgO, fused-quartz, sapphire, and KTaO$_3$. Raman spectra from these heterostructures are presented. All of the films exhibited the strong, narrow spectral features characteristic of PbTiO$_3$ perovskite-oxide crystals and indicative of high crystalline quality. Temperature behaviors of the Raman modes, including the so-called "soft-mode", were studied. A "difference-Raman" technique was used to distinguish the contributions of the PbTiO$_3$ film and the KTaO$_3$ single-crystal substrate.

INTRODUCTION

Lead titanate (PbTiO$_3$) is an important ferroelectric material for applications in electronic and optical devices. In recent years, the epitaxial growth of PbTiO$_3$ thin films by various techniques and on different substrate materials has attracted a great deal of attention. Metalorganic chemical vapor deposition (MOCVD) is a powerful technique for the growth of PbTiO$_3$ thin films [1-4]. We have grown various PbTiO$_3$ thin films on a variety of substrates by using the MOCVD technique. The substrates employed included Si, GaAs, MgO, sapphire, fused-quartz, and KTaO$_3$.

The selection of these substrates is very significant. The potential combination of ferroelectric devices with Si-based integrated circuits and GaAs-based optoelectronics may open a new field for the applications of ferroelectric thin films. It may also be possible to use ferroelectrics with high dielectric constants to replace SiO$_2$ in dynamic random access memories.

MgO, sapphire or a-Al$_2$O$_3$, and fused quartz are used because of their popular use as substrate materials for various devices and thin film structures, including superconductors. A perovskite oxide, potassium tantalate (KTaO$_3$), was chosen to investigate the combination of two different perovskite oxides.

In the present study, Raman scattering was employed to characterize the resulting PbTiO$_3$ thin films. Raman spectra from these heterostructures, taken at both room and liquid nitrogen temperature and under different excitations, are presented. All of the films exhibited the strong, narrow spectral features characteristic of PbTiO$_3$ perovskite-oxide crystals, indicative of high crystalline quality.

EXPERIMENT

The PbTiO$_3$ depositions were performed in an inverted, vertical, warm-wall reactor with a resistively heated susceptor. Tetraethyl-lead [Pb(C$_2$H$_5$)$_4$], and titanium isopropoxide [Ti(OC$_3$H$_7$)$_4$], were chosen as the carrier gas. Argon was used as the carrier gas. The detailed growth conditions used in forming the lead titanate films have been described elsewhere [1,3,4]. The film thickness in this study ranged between one and several thousand angstroms, as determined by Rutherford backscattering spectroscopy.

All PbTiO$_3$ thin films were first examined by x-ray diffraction (XRD) technique in a digitized horizontal diffractometer from Siemens with monochromatized CuKa radiation source. The sample stage was equipped with rotational and rocking capabilities to insure statistically correct averaging over the reciprocal lattice points.

Raman scattering measurements were performed in a near-backscattering geometry with the excitation source of different lines between 4579 and 5145 Å from an Ar$^+$ laser. The samples were mounted on a Cu block in a liquid nitrogen vacuum dewar, which was set on a micrometer-adjustable stage to allow precise positioning of the samples. The scattered light from the sample was dispersed by a triple spectrometer and subsequently detected by an optical multichannel analyzer (OMA) using a Si detector array. The signals were accumulated under computer control to achieve a high signal-to-noise ratio.

RESULTS AND DISCUSSION

PbTiO$_3$ on Si and GaAs

Figure 1 shows the Raman spectra for a PbTiO$_3$/Si(100) samples, measured at 300 and 80 K. The lines at 520 cm^{-1} (300 K) and 522 cm^{-1} (80 K) which are from Si substrate are the strongest features and not fully displayed in the figure. The broad bands between 900 and 1150 cm^{-1} are due to the second order scattering of the Si phonon modes, which shift up slightly as the temperature is decreased from 300 K to 80 K.

Figure 2 shows the spectra of a MOCVD-grown PbTiO$_3$/GaAs, at 300 and 80 K. The bands at 258 cm^{-1} (300 K) and 262 cm^{-1} (80 K) are due to the GaAs TO mode. The GaAs LO appears as shoulders at the high frequency side of the silent mode, which are located at 290 cm^{-1} (300 K) in Fig. 2a and 293 cm^{-1} (80 K) in Fig. 2b, and clearly splitted out from the silent band at low temperature than at the room temperature measurement.

All other features in Figs. 1 and 2 are characteristic of the bulk PbTiO$_3$ with the assignment and notations followed those of Burns and Scott [5]. Below 800 cm^{-1}, there exist several longitudinal optical (LO) and transverse optical (TO) phonon modes with E or A$_1$ symmetries. To label the modes, we use the numbers, 1, 2 and 3, placed prior to TO or LO inside the parenthesis following the symmetry notation E or A$_1$, as shown in the figures. The lowest bands in Figs. 1a and 1b are the so-called soft mode [6] with the E(TO) symmetry, E(1TO), which are located at 84 cm^{-1} (300 K), and 87 cm^{-1} (80 K), respectively. Near 150 cm^{-1}, an E symmetric LO mode, E(1LO), can be seen in all the four spectra with the band intensity relatively strong at low temperature than at high temperature. Features between E(1TO) and E(1LO) are not well resolved for

PbTiO3/Si in Fig. 1, better distinguishable in Fig. 2a and well resolved with a clear A_1(1TO) peak in Fig. 2b for PbTiO3/GaAs. This indicates that the quality of PbTiO3 film on GaAs is better than that on Si.

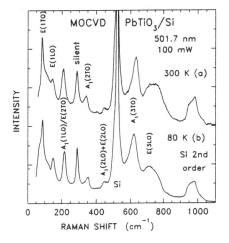
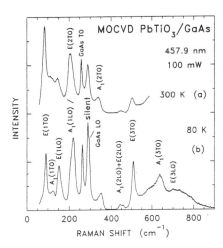

Figure 1   Raman spectra of MOCVD-grown PbTiO3/Si taken at 300 and 80 K under excitation of 501.7 nm and 100 mW.

Figure 2   Raman spectra of MOCVD-grown PbTiO3/GaAs taken at 300 and 80 K under 457.9 nm and 100 mW.

The mixture of A_1(1LO) and E(2TO) are located at just above 200 cm^{-1}. The "silent" (or B_1+E) mode [5], located at 287-288 cm^{-1} for all four spectra, exhibits an unique temperature dependence where the position almost does not vary with the temperature. Figs. 1 and 2 show also other PbTiO3 characteristic modes, A_1(2TO) near 350 cm^{-1}, A_1(2LO)+E(2LO) near 450 cm^{-1}, E(3TO) just above 500 cm^{-1} (unrecognizable in Fig. 1 due to strong Si phonon mode near 520 cm^{-1}), A_1(3TO) close to 550 cm^{-1}, and E(3LO) above 700 cm^{-1}. The comparison between Figs. 1 and 2 shows that E(3TO) and A_1(3TO) bands for PbTiO3/Si in Fig. 1 are much broader than that for PbTiO3/GaAs in Fig. 2. This broadness should be caused by disorders, and also provides evidence that the film quality of PbTiO3/Si is poorer than that of PbTiO3/GaAs.

PbTiO3 on oxides of MgO, fused-quartz and sapphire

Figure 3 shows the Raman spectra for a PbTiO3 film on MgO with all the PbTiO3 characteristic features mentioned above. The rising backgrounds with increasing the wavenumber in the figure are due to the MgO substrate.

Figure 4 shows three Raman spectra from (a) PbTiO3/quartz, (b) bared quartz substrate and (c) their difference. Once again, the spectra clearly show all the PbTiO3 characteristic bands as described above. One can observe that most of PbTiO3 characteristic Raman lines for PbTiO3/MgO in Fig. 3 are narrower than that for PbTiO3/quartz in Fig. 4. This reflects that the film quality of PbTiO3 grown on MgO is better than that of PbTiO3 grown on quartz.

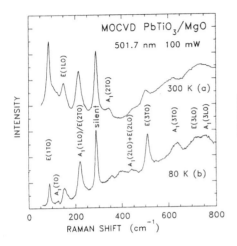

Figure 3 Raman spectra of MOCVD-grown PbTiO$_3$/MgO taken at 300 and 80 K under excitation of 501.7 nm and 100 mW.

Figure 4 Raman spectra of (a) PbTiO$_3$/quartz, (b) quartz, and (c) their difference.

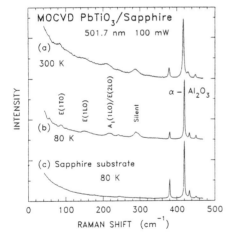

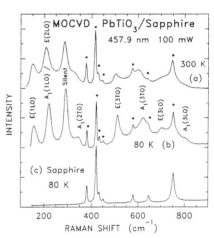

Figure 5 Raman spectra of MOCVD-grown PbTiO$_3$/sapphire taken at (a) 300 K, (b) 80 K, and (c) α-Al$_2$O$_3$ under excitation of 501.7 nm and 100 mW, in the range of 40-470 cm^{-1}.

Figure 6 Raman spectra of MOCVD-grown PbTiO$_3$/sapphire taken at (a) 300 K, (b) 80 K, and (c) α-Al$_2$O$_3$ under excitation of 457.9 nm and 100 mW, in the range of 150-900 cm^{-1}. Asterisks mark the features from α-Al$_2$O$_3$.

Figures 5 and 6 exhibit Raman spectra of MOCVD-grown PbTiO3 on sapphire, i.e. α-Al2O3, under excitation of 501.7 nm in the range of 40-470 cm^{-1} and under excitation of 457.9 nm in the range of 150-900 cm^{-1}. Comparative Raman spectra from sapphire substrate with seven typical Raman lines from α-Al2O3 [6] are shown in Figs. 5c and 6c. These seven α-Al2O3 lines also appear clearly in the spectra for PbTiO3/sapphire measured at 300 and 80 K. The comparison between Figs. 5 and 6 shows that under excitation of 501.7 nm, the intensities of PbTiO3 characteristic Raman lines are much weaker than that of α-Al2O3 lines, while under excitation of 457.9 nm, the intensities of the PbTiO3 lines with respect to the α-Al2O3 lines become much stronger than the case of 501.7 nm excitation. Similar phenomena were observed for PbTiO3 films grown on other substrates. This may be due to resonant enhancement since the high excitation approachs towards the energy gap of PbTiO3. Further investigation is needed to explain these observations.

PbTiO3 on KTaO3

We have grown PbTiO3 on another perovskite oxide KTaO3. But, the Raman spectrum from PbTiO3/KTaO3 is very similar to that from pure KTaO3 [7], shown as in Figs. 7a and 7b. By a direct examination, one cannot find any obvious difference between these two spectra. However, their difference spectrum (Fig. 7c) shows the major features of PbTiO3 as observed in Figs. 1-6 for PbTiO3 grown on other substrates. Particularly, the first four modes in the low frequency range below 300 cm^{-1}, show up clearly. Other PbTiO3 features between 400 and 800 cm^{-1} are also seen. A strong band near 600 cm^{-1} in Fig. 7c is due to KTaO3, which may be caused by the different dispersion properties between PbTiO3 and KTaO3.

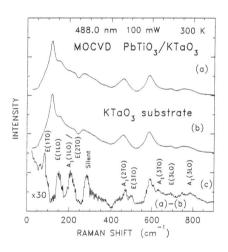

Figure 7   Raman spectra from (a) PTiO3/KTaO3, (b) KTaO3 substrate, taken at 300 K under an excitation of 488.0 nm and 100 mW, and (c) their difference.

Different temperature behaviors of Raman line shifts

As mentioned above, the "silent mode" has its frequency almost unchanged with temperature. All the other PbTiO3 modes, including the soft mode E(1TO), vary their frequencies with temperature. The temperature T dependence of the mode frequency w can be described by the soft mode theory with a relationship [7,8]

$$w = Aw_0|T - T_0|^{1/2} = Aw_0(T_0 - T)^{1/2}, \qquad (1)$$

where A is a constant and w_0 is the mode frequency at the transition temperature T_0. The second equality is valid for T between 80 and 300 K because the T_0 of PbTiO3 is near 500°C [9]. Thus, the frequency of the E(TO) soft mode increases in wavenumber with a decrease of T from 300 K to 80 K. This is also true for other non-silent modes.

The observed Raman line frequencies of $PbTiO_3$ thin films are also different when compared to bulk values. This presumably can be related to substrate-dependent strain-relaxations [10] in the $PbTiO_3$ layers.

CONCLUSION

In summary, we have grown lead titanate (PT) thin films on a variety of substrates, including the most popular semiconductors Si and GaAs, widely-used oxides of MgO, fused quartz and sapphire, and a perovskite oxide $KTaO_3$. Films were characterized by Raman spectroscopy, showing all the $PbTiO_3$ characteristic modes. Comparative Raman spectra indicate that the film quality is better for $PbTiO_3$ grown on GaAs than on Si, better on MgO than on quartz. The "difference-Raman" technique was used successfully to distinguish the contributions of the $PbTiO_3$ film and the $KTaO_3$ single-crystal substrate. The temperature behaviors of the $PbTiO_3$ Raman lines were explained by the soft mode theory. The success of the growth of the $PbTiO_3$ and other complex ferroelectric films on various common wafers by MOCVD growth technique may enhance the development and application of ferroelectric thin films in integrated electronics and ferroelectric devices.

ACKNOWLEDGMENT

We acknowledge Profs. S. Perkowitz and S. H. Tang for their support and help in this work. This work was partially (B.S.K. and A.E.) supported by the Division of Materials Sciences, U.S. Department of Energy under the contract no. DE-AC05-84OR21400 with Martin Marietta Energy Systems, Inc.

REFERENCES

1. B. S. Kwak, E. P. Boyd and A. Erbil, Appl. Phys. Lett. **53**, 1702, (1988).
2. M. de Keijser, G. J. M. Dormans, J. F. M. Cillesen, D. M. de Leeuw and H. W. Zandbergen, Appl. Phys. Lett. **58**, 2636 (1991).
3. B. S. Kwak, K. Zhang, E. P. Boyd, A. Erbil, and B. J. Wilkens, J. Appl. Phys. **69**, 767, (1991).
4. B. S. Kwak, K. Zhang, A. Erbil, B. J. Wilkens, J. D. Budai, M. F. Chisholm, and L. A. Boatner, in *Ceramics Transaction Vol. 25: Symposium on Ferroelectric Films*, ed. by A. F. Bhalla and A. S. Nair (American Ceramic Society, Westerville,1992) p. 203.
5. G. Burns and B. A. Scott, Phys. Rev. B**7**, 3088 (1973).
6. S. P. S. Porto and R. S. Krishnan, J. Chem. Phys. **47**, 1009 (1967).
7. Z. C. Feng, B. S. Kwak, A. Erbil, and L. A. Boatner, Appl. Phys. Lett., **62**, 349 (1993).
8. J. F. Scott, Rev. Mod. Phys. **46**, 83 (1974).
9. N. E. Tornberg and C. H. Perry, J. Chem. Phys. **53**, 2946 (1970).
10. B. S. Kwak, A. Erbil, B. J. Wilkens, J. D. Budai, M. F. Chisholm, and L. A. Boatner, Phys. Rev. Lett. 68, 3733 (1992).

Deposition of Dielectric Thin Films by Irradiation of Condensed Reactant Mixtures

J.F. Moore[*], D.R. Strongin[*], M.W. Ruckman[+], and M. Strongin[+]
* State University of New York, Stony Brook, New York, 11794
+ Physics Department, Brookhaven National Laboratory, Upton, New York, 11973

Abstract

Results of ongoing investigations of a new method for processing molecular solids to form refractory thin films are presented. We have extended past work on boron nitride and alumina growth from molecular solids induced by synchrotron radiation (SR), to silica growth from tetramethylsilane and water at 80 K. Also reported are experiments where alumina is formed using 4.64 eV laser irradiation. The majority of the results suggest that with SR, the film growth is induced by secondary electrons from the substrate, whereas laser light excites the film growth by direct photolysis. Implications of this will be discussed.

Introduction

Much research has recently been devoted to understanding the growth of dielectric thin films, particularly in chemical vapor deposition (CVD) processes. Limitations in current techniques include relatively high processing temperatures, low growth rates, and the need for lithographic processing for patterning. Methods based on volatile reactants in the solid state offer a possible alternative. Since the density of molecules is higher, the rate of growth should follow commensurately. Also, if an excitation source is used to drive the reaction, the processing temperature can be kept low, protecting sensitive substrates and limiting diffusion. Finally, the growth can be initiated selectively, allowing direct write methods to be used.

Several researchers have studied decomposition and deposition from molecular solids at low temperatures.[1,2,3] However, none have attempted dielectric film growth, which generally requires two reacting species, and presents more complex chemistry issues (such as removal of contaminants). We have successfully deposited thin film dielectrics from condensed reactants, particularly BN from diborane and ammonia[4], and alumina from trimethylaluminum (TMA) and water[5]. This paper addresses the most recent work, in which we strive not only to extend the application of this method, but also to understand fundamentally how the film is grown.

Experimental

All experiments were carried out in one of several ultrahigh vacuum (UHV) systems, each having a base pressure below 1×10^{-9} Torr. The synchrotron radiation studies were done at beamlines U7a and U7b at the National Synchrotron Light Source, which provided light having energies from 30 to 1200eV and a flux of $\sim 10^{10}$ photons/cm^{2}sec. The ultraviolet deposition studies used the 4th harmonic of a Nd:YAG laser and a UHV system at the IBM Almaden Research Center[6].

Substrates were formed by evaporation of either silver or gold onto a cooled cryostat, which reached 78 K when filled with liquid nitrogen. Following evaporation, a mixture of gases was introduced through separate stainless steel dosing tubes, forming multilayers of condensed reactants on top of the substrate. Exposure to the excitation source was then carried out, generally for several minutes, followed by warming the substrate to room temperature.

The film was monitored during each of the above steps with a variety of surface sensitive techniques. Soft x-ray photoelectron spectroscopy (SXPS) allows sensitive measurement of core level binding energy shifts and intensities of the reacting species, indicating changing oxidation state and concentration, respectively. Near-edge x-ray absorption fine structure (NEXAFS) gives spectra characteristic of chemical environment for a given species. Auger electron spectroscopy (AES) was used in the laser experiments to check the product film stoichiometry.

Recently, we have added photon-stimulated desorption (PSD) in an effort to elucidate the mechanism behind growth. Following the method of previous workers in this field[7], a detector consisting of a short (6 cm long x 1 cm diameter) drift tube and a pair of channelplates was held a few millimeters from the substrate. The tube was generally biased from -1.0 kV to -1.9 kV, and the substrate was grounded. Light pulses from the NSLS storage ring illuminate the surface for ~2 ns and occur every 170 ns. The time delay between an ion signal in the detector and the next photon pulse was measured with a time-to-amplitude converter and stored in a computer with a pulse height analyzer card.

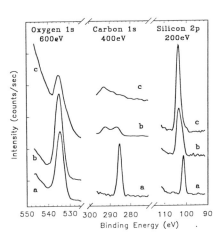

1 SXPS spectra of: a) TMS and H_2O, b) after SR exposure, and c) warmed to 300K.

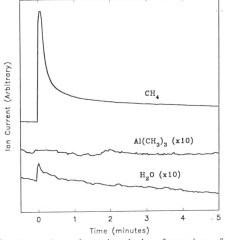

2 Desorption of species during formation of alumina from TMA and H_2O with SR.

Results and Discussion

Previous and forthcoming publications include extensive detail, so we provide only a general overview here. A series of SXPS spectra taken during silica film growth are shown in Figure 1. Clearly there are significant compositional and chemical changes taking place as a condensed tetramethylsilane (TMS) and water mixture is irradiated with SR. The Si 2p core level increases in intensity and shifts to higher binding energy, indicating oxidation. The Si binding energy position in the final product is close to values reported for SiO_2 [8]. The carbon signal decreases dramatically, leaving behind a relatively pure silica film of ~25 Å thickness.

These results are sililar to our previous study of alumina formation from TMA and water, where it was observed that the carbon leaving the surface is primarily in the form of methane. Figure 2 shows the neutral desorption yield from the surface as a function of time, monitored by a quadrupole mass spectrometer. Although there is some desorption of water, the intensity of methane (verified by the fragmentation pattern) is clearly higher and indicates a reaction as the major pathway (over desorption).

We have also attempted film growth by substituting pulsed laser light (4.64 eV) for the SR (150-650 eV). As before, relatively pure alumina films from TMA and water were obtained (representative AES shown in figure 3), but the obtainable thickness is unlimited. It was postulated that the mechanism proceeds through a direct photolytic channel, unlike the SR induced growth, which is primarily secondary-electron mediated. Additionally, it is important to mention that scanning electron micrographs of the laser-deposited films showed a very porous morphology, presumably due to loss of methane from the growing film.

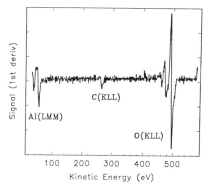

3 AES of alumina formed from condensed TMA and H_2O with a 4.64eV laser.

The photon-stimulated desorption study has given us some insight into possible mechanisms. It is important to point out that this technique uses monochromatic SR, unlike the data from figure 2 which was from "white", or broadband, SR (about 3 orders of magnitude more intense). It would be exceedingly difficult to measure the neutral yield with the monochromatized flux, since the background in the UHV chamber is substantial. Also, the yields observed from photoions are expected to be more representative of the species leaving the surface, since there is no further ionization and fragmentation.

The most intriguing measurement PSD allows is the monitoring of a given mass as a function of photon energy. An example of this, from a condensed mixture of TMA and water on a gold substrate, is shown in figure 4. The H$^+$ intensity (96% of the total ion yield) is measured as the photon energy increases through the Al and Au absorption edges. Although there is a slight (50%) increase at the Al resonance, the broad increase in intensity from 90 to 150 eV (only partially seen in the figure) strongly suggests that secondary electron emission from the gold 4f states (88 and 92 eV) is the major pathway to desorption.

One feature of the growth induced by SR is the exponentially decaying film growth rate with irradiation time. This leads to self-limiting growth at 20-100 Å, depending on conditions, which may be important for applications requiring uniform ultrathin oxides, for example in microelectronic devices or optical nanostructures. This effect is intrinsic to the method, apparently because the growth is being initiated by substrate secondary electrons. We hope to explore this further by extensive PSD studies using varying thicknesses of reactants on a number of different substrates.

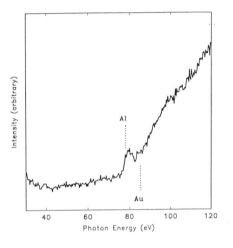

4 Total ion yield from a condensed mixture of TMA and H$_2$O. Absorption edges are indicated.

Acknowledgements

We appreciate the invaluable assistance and equipment from E.D. Johnson of the NSLS, who made the PSD measurements possible. P.B. Comita was our collaborator and host for the laser studies. Fran Loeb and Ray Raynis of BNL provided much help in cryostat design and construction, beamline maintenance and upgrading, and advice.

This work was performed under contract DE-AC02-76CH00016 with the Division of Materials Sciences, U.S. Department of Energy. D.R.S. appreciates support from the N.S.F. (#DMR-9258544).

References

1. P.J. Love, R.T. Loda, R.A. Rosenburg, A.K. Green, and Victor Rehn, SPIE vol.459 Laser Assisted Deposition, Etching, and Doping, 25 (1984).

2. Jun-Ying Zhang and Hilmar Esrom, Applied Surface Science 54, 465 (1992).

3. C.D. Stinespring and A. Freedman, SPIE vol.1190 Laser/Optical Processing of Electronic Materials, 35 (1989).

4. D.R. Strongin, J.K. Mowlem, M.W. Ruckman, and Myron Strongin. Appl. Phys. Let. 60, 2561 (1992).

5. D.R. Strongin, J.F. Moore and M.W. Ruckman. Appl. Phys. Let. 61, 729 (1992).

6. J.F. Moore, P.B. Comita, D.R. Strongin, M.W. Ruckman, and Myron Strongin, forthcoming publication.

7. M.L. Knotek, Springer Series in Chemical Physics v.24, 139 (1983).

8. Tery L. Barr. Appl. Surf. Sci. 15 (1983) 1-35.

INVESTIGATION OF SECOND HARMONIC GENERATION IN MOCVD GROWN BARIUM TITANATE THIN FILMS

BIPIN BIHARI,* XIN LI JIANG,* JAYANT KUMAR,* GREGORY T. STAUF** AND PETER C. VAN BUSKIRK**
*Department of Physics, University of Massachusetts Lowell, One University Avenue, Lowell, MA 01854
**Advanced Technology Materials Inc., 7 Commerce Drive, Danbury, CT 06810

ABSTRACT

Ferroelectric Barium titanate ($BaTiO_3$) thin films have potential applications in optical devices such as frequency doublers. Second harmonic generation (SHG) in MOCVD grown epitaxial thin films of $BaTiO_3$ on MgO substrates is reported at an incident wavelength of 1.064 microns. A number of as-grown films show substantial second harmonic signal. Angular dependence of second harmonic intensity from as-grown films reveals that the orientation of the c-axis in these films are dependent on the growth conditions. We have examined the effect of heating and corona poling on the second harmonic generated by the samples. A higher level of SHG was observed during application of the electric field. On the removal of the poling field the second harmonic signal relaxes back. This behavior may indicate substrate-induced strain stabilization of domains.

INTRODUCTION

Bulk $BaTiO_3$ in the tetragonal phase possesses large optical nonlinearities and has the potential to be exploited for thin film electrooptic-modulation and wave guide frequency doubler applications. Device applications of ferroelectric films require that properties similar to those found in bulk be maintained in deposited films. Thus film characteristics such as stoichiometry, crystallinity, density, micro-structure and crystallographic orientations are important. Recently second harmonic generation from $BaTiO_3$ thin films grown on $LaAlO_3$ substrate have been reported by Lu et al [1]. The authors have shown an improvement in the second order nonlinearity by corona poling although the increased nonlinearity decays somewhat after removal of the poling field. We report second harmonic generation in MOCVD grown $BaTiO_3$ thin films on MgO substrates. Effective d values have been calculated for as-grown films. Effect of in-situ corona poling and temperature are also investigated.

EXPERIMENTAL

Our investigation of second harmonic generation from $BaTiO_3$ thin films was motivated by its possible applications in the integrated optics. Design aspects of the wave guide device and procedure for thin film growth have been given elsewhere [2,3]. Thin films were grown in a inverted vertical MOCVD reactor using Titanium bis-isopropoxide bis(thd) and Ba(thd)$_2$-tetraglyme adduct as the source reagents. The substrate temperature was kept at about 840 °C. Details of the growth conditions are geven in reference 2. MgO as a substrate was selected in order to minimize lattice mismatch between the substrate and film, and at the same time achieve reasonable refractive index difference (minimizing required film thickness to achieve efficient wave guiding). Films deposited were typically smooth and featureless when imaged by scanning electron microscope (SEM) at a magnification of 10^4. X-ray diffraction (XRD) revealed films deposited on MgO [100] to be essentially [100] orientation with trace levels of other orientations [2].

Figure 1 shows the schematic of the second harmonic measurements and in-situ corona poling setup. A fundamental beam from a Q-switched Nd-YAG laser (Quantel 660A) at a wavelength of 1.064 micron was incident on the film mounted on a rotation stage. Fundamental wavelength was removed from the second harmonic signal by using $CuSO_4$ solution and a 532 nm interference filter. The second harmonic signal was detected by a photomultiplier tube, amplified and averaged in a boxcar integrator. In order to make measurements with different polarization of the fundamental radiation a rotatable half-wave plate was inserted between a high power polarizer and sample stage. The p-polarized second harmonic was selected by an analyzer placed before the detector. The second harmonic signal from a Y-cut quartz crystal (d_{11}=0.34 pm/V) was used as a reference to estimate the effective values of d-coefficients for the films. In-situ corona poling was performed using a tungsten needle about 2.5 mm from the film surface at a potential of 5-10 kV.

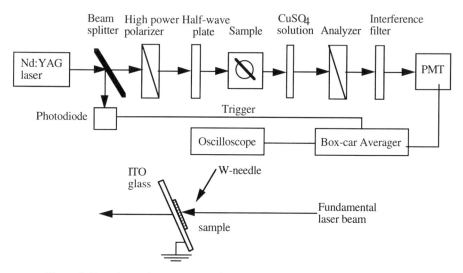

Figure 1. Experimental arrangements for second harmonic measurements and in-situ corona poling

The thicknesses of barium titanate films were almost an order of magnitude less than the coherence length (~3 micron) of bulk barium titanate. For film thicknesses much smaller than the coherence length, the effective values of d-coefficients for $BaTiO_3$ films can be given to a good approximation by [1,4]

$$d_{eff} \sim \frac{l_{c,q}}{l_s} \left[\frac{n(2\omega) \; n^2(\omega) \; I_s}{n_q(2\omega) n_q^2(\omega) T I_q} \right]^{1/2} d_{11,q} \qquad (1)$$

where $l_{c,q}$ (~20 μm) is the coherence length of quartz, l_s is the film thickness, $n(2\omega)$ and $n(\omega)$ are the refractive indices of the film at the second harmonic and fundamental frequencies respectively, $n_q(2\omega)$ and $n_q(\omega)$ are that of the quartz, T is the transmission factor of the film/substrate. Refractive indices of the film were taken as that of the bulk $BaTiO_3$ [5].

OBSERVATIONS AND RESULTS

Effective d-coefficients for the as-grown films, were calculated using expression 1. There was considerable variation in effective d-values from one film to another. The largest values were as high as 2.13 pm/V. These values are much lower than the d-coefficients for the bulk barium titanate ($d_{15} \approx 17.0$, $d_{31} \approx 15.7$ and $d_{33} \approx 6.8$) [6]. The low values of the effective d-coefficients for the thin films can be attributed to the fact that, CVD deposited films form microcrystals that have reduced long range order in the film and if the average microcrystal size is much smaller than the wavelengths of the light in the medium, the average value of d coefficient will be considerably smaller than the bulk single crystal. If the average microcrystal size approaches zero, one would expect a completely isotropic material resulting in no second order response ($d \approx 0$) in the material. Our inability to observe any microstructure at micron levels seems to support the conjecture that the average microcrystal size is much smaller than the wavelength of the Nd:YAG laser in the medium. We assume the second order susceptibility tensor to have the same form as the bulk crystal. The validity of this assumption may be verified by investigating the angular and polarization dependencies of second harmonic signal from these films. Bulk BaTiO3 at room temperature has perovskite type tetragonal structure with point group symmetry 4mm (C_{4v}), which is responsible for its ferroelectric and second order optical nonlinear properties. The second order nonlinear polarization for this symmetry can be expressed as [7,8]

$$P^{NL} = 2 \begin{bmatrix} 0 & 0 & 0 & 0 & d_{15} & 0 \\ 0 & 0 & 0 & d_{24} & 0 & 0 \\ d_{31} & d_{32} & d_{33} & 0 & 0 & 0 \end{bmatrix} \begin{bmatrix} E_x^2 \\ E_y^2 \\ E_z^2 \\ 2E_yE_z \\ 2E_xE_z \\ 2E_xE_y \end{bmatrix} \quad (2)$$

We assume that Klienmann symmetry is approximately valid for barium titanate thin films ($d_{15}=d_{24} \approx d_{31}=d_{32}$). Angular dependence of the second harmonic signals from as-grown samples indicates two kinds of growth on the MgO substrates. Figure 2 shows the angular dependence of second harmonic (p-polarized) intensities from two such samples for incident p-polarized fundamental beam.

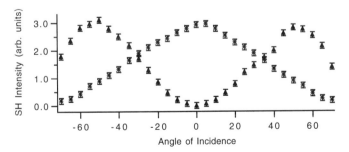

Figure2. Angular dependence of second harmonic intensity for as-grown a-textured and c-textured BaTiO3 films on MgO.

From expression (2) we do not expect any second harmonic signal from a c-textured film (having c-axis perpendicular to the substrate surface), when the fundamental beam is incident normal to the film, whereas it will be maximum for an a-textured film (having c-axis in the plane of the film). Effective d-coefficients for the c-textured films can be given by the expression

$$d_{eff} = 2d_{15}\cos\theta_\omega \sin\theta_\omega \cos\theta_{2\omega} + (d_{31}\cos^2\theta_\omega + d_{33}\sin^2\theta_\omega)\sin\theta_{2\omega} \qquad (3)$$

where θ_ω are given by $n(\omega)\sin\theta_\omega = \sin\theta = n(2\omega)\sin\theta_{2\omega}$, $n(\omega)$ and $n(2\omega)$ are the refractive indices of the NLO material at fundamental and second harmonic frequencies respectively. Effective d-coefficients for an a-textured film for the fundamental beam incident normal to the film can be given by

$$d_{eff} = 2d_{15}\cos\phi_\omega \sin\phi_\omega \cos\phi_{2\omega} + (d_{31}\cos^2\phi_\omega + d_{33}\sin^2\phi_\omega)\sin\phi_{2\omega} \qquad (4)$$

where ϕ_ω and $\phi_{2\omega}$ are the angles between y-axis (b-axis) and the electric field vectors of the fundamental and second harmonic radiation respectively.

Figures 3 shows the p-polarized second harmonic intensity as a function of the polarization of the incident fundamental beam for an a-textured film. The fundamental beam was incident normal

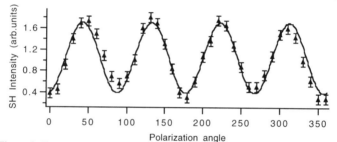

Figure 3. Fundamental polarization dependence of p-polarized second harmonic intensity for as-grown a-textured film.

to the film. A fairly good fit to the function $[A\sin^2(\phi) + B\cos^2(\phi) + C\sin 2(\phi)]^2$, with non-zero A, B and C indicates that indeed, there is an in-plane anisotropy in these films. Figure 4 shows dependence of second harmonic signal on the polarization of the incident fundamental beam for

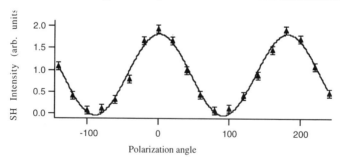

Figure 4. Fundamental polarization dependence of p-polarized second harmonic intensity for as-grown c-textured film.

an as-grown film having no second harmonic signal when fundamental beam was incident normal to the substrate surface. In this case angle of incidence for fundamental beam was kept at $40°$. A function of type $(A\sin^2\phi+B\cos^2\phi)^2$ in this case gives a nearly perfect fit, indicating that these films are isotropic in the plane of the substrate.

In Figure 5 the real time second harmonic signal monitored during heating and corona poling process is shown. An as-grown c-textured film was heated to 116 °C. During heating process a significant reduction in the initial signal was observed. When temperature was stabilized at 116 °C, a negative voltage of about 6.2 kV and 10 µA corona current was applied on the needle for

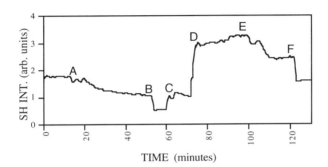

Figure 5. In-situ second harmonic intensity measurement during heating and poling: From A to B temperature was raised to 116 °C; at B negative field (6.2 kV/10 µA) was applied; at C negative field was removed; at D positive field (5.3 kV/2 µA) was applied; E to F temperature was reduced to room temperature in presence of positive field on and at F field was removed as room temperature was achieved.

about 10 minutes. Application of the negative field made the second harmonic signal drop sharply to a lower value, and it remained constant at that level while the negative field was kept on. On removal of the negative field, the second harmonic signal recovered to the value before the application of the field. When a positive voltage of 5.2 kV and corona current of 2 µA was applied, there was a sharp increase in the second harmonic signal. There was no further appreciable increase in the second harmonic intensity for about 30 minutes while this field was kept on. Finally, the sample was allowed to cool down to room temperature in the presence of the positive poling field. On removal of the field the second harmonic signal came down to about its original value. This process was repeated for several samples with some variation in final temperatures and applied electric fields, but no appreciable improvement in the final second harmonic signal was observed. When a similar in-situ poling experiment was performed on an a-textured film, there was a decrease in the second harmonic signal because of the application of electric field, on removal of the field second harmonic signal recovered back to its original value. In Figure 6 effect of heating on a c-textured film is shown. The sample was heated upto 150 °C and then cooled down to room temperature. The second harmonic signal reduced appreciably while heating and increased again on cooling, but did not go to its original value. While, heating the film to 150 °C, second harmonic signal did not reduce to zero, on the other hand, the bulk Barium titanate under goes a phase transition from tetragonal to cubic at about 128 °C and loses its non-centrosymmetric nature. This indicates a contribution of the strain induced second harmonic generation in these films. The observed temperature dependence in $BaTiO_3$ thin films are presumably due to temperature induced changes in the stresses and optical constants of film/substrate system.

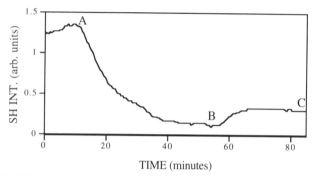

Figure 6. Effect of heating process on the second harmonic intensity: From A to B sample was heated from room temperature to 150 °C; From B to C it was cooled down to room temperature.

In conclusion, we have observed that as-grown BaTiO3 films on MgO substrate have a preferred orientation either in the plane of the film or normal to the film depending upon substrate and growth conditions. Effective d-coefficients are lower than the bulk BaTiO3. Corona poling of these films increases the second harmonic signal only when the field is on, indicating that this increase is field induced. Substrate induced strains seems to determine orientations for this film/substrate system.

REFERENCES

1. H. A. Lu, A. Wills, B. W. Wessels, W. P. Lin, T. J. Zang, G. K. Wong, D. A. Neumayer and T. J. Marks, Appl. Phys. Lett., **62** 1314 (1993).
2. P. C. Van Buskirk, G. T. Stauf, R. Gardiner, P. S. Kirlin, B. Bihari, X. L. Xiang, J. Kumar and G. Gallatin (Mater. Res. Soc. Proc. **310,** Pittsburgh, PA, 1993) pp 119-124.
3. P. C. Van Buskirk, R. Gardiner, P. S. Kirlin and S. Krupanidhi J Vac. Sci. Tech. **A 10**(4) 1578 (1992).
4. Y. R. Shen, The Principles of Nonlinear Optics (J. Wiley and Sons Inc.1984).
5. C. Wong, Y. Y. Teng, J. Ashok, and P. L. H. Varaprasad, Hand Book of Optical Constants of Solids II (Academic Press, 1991) p. 789-803.
6. Hand Book Of Laser Science and Technology, edited by S. Singh (Chemical Rubber, Boca Raton, Fl, 1986) Vol. III. pp.75
7. F. Zernike and J. E. Midwinter Applied Nonlinear Optics (J. Wiley and Sons Inc. 1973).
8. A Yariv, Quantum Electronics (J. Wiley and Sons Inc. 1989).

LEAD TITANATE THIN FILMS PREPARED BY METALLORGANIC CHEMICAL VAPOR DEPOSITION (MOCVD) ON SAPPHIRE, Pt, AND RuO_x SUBSTRATES

WARREN C. HENDRICKS, SESHU B. DESU, AND CHIEN H. PENG
Department of Materials Science and Engineering, Virginia Polytechnic Institute & State University, Blacksburg, VA

ABSTRACT

Transparent and highly specular $PbTiO_3$ thin films were deposited on sapphire, platinum and ruthenium oxide-coated silicon wafers by hot-wall metallorganic chemical vapor deposition (MOCVD). Lead bis-tetramethylheptadionate and titanium ethoxide were used as chemical precursors. Films were deposited over a range of experimental conditions. X-ray diffraction (XRD) was used to determine the phases present in the films; Scanning Electron Microscopy (SEM) was used to examine the surface morphology and Energy Dispersive Spectroscopy (EDS) was used to determine the composition. Optical spectra were obtained to confirm the highly dense and transparent nature of the films. The chemical stability of the ruthenium oxide substrates in the MOCVD environment as well as the existence of a high-temperature deposition regime for composition control are also discussed.

INTRODUCTION

Lead titanate ($PbTiO_3$) based ceramics have been well known for their interesting piezoelectric, pyroelectric, ferroelectric, and electro-optic properties. In recent years, the thin film forms of $PbTiO_3$ (PT) and PT derived materials such as PZT and PLZT have generated considerable interest due to their potential application in nonvolatile ferroelectric RAMs (FRAMs). Undoped lead titanate thin films also find applications in optoelectronic devices, sensors and transducers.

While many processing techniques are available to synthesize lead titanate thin films, metallorganic chemical vapor deposition (MOCVD) is a promising technique for several reasons: the equipment is relatively simple (i. e. no high vacuum equipment or plasma generators are necessary), step-coverage is superior to all other processes, compositional control is extremely flexible and the process can be incorporated into large-scale industrial processing.

In this study, the deposition behavior of lead titanate was studied using the metallorganic precursors, lead bis-tetramethylheptadionate ($Pb(thd)_2$) and titanium ethoxide ($Ti(OEt)_4$). The structure, composition and thicknesses of the resulting films were studied. Several important processing issues are also addressed including the existence of a high temperature self-limiting reaction regime for high reproducibility of film stoichiometry, and also the stability of the ruthenium oxide-coated silicon substrates in the MOCVD process environment.

EXPERIMENTAL PROCEDURE

The metallorganic precursors chosen for this study were lead bis-tetramethylheptadionate ($Pb(thd)_2$) for the lead source, and titanium ethoxide ($Ti(OEt)_4$) for the titanium source. The substrates used in this study include sapphire, as well as multilayer structures consisting of Pt/Ti/SiO$_2$/Si and RuO$_x$/SiO$_2$/Si, which will be referred to hereafter as platinum (Pt) and ruthenium oxide (RuO$_x$) substrates, respectively.

Ruthenium oxide has recently gained recognition as having superior properties to platinum electrodes, especially regarding the fatigue properties of ferroelectric thin films [1,2]. Ruthenium oxide was processed in-house by oxidizing silicon wafers (both (100) and (111) oriented) at 950 °C in a wet oxygen environment, followed by sputter coating the ruthenium oxide using a ruthenium target in a reactive oxygen atmosphere. Finally, the substrates are annealed in air for one hour at 600 °C to relieve residual stress and complete the oxidation of the films.

The hot-walled reactor used consisted of a 2" (5.08 cm) diameter stainless steel tube which was put inside a resistively heated tube furnace resulting in a 30.5 cm heated zone. A vacuum pump and liquid nitrogen cooled cold trap were used to provide a reduced pressure. A bypass line was used to allow the system to equilibrate before beginning the actual deposition process; this has been shown to improve the compositional uniformity[3].

The precursors are stored in stainless steel vacuum-sealed bubblers 1" in diameter. Dry nitrogen was used as the carrier gas and dry oxygen was used as the diluent gas; both were controlled with mass flow controllers. The total pressure inside the reactor was monitored with a pressure sensor and was controlled using a valve installed between the vacuum pump and the cold trap. Substrates were placed on a tilted susceptor at positions of 9.0, 13.0, 17.0, and 21.0 cm from the beginning of the heated zone where the gas stream first enters; each will be referred to subsequently as reactor position #2, #3, #4, and #5, respectively. Deposition was allowed to continue for 30 minutes.

In order to determine the resulting thicknesses of the deposited films, the substrates were weighed both before and after deposition. X-ray diffraction was performed on all of the samples deposited in order to determine the phases present and the crystal structure of the phases. SEM was used to investigate the surface morphology and EDS was used to determine the Pb/Ti ratio of the deposited films. The optical properties of the films on sapphire were studied using UV-VIS-NIR spectrophotometry.

RESULTS AND DISCUSSION

By a largely trial and error process, it was found that reproducible lead titanate films of excellent composition and structure which were optically specular and transparent could be produced using a deposition temperature, T_d, of 550 °C and the following experimental parameters:

TABLE 1: Optimum PbTiO$_3$ MOCVD Parameters for T_d = 550 °C

Total Pressure, p	6 torr
Dilute Gas Flow Rate, f_{O2}	550 sccm
Ti Bubbler Temp., $T_{h,\ TiO2}$	114 °C
Pb Bubbler Temp., $T_{h,\ Pb}$	140 °C
Pb Carrier Gas Flow Rate, $f_{N2,\ Pb}$	50 sccm
Ti Carrier Gas Flow Rate, $f_{N2,\ Ti}$	5 sccm

Typical deposition rates were between 3 and 30 nm/min depending on the reactor location; those films on substrates placed closer to the entrance were invariably thicker. Once these guidelines were established the effects resulting from slight changes in the various experimental parameters regarding the composition, deposition rate (thickness) and crystal structure were examined. Experimental parameters including individual bubbler temperatures, pressure and total gas flow rate were varied independently to determine their respective effects on the deposition behavior. Since the ability to achieve stoichiometric films of reproducible thickness is probably the most important desired characteristic of the process, all of the films were characterized by EDS for compositional analysis and were weighed to determine their thicknesses. The single most important variable, using this criteria, is undoubtedly the titanium bubbler temperature; the effect of small variations in this parameter are shown in Fig. 1.

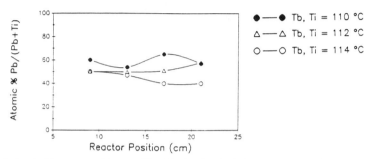

Figure 1: Variation in film composition profile with titanium bubbler temperature for Td = 550 C; p = 6 torr; f_{O2} = 550 sccm

The high sensitivity of the film composition to precursor bubbler temperatures is an indication of a strong dependence of the film composition on the source gas composition. However, in a paper by de Keijser, et al., 1991[4], it was shown that for the MOCVD of lead zirconate titanate (PZT), as the deposition temperature is increased to 600 °C, a process window develops where a range of gas stream compositions will result in stoichiometric PZT. For PZT, stoichiometric is defined as having the ratio of A-site cations to B-site cations equal to 1:1; in other words, the films are 50 mol% PbO; oxygen is assumed to exist in the proper ratio as is generally the case for reasonably low deposition rates. As the deposition temperature increases the width of the process window increases.

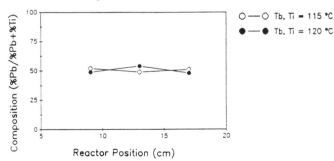

Figure 2: Variation in film composition profile with titanium bubbler temperature for Td = 650 C; p = 6 torr; f_{O2} = 850 sccm

The existence of a process window was confirmed for lead titanate also in the hot-walled reactor used in this study. PbTiO$_3$ was deposited on sapphire using a lead bubbler temperature of 150 °C and a titanium bubbler temperature of 115 °C. The reactor temperature was increased to 650 °C and the oxygen diluent gas flow rate was increased to 850 sccm to account for the higher depletion rate at the higher temperature. Each of the three samples were placed in the reactor at different positions and hence different temperatures, yet each was found to contain exactly 50% PbO and 50% TiO$_2$ within the experimental accuracy of the EDS detector as can be seen in Fig. 2. The only effect was a substantial increase in the deposition rate. To confirm that the process is insensitive to disturbances in the experimental conditions, the titanium bubbler temperature was increased to 120 °C, an increase of 5 °C, which is more than the entire range of titanium bubbler temperatures studied previously; these films were also found to be nearly stoichiometric.

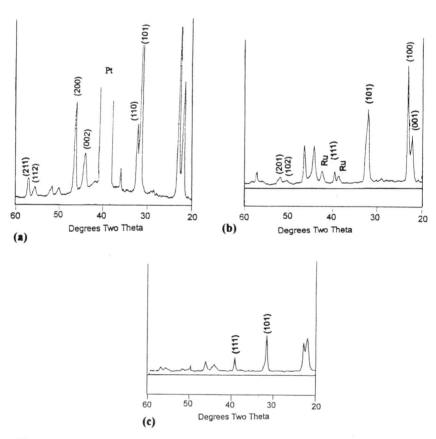

Figure 3: XRD of PbTiO$_3$ deposited on (a) platinum; (b) ruthenium oxide; and (c) sapphire

XRD was used to identify the crystalline phases present; pure perovskite could be produced on all three substrates used as can be seen in Fig. 3. Typical surface morphologies for $PbTiO_3$ films grown on ruthenium oxide at 550 °C and 650 °C, respectively, are given in Figs. 4(a) and (b). The surface morphology of the films deposited at the higher temperature shows an extremely fine-grained microstructure characteristic of a high nucleation rate resulting from the increased deposition temperature. The films deposited at either temperature are generally smooth and fine grained and appear mirror-like and transparent to the eye, as demonstrated by the specular transmission spectrum obtained by UV-VIS-NIR spectrophotometry shown in Fig. 5.

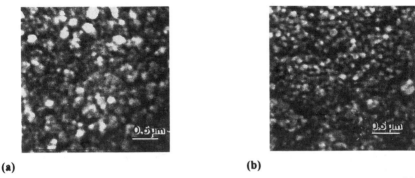

(a) (b)

Figure 4: SEM micrographs of $PbTiO_3$ on ruthenium oxide deposited at (a) 550 C and (b) 600 C

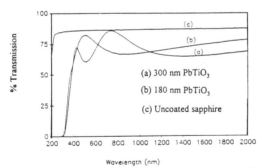

Figure 5: Specular transmission curves for the films on sapphire

The process window exists for the CVD of these lead-based compounds because PbO is much more volatile than either lead titanate or PZT, especially as the temperature exceeds 600 °C. As long as a sufficient quantity of lead precursor is available such that PbO deposits in excess of that required to form the stoichiometric compound, any excess PbO that is deposited is desorbed before it can be incorporated into the film. Thus, only the PbO that actually reacts with the TiO_2 to form $PbTiO_3$ will actually remain. Although it has not been shown, it is suspected that this self-limiting behavior will not occur for extremely high deposition rates as the PbO will not have sufficient time to desorb before the film begins to grow over it.

While successful PbTiO$_3$ films were deposited on ruthenium oxide for temperatures as high as 550 °C, at 650 °C significant problems had occurred with the film peeling. For some wafers, this was even a problem at low temperatures (550 °C). Several observations were made at this time: (1) the properties of substrates cut from the same wafer were generally consistent from one CVD run to the next; (2) film peeling was generally found to occur between the RuO$_2$ layer and the SiO$_2$ layer which could be confirmed by the metallic luster of the underside of the peels; (3) the XRD pattern of peeled and partially peeled films showed the existence of either ruthenium metal and no ruthenium oxide or a mixture of ruthenium metal and ruthenium oxide. Observation #3 was especially noteworthy because in no case was ruthenium metal observed for a film that had not peeled or cracked.

Based on the XRD patterns, it seems that the cause of the film stress is the chemical reaction of the ruthenium oxide electrode material. This is very surprising since ruthenium oxide is the stable phase at the temperature and oxygen partial pressures that are used in the MOCVD system [5,6]. Regardless of whether the reaction occurs by reduction or disproportionation, there will be a large decrease in volume associated with each reaction and consequently a large tensile stress will be generated in the ruthenium/ruthenium oxide film.

To determine the influence of the CVD process on the reduction reaction, the CVD run was performed with the ruthenium oxide substrates as usual but without sending precursor vapors, and hence no film formation resulted. These substrates were completely untransformed after removing them from the reactor; this suggests that either the presence of the film or the reactive species in the CVD gas stream play a role in the ruthenium formation reaction. However, using a nitrogen atmosphere of 6 torr at 550 °C, it was possible to form the Ru even in the absence of precursor species or film. The reason for the variability in the peeling behavior may be related to variations in the ruthenium oxide deposition process; however, further investigations are required to solve this problem. Optical microscopy was used at relatively low magnification to show the cracked and pock-marked surface of the lead titanate coated substrate that had been reduced during processing, seen in Fig. 6.

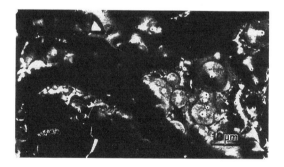

Figure 6: Optical micrograph of cracked surface of PbTiO$_3$-coated RuO$_x$ substrate

SUMMARY AND CONCLUSIONS

Lead titanate was deposited using a hot-walled MOCVD process with $Pb(thd)_2$ as the precursor for the lead source and $Ti(OEt)_4$ as the precursor for the titanium source. The films were deposited under a variety of conditions; the composition of the films was found to be extremely sensitive to the source gas composition at the low temperatures (550 °C and below), while at the higher temperatures (greater than 600 °C) it was found that the composition was insensitive to experimental parameters. XRD was used to study the constituent phases in the films which were found to consist always of the perovskite phase of lead titanate with lead oxide appearing occasionally. Some problems were encountered with the peeling of the ruthenium oxide substrates in the MOCVD environment; this problem is believed to be caused by the formation of ruthenium metal from the ruthenium oxide and the large tensile stress associated with the transformation.

ACKNOWLEDGEMENTS

Jie Si and Chen C. Li deserve great recognition for their many hours spent performing EDS analysis. I would like to give special thanks to Steve Mc Cartney and Chien Chiu and for help in taking the SEM photographs. Wei Pan, Dilip Vijay and Jimmy Xing deserve my appreciation for depositing the ruthenium oxide coated substrates.

REFERENCES

1. C. K. Kwok, D. P. Vijay, S. B. Desu, N. R. Parikh, and E. A. Hill, Proceedings of the 4th International Symposium on Integrated Ferroelectrics, IEEE, 412 (1992).

2. I. K. Yoo, S. B. Desu, Mat. Sci. and Eng. B, **13** (4), 319-322 (1992).

3. C. H. Peng, PhD thesis, Virginia Polytechnic Institute & State University, 1992.

4. M. De Keijser, G. J. M. Dormans, P. J. Van Veldhoven, and P. K. Larsen, Proceedings of the 1992 IEEE ISIF Conference, 243 (1992).

5. E. A. Seddon and K. R. Seddon, Chemistry of Ruthenium, Elsevier, New York, 1984.

6. V. K. Tagirov, D. M. Chizhikov, E. K. Kazenas, and L. K. Shubochkin, Zhurnal Neorganicheskoi Khimii, **20**, 1135 (1975).

DEPOSITION AND PROPERTIES OF ALN FILMS PREPARED BY LOW PRESSURE CVD WITH A METALORGANIC PRECURSOR

M.J. COOK, P.K. WU, N. PATIBANDLA, W.B. HILLIG AND J.B. HUDSON
Materials Engineering Department, Rensselaer Polytechnic Institute, Troy, NY 12180

ABSTRACT

Aluminum nitride films were deposited on Si (100) and sapphire ($1\bar{1}02$) substrates by low pressure chemical vapor deposition using the metalorganic precursor tris-dimethylaluminum amide, $[(CH_3)_2AlNH_2]_3$. Depositions were carried out in a cold wall reactor with substrate temperatures between 500 and 700 °C and precursor temperatures between 50 and 80 °C. The films were analyzed by X-ray photoelectron spectroscopy, X-ray diffraction, optical microscopy and scanning electron microscopy. The films were generally smooth and adherent with colors ranging from transparent to opaque grey. Cracking and spallation were seen to occur at high film thickness. Deposition rates ranged from 20 to 300 Å/min and increased with both precursor and substrate temperature. Carbon concentrations were small, < 5 at. %, while oxygen concentrations were higher and showed a characteristic profile versus depth in the film. High temperature compatibility testing with sapphire/AlN/$MoSi_2$ samples was carried out to determine film effectiveness as a fiber coating in a composite.

INTRODUCTION

Aluminum nitride has received much attention in recent years due to its unique properties. The combination of a high band gap (6 eV), high thermal conductivity (200 W/(m·K)) and close match to silicon in thermal expansion coefficient make it favorable for use in electronic packaging applications, while its piezoelectric nature with a high acoustic wave velocity make it attractive for surface acoustic wave devices. Aluminum nitride also sees use in refractory applications due to its high thermal and chemical stability. This refractory nature and good mechanical properties (elastic modulus of 300 kN/mm² and bending strength of 300 N/mm²) make AlN an option in high temperature structural applications, including as a component in composite materials[1].

Though several techniques have been used to produce AlN films, such as reactive sputtering[2], reactive evaporation[3], and organometallic surface chemical adsorption deposition[4], chemical vapor deposition (CVD) is currently the most widely used. Many precursor systems have been used to produce CVD AlN including: Trimethylaluminum (TMA) and ammonia[5], aluminum trichloride and ammonia[6,7], hexakis(dimethylamido)-dialuminum and ammonia[8,9] and aluminum monoselenide and nitrogen[10]. Organometallic azides[11] and amides[12,13] have been used as single component precursors which decompose to produce AlN. The use of a single precursor has been shown to allow easier film stoichiometry control, allow deposition at lower temperatures and avoid the production of harmful by-products, such as HCl, which can result from other processes.

We report here the production of CVD AlN films using the organometallic precursor Tris-dimethylaluminum amide (TDAA), $[(CH_3)_2AlNH_2]_3$. In an earlier work, Interrante, et al[13], grew smooth, fine-grained films on Si(100) at temperatures between 400 and 800°C in a hot wall reactor. The films adhered well, had good stoichiometry and had a preferred (100) orientation. Oxygen was the main impurity in the films, usually present in amounts between 5 and 10 %. In this work we report the results of investigations of films produced from this precursor on silicon and sapphire substrates in a horizontal, cold wall reactor. The eventual goal of this work is to produce protective coatings on sapphire fibers for use in a $MoSi_2$ matrix composite.

EXPERIMENTAL

The silicon substrates were cleaved from single crystal wafers provided by NBK and had a (100) orientation. The sapphire substrates were 1 cm x 1 cm x 1 mm thick, with a [1$\overline{1}$20] orientation provided by Saphikon, Inc. or Meller Optics, Inc. All substrates were ultrasonically cleaned for five minutes in acetone followed by five minutes in methanol. They were then clipped onto a Mo boat which was in contact with a K-type thermocouple on the underside for temperature measurements. Substrates were annealed in vacuum at 800°C for 30 minutes prior to deposition.

The TDAA precursor was synthesized by Y. Wang in the laboratory of Professor L. Interrante at the RPI chemistry department, using the procedure described in Ref. 12. The final product is a white powder which was loaded into an air tight bubbler inside a dry box to prevent reaction with water vapor. The bubbler was incorporated directly into the deposition system. Before heating the substrate, the bubbler was opened to vacuum for 20 min. prior to deposition to allow any trapped gases to escape.

A schematic of the deposition system is shown in figure 1. Resistance heating of the Mo brought the substrate to its deposition temperature. The Mo boat was held at an angle to the horizontal with the precursor gas directed at its lower end. Due to uneven heating of the boat, higher temperatures were achieved at the lower end. The thermocouple gave temperature readings from a point near the middle of the substrate. Depositions occurred in a 2.5 in. diameter quartz tube which allowed viewing of the depositions. System pressure was maintained by a Balzers TPU 240 turbomolecular pump, which gave a background pressure of about 8 x10^{-8} torr.

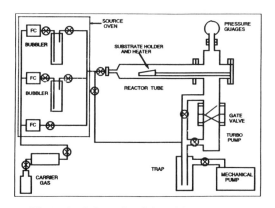

Figure 1 - Schematic of deposition system

The bubbler was held inside a water-filled insulated beaker attached to a Haake D1-L heating bath / circulator which heated the precursor to cause sublimation. An oven containing all gas lines and the bubbler was kept at the precursor temperature to insure that no precursor condensation occurred prior to reaching the sample surface.

Films were deposited through the sublimation and subsequent decomposition of the precursor on contact with the heated substrate. Precursor temperatures (T_p) of 50, 65 and 80°C were used at substrate temperatures (T_s) between 500 and 700 °C. The precursor flow was directed onto the substrate by a tube, attached to the precursor entry port, extending into the reactor tube. Deposition time was normally 30 minutes. Some longer deposition times were used to study thickness effects. After the completion of a deposition, the bubbler valve was closed, the precursor heaters shut off, and the substrate allowed to cool slowly in vacuum to minimize thermal stresses.

Sapphire substrates with AlN films and subsequently sputter deposited $MoSi_2$ were used for chemical compatibility testing at 1200°C. Annealing was done in a tube furnace in flowing N_2.

The films were analyzed to determine structure, composition and deposition rate using X-Ray Diffraction (XRD) and X-ray Photoelectron Spectrometry (XPS). XRD was done using a Philips APD 3520. XPS measurements were made with a Perkin-Elmer 5500 multitechnique analysis system equipped with a sputtering gun for depth profiling. Measurements were made on samples immediately after removal from the deposition system as well as after exposing samples to air for long periods of time (> 24 hrs) to study the oxidation characteristics of the films. Film thicknesses were measured with a profilometer, after the films had been sputtered through, to determine the deposition rate. Note that this procedure yields an average deposition rate, and does not account for the possible presence of an initial nucleation period. Film adherence was tested by scotch tape peeling.

RESULTS AND DISCUSSION

The films deposited on silicon were smooth, transparent and adherent. Cracking, visible under an optical microscope, was encountered when film thickness exceeded approximately 0.4 μm. Close examination of the films during their formation showed that the cracking occurred during growth. When deposition was allowed to continue after cracking started, the film began to take on a cloudy appearance with the cloudy area increasing with time. Eventually, a tan colored powder could be seen forming on the film surface. The cracking occurred as a result of growth stresses in the film, as the amount of cracking increased with film thickness. The scotch tape test showed that these films were highly adherent with only the material that had flaked up being removed.

Two of the films deposited on sapphire (T_s = 700 and 750 °C and T_p = 80°C) showed cracking, visible with the naked eye. The cracking seemed more directional than on the silicon substrates. Again, the cracking was seen to be occurring during film growth. In one case, where deposition time exceeded two hours, ribbons of film spalled from the surface. In another extended deposition, a tan powder appeared on top of the cracked film. Both the "ribbons" and powder came off very easily, while the remaining films underneath were cracked but showed good adherence. Two films deposited on Si with T_s = 600°C and T_p = 50 and 65°C and shorter deposition times (30 and 49 min.) were adherent with no cracking.

XPS results showed that in all cases the major impurity present was oxygen, with the relative concentrations ranging from 3 to over 50 %. Large amounts of oxygen and carbon were detected at the free surface owing to the oxidation characteristics of AlN[14] and the introduction of hydrocarbons through handling. Higher concentrations of oxygen were also observed close to the film-substrate interface as shown in figure 2, a depth profile of a film grown on Si. The lowest atomic percentage of oxygen, ≈2%, was found in the upper half of a 1 μm thick film. The reduction in oxygen concentration with film growth suggests that the amount of contaminating species present in the system decreased with deposition time. The oxygen content can be due to either a high background concentration of water vapor or contaminated precursor.

Testing of samples after different amounts of exposure time to the atmosphere showed that the oxygen atomic concentration increased at the surface with time. This effect was minimal at depths greater than 200 Å, suggesting the oxygen penetration was slowed by the presence of a protective layer, in agreement with the results of Sternitzke[14]. Al/N ratios were usually between 1.2 - 1.5 in the films. The surplus aluminum can be explained by the formation of amorphous aluminum oxides and/or hydroxides at the surface as was witnessed for the oxidation of AlN by Sternitzke[14]. The presence of oxidation in the bulk of the films may be as Al-O-N polytypoids, resulting in a total composition showing excess aluminum. The general structure is described in more detail in Krishnan, et al[15].

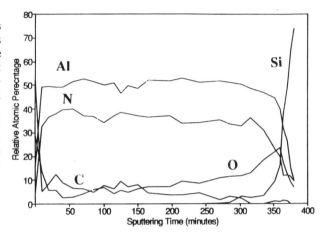

Figure 2 - XPS depth profile of an AlN film on Si. Substrate

X-ray diffraction was used to determine the phases present and the extent of any preferred orientation. Figure 3 shows an XRD pattern for a 1 μm thick film deposited on silicon (100).

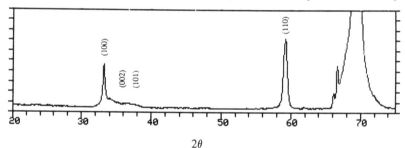

Figure 3. XRD pattern for a 1 μm thick film.

The (100), $2\theta = 33°$, and (110), $2\theta = 59°$, peaks of AlN are visible. A low, broad hump evident on the trailing edge of the (100) peak is due either to the (101) $2\theta = 37.9°$, and/or the (002), $2\theta = 36°$. Films less than 1000 Å thick showed only the (100) peak, indicating a strong preferred orientation with (100) planes parallel to the surface, in agreement with the results of Interrante, et al[13]. The other peaks seen in Figure 3 grew as a function of film thickness, indicating the evolution to a more random orientation.

XRD study of films deposited on sapphire produced no discernable peaks other than those for the substrate. The fact that highly directional cracking was seen on these films suggests that there is some crystallinity present.

Figure 4 is an Arrhenius type plot of deposition rate as a function of reciprocal temperature. Arrhenius behavior is followed at all three precursor temperatures, with apparent activation energies 16.8 kcal/mol at 80°C, 14.5 kcal/mol at 65°C and 8.0 kcal/mol at 50°C. As expected, higher precursor temperatures lead to higher deposition rates due to increased precursor vapor pressure. Higher substrate temperatures also lead to higher deposition rates.

Interpretation of the growth rate data is complicated by the fact that the precursor used has been shown previously to decompose on vaporization to a mixture of the original trimer plus both lower (dimer) and higher (pentamer) molecular weight species [16]. The apparent activation energy for the deposition process will thus contain contributions from both the disproportionation during the vaporization process and the energetics of the decomposition of the various vapor species on the growth surface. More detailed studies of the vaporization and surface decomposition processes are required for mechanistic interpretation of the overall growth kinetics.

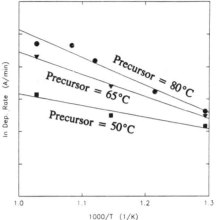

Figure 4. Arrhenius plot of AlN dep. rate vs. temperature.

Two AlN films deposited on sapphire were sputter coated with 1 μm thick layer of $MoSi_2$. One of these sample was tested for adherence using Scotch tape. The silicide as well as the AlN were completely lifted off the sapphire by the tape, as shown by XPS measurements subsequent to the adhesion test. Thus AlN is more adherent to the $MoSi_2$ than to the sapphire. This should not be a result of thermal stresses since the expansion coefficients of $MoSi_2$ and sapphire are closely matched. The adherence may be the result of a better lattice match between the AlN and $MoSi_2$, or to the presence of impurities at the AlN-Al_2O_3 interface.

The other sample was annealed at 1200°C for 24 hours in a flowing N_2 atmosphere; most of the $MoSi_2$ layer flaked off during the annealing, with only a small amount remaining at one edge. XPS investigation of the annealed sample showed that no N could be detected where the film had flaked off. Analysis of the piece of $MoSi_2$ that remained showed that it had oxidized to form a SiO_2 layer at the outer surface, which is characteristic of this material upon heating to 1200°C. Underneath this layer only molybdenum, aluminum, silicon and oxygen were found, suggesting that the aluminum nitride had oxidized.

CONCLUSIONS

Polycrystalline aluminum nitride films can be grown from the organometallic precursor tris-dimethylaluminum amide in a cold-wall reactor. Smooth adherent films were grown at substrate temperatures between 500 and 700 °C over the precursor temperature range 50 to 80 °C. The films contained oxygen, most likely due to contamination of the deposition system or precursor material. On Si(100) substrates the films grew with a (100) orientation at first, switching to a more random orientation at film thickness >0.6 μm. Cracking and flaking of the films occurred during deposition as a result of growth stresses. Deposition rates ranged from 350 Å/min with $T_s = 700°C$ and $T_p = 80°C$ to <50 Å/min at lower temperatures. It was found that the AlN is more adherent to a layer of $MoSi_2$ sputter deposited on it than to the sapphire it was deposited on. High temperature annealing of a sapphire/AlN/$MoSi_2$ sample showed that the AlN oxidized, with no nitrogen being detected in the remaining structure. The deposition rate increased with increased substrate temperature and with increasing precursor source temperature.

ACKNOWLEDGEMENT

This project is supported by funding from ONR/DARPA. Construction of the deposition system was funded by NYSERDA.

REFERENCES

1. W.J. Lackey, G.B. Freeman, et al, "Ultrafine Microstructure Composites Prepared by Chemical Vapor Deposition", Final Report A-4699-3, Georgia Tech Research Institute, Georgia Institute of Technology, Atlanta, GA.
2. Siettman and Aita, J. Vac. Sci. Tech., **A6**, 1712 (1988).
3. Itoh and Misawa, Jpn. J. Appl. Phys., **Suppl 2**, 467 (1974).
4. Z.J. Yu, J.H. Edgar, A.U. Ahmed and A. Rys, J. Electrochem. Soc., **138**, 196 (1991).
5. Y. Chubachi, K. Sato and K. Kojima, Thin Solid Films, **122**, 259 (1984).
6. T.L. Chu, D.W. Ing and A.J. Noreika, Sol. St. Elec., **10**, 1023 (1967)
7. Y.G. Roman and A.P.M. Adriaansen, Thin Solid Films, **169**, 241 (1989)
8. R.G. Gordon, D.M. Hoffman and U. Riaz, J. Mater. Res., **6**, 5 (1991)
9. R.G. Gordon, U. Riaz and D.M. Hoffman, J. Mater. Res., **7**, 1679 (1992)
10. P.M. Dryburgh, J. Cryst. Gr., **94**, 23 (1989)
11. D.C. Boyd, R.T. Haasch, D.R. Mantell, R.K. Schulze, J.F. Evans and W.L. Gladfelter, Chem. of Mater., **1**, 119 (1989)
12. L.V. Interrante, L.E. Carpenter II, C. Whitmarsh and W. Lee in Better Ceramics Through Chemistry II, edited by C.J. Brinker, D.E. Clark and D.R. Ulrich (Mater. Res. Soc. Proc. **73**, Pittsburgh, PA, 1986)pp. 359-366.
13. L.V. Interrante, W. Lee, M. McConnell, N. Lewis and E. Hall, J. Electrochem. Soc., **136**, 472 (1989).
14. M. Sternitzke, J. Am. Ceram. Soc., **76**, 2289 (1993).
15. K.M. Krishnan, R.S. Rai, G. Thomas, N.D. Corbin and J.W. McCauley in Defect Properties and Processing of High-Technology Nonmetallic Materials, edited by Y. Chen, W.D. Kingery and R.J. Stokes (Mater. Res. Soc. Proc. **60**, Pittsburgh, PA, 1986) p. 211.
16. C.C. Amato-Wierda, PhD thesis, Rensselaer Polytechnic Institute, 1993.

ELECTRICAL AND DIELECTRIC CHARACTERISTICS OF SrTiO$_3$ THIN FILMS GROWN BY PE-MOCVD TECHNIQUE

Z.Q. Shi, C. Chern and S. Liang, EMCORE Corporation, Somerset, NJ 08873; Y. Lu and A. Safari, College of Engineering, Rutgers University.

Abstract

Epitaxial strontium titanate (SrTiO$_3$) thin films have been grown by plasma enhanced metalorganic chemical vapor deposition (PE-MOCVD) technique on different bottom electrode materials, such as Pt/MgO and YBCO/LaAlO$_3$. The as grown SrTiO$_3$ film exhibited an epitaxial structure with <100> orientation perpendicular to the substrates as examined by X-ray diffraction (XRD). The electrical and dielectric properties of the films were investigated by capacitance-voltage (C-V) and current-voltage (I-V) measurements in a temperature range from 80K to 300K. The dielectric constant and dielectric loss were found to be 320 and 0.08 at 100 kHz and room temperature. The dc leakage current density and breakdown voltage were strongly dependent on the choice of the bottom electrode materials. For the films grown on YBCO/LaAlO$_3$, the leakage current density is 8.8×10^{-7} A/cm^2 at 200 kV/cm and the breakdown voltage is about 2.0 MV/cm. These results indicated that the SrTiO$_3$ films are suitable for many devices applications. The frequency dependence of the dielectric constant and dielectric loss were studied in the range of 12 Hz to 1 MHz at room temperature. With the increase of the frequency, the dielectric constant showed a little decrease, while the dielectric loss exhibited a sharp increase which could be attributed to the electrode resistance loss.

Introduction

SrTiO$_3$ is a promising dielectric material for future high density DRAMs applications because of its high dielectric constant, good chemical stability and good insulating properties. Its dielectric constant does not sharply decrease at high frequency as encouraged in ferroelectric PZT material[1]. Chemical vapor deposition[2,3] and other deposition techniques such as RF sputtering[4], pulsed laser ablation[5] have been used to grow SrTiO$_3$ thin films. For practical applications of the SrTiO$_3$ thin films, there are special requirements for its dielectric constant, capacitance per unit area, dielectric loss, dc leakage current and dielectric breakdown strength. In addition, the frequency dependence, temperature dependence and applied voltage dependence of its electrical and dielectric properties are also very important for high frequency applications such as ULSI DRAMs. The growth conditions and the microstructures of the SrTiO$_3$ films have been reported elsewhere[6]. In this paper, the electrical and dielectric properties of the SrTiO3 films deposited by MOCVD technique are reported.

Experimental

SrTiO$_3$ (STO) thin films were grown on YBCO/LaAlO$_3$ substrate in an EMCORE oxide PE-MOCVD system. The β-diketonate complex of Sr(dpm)$_2$ and titanium isopropoxide (TIP) were used as strontium and titanium sources. Ultra high purity (99.995%) oxygen, with a flow rate of 300 sccm, was used as the oxidant. Argon was used as carrier gas at a flow rate of 100 sccm for Sr(dpm)$_2$ and 70 sccm for TIP, respectively. The bubble temperature for Sr(dpm)$_2$ and TIP were maintained at 220°C and 23.5 °C, respectively. The TIP bubbler pressure was kept at 450 Torr to provide stabilized TIP transport flux and to match that of Sr(dpm)$_2$. Reactor chamber pressure during the deposition was 10 Torr with a nominal oxygen partial pressure of about 5 Torr. The substrate temperature was controlled at 680 - 700°C.

To measure the electrical and dielectric properties of the thin films, an Au pad with a diameter of 0.6 mm was sputtered on the top of the STO film to form a metal-insulator-metal (MIM) capacitor. As shown in Figure 1, the YBCO serves not only as a lattice-matched template layer but also as a bottom electrode of the capacitor. All of the tests are measured by both single-dot (SD) and double-dot (DD) configurations.

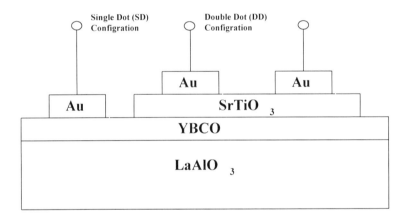

Figure 1: MIM capacitor structure used for I-V, C-V, C-f, C-T measurements.

The current-voltage (I-V) characteristics of the capacitors were measured at room temperature using a Keithley model 614 electrometer and 230 programmable voltage source. The dc leakage current density of the thin film was determined from I-V curve at a voltage of +2.0V. The capacitance-voltage (C-V) characteristics of the capacitor were obtained using a HP 4194A Impedance/Gain-Phase Analyzer in the frequency range of 10 Hz to 10 MHz. The relative permittivity, ε_r of the film was calculated from the C-V curve and by knowing the film thickness as well as the capacitor area. The capacitance-temperature (C-T) and dissipation factor-temperature (tanδ-T) measurements at frequencies of 1, 10 and 100 KHz were obtained by using HP 4194A Impedance /Gain-Phase Analyzer in the temperature range of -170 to 50°C. The Curie temperature (Tc) of the STO films was proved not in this range from the C-T measurement (Tc of STO is at about -241°C). The frequency dependence of the relative permittivity and dissipation factor of the films were measured at room temperature from 10 Hz to 10 MHz.

Results and Discussion

Figure 2 shows the current-voltage (I-V) characteristics of the films with a thickness of 1500Å. The dc leakage current densities are less than $1\times10^{-6} A/cm^2$ at 2.0V. This indicates that the films have a very good insulating property and can meet the requirement for the practical device application.

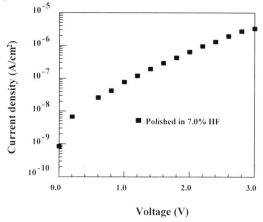

Figure 2 : Current-voltage (I-V) characteristics of a STO film

Figure 3 shows the voltage dependence of the relative permittivity (ε_r) for the STO films. The test was performed at frequencies of 100KHz and 1MHz. The relative permittivity decreased a little at high applied voltage due to the increase of dc leakage current as shown in Figure 2.

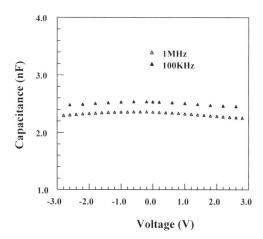

Figure 3 : The voltage dependence of the relative permittivity (ε_r) for the STO films.

The frequency dependence of the relative permittivity (ε_r) and dissipation factor (tanδ) for the STO films are shown in Figures 4 and 5. The ε_r exhibits a monotonic decrease with increasing frequency. The ε_r values decreased with increasing frequency from 10KHz to 1MHz. This frequency range is just located in the ε_r transition region.

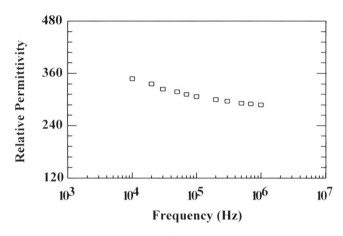

Figure 4 : Frequency dependence of the relative permittivity of STO thin film.

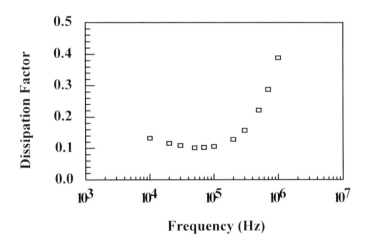

Figure 5. Frequency dependence of dissipation factor of STO thin film.

The tanδ, however, shows a minimum value of 0.10 at a frequency of around 100 kHz. At frequency lower than 100KHz, tanδ increased with increasing frequency. This phenomenon can be explained by the loss mechanism from the polarization relaxation. At frequency higher than 100KHz, tanδ increased with decreasing frequency. This is due to the series resistance effect from the bottom electrodes.

Conclusion

High quality $SrTiO_3$ thin films have been successfully grown by plasma enhanced metalorganic chemical vapor deposition (PE-MOCVD) technique. The dielectric constant and dielectric loss were found to be 320 and 0.08 at 100 kHz and room temperature. The dc leakage current density of the films grown on $YBCO/LaAlO_3$ is 8.8×10^{-7} A/cm^2 at 200 kV/cm. The breakdown voltage is about 2.0 MV/cm. At higher frequency, the dielectric constant showed a little decrease, while the dielectric loss exhibited a sharp increase which could be attributed to the electrode contact loss. The dc leakage current density and breakdown properties were strongly dependent on the microstructure of the films. The dielectric constant, however, was not only dependent on the crystal structure but also the interface layer between the film and electrodes.

References

1. J.F. Scott, Proceedings of the 8th ISAF, 356, August 31-September 2, 1992.
2. A. Greenwald, Mat. Res. Soc. Symp. Proc., Vol. 243, 457, (1992).
3. L.A. Wills, J. Crystal Growth, Vol. 107, 712, January (1991)
4. T. Sakuma, Appl. Phys. Lett., 57 (23), 2431, December (1990)
5. R.F. Pinizzotto, Mat. Res. Soc. Symp. Proc., Vol. 243, 463 (1992)
6. S. Liang, C. Chern, and Z.Q. Shi, will be published in Mat. Res. Soc. Symp. Proc., (1993)

MOCVD GROWTH OF EPITAXIAL $SrTiO_3$ THIN FILMS ON $YBa_2Cu_3O_{7-x}$/$LaAlO_3$

S. Liang, C. Chern, and Z.Q. Shi, EMCORE Corporation, Somerset, NJ 08873; Y. Lu, P. Lu, B.H. Kear and A. Safari, College of Engineering, Rutgers University, Piscataway, NJ 08855-0909.

Abstract

Strontium titanate ($SrTiO_3$) thin films have been epitaxially grown on $YB_2Cu_3O_{7-x}$ (YBCO)/$LaAlO_3$ at substrate temperatures of 660 to 700 °C. X-ray diffraction (XRD) results indicated that single crystalline $SrTiO_3$ thin films were epitaxially grown on the substrate with <100> orientation perpendicular to the substrates. The compositions of the films with different growth conditions were examined by Rutherford backscattering spectroscopy (RBS) and energy dispersive x-ray spectroscopy (EDX). The ratio of Sr/Ti is in the range of 0.9 to 1.1 for the films with a thickness of 1000-2000Å. The surface morphology of the films and the interfaces of the $SrTiO_3$/YBCO structure were examined by scanning electron microscopy (SEM) and high resolution transmission electron microscopy (HRTEM). Very smooth surface and sharp interface were observed. The superconducting property of the YBCO layer, as measured by ac susceptibility, did not degrade after growth of $SrTiO_3$ film. The dielectric constant as high as 320 was obtained at 100KHz. The leakage current density is less than $1 \times 10^{-6} A/cm^2$ at 3V operation.

Introduction

$SrTiO_3$ (STO) thin films have variety of applications in multi-chip modules (MCMs), DRAMs and other storage capacitors. Various deposition techniques have been explored to grow high quality STO thin films. These techniques include RF sputtering [1], co-evaporation [2-3], laser ablation [4], and chemical vapor deposition (CVD) [5-6]. Among these methods, CVD technique was found to be the most promising approach because of its capability of step coverage, large area uniform deposition, and easy control of film composition. For device applications, the co-relationship between growth conditions and microstructures of the films is important for understanding the growth mechanism and obtaining the desirable material properties. So far, however, no details about the correlation of growth and microstructures have been reported. In this paper, we will report the process of the thin film growth and the results of the microstructural analysis

Experimental

A commercial PE-MOCVD system has been used for $SrTiO_3$ thin film growth. The information of the system was reported elsewhere [7]. Ti(i-propoxide)$_4$ and Sr(thd)$_2$ were chosen as Ti and Sr metalorganic sources, respectively. The gas delivery system is a standard gas panel for most MOCVD systems except that the Sr transport lines need heating control. All transport lines involving Sr transportation were maintained at about 30°C above the Sr bubbler temperature. A pressure balance setup was applied to control the Ti bubbler pressure to provide stabilized TIP transport flux and to match that of Sr(thd)$_2$ since the TIP is too volatile. Deposition was carried out at a system pressure of 10-30 Torr and at substrate temperature of 650-750°C with rotation speed of 900~1400 rpm. A microwave oxygen plasma (2.45GHz, 300 Watt) was used in order to reduce the substrate temperature required for epitaxial growth of STO thin films. The substrates used were $YB_2Cu_3O_{7-x}$ (YBCO) thin films on $LaAlO_3$. The YBCO films were mechanically and chemically polished in 7% HF solution (The details of the substrate preparation will be published elsewhere [8]). The growth rate for large-area deposition is controlled in the range of 200A to 500A/hour.

EDX and RBS were utilized to determine the compositions of the STO films. XRD θ-2θ and ϕ scan were used to confirm the film orientation and phase. The film crystallinity was tested by the XRD ω scan. The microstructures and interface properties were investigated by HRTEM. The electric properties were evaluated by C-V and I-V measurements. Also, the growth effects on the substrate YBCO were tested by AC susceptance measurements.

Results and Discussions

EDX and RBS were used to determine the composition (Sr/Ti) of the STO thin films. Two percent variation was estimated in measurement accuracy. Figure 1 shows the Sr bubbler temperature dependence of the absolute metal atomic ratio of the STO films. The growth conditions were fixed at a substrate temperature of 700°C, a TIP bubbler temperature of 23.9°C, and a TIP bubbler pressure of about 500 Torr in Figure 1.

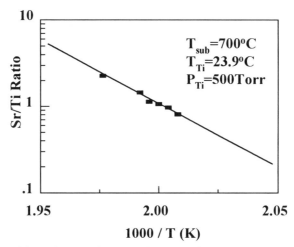

Figure 1. Sr/Ti atomic ratio of the STO films vs. Sr bubbler temperature.

XRD θ–2θ scan was used to determine the crystal phase and preferred orientation of the grown films. From the XRD θ–2θ pattern, it was confirmed that the STO films with (001) orientation were deposited. Figure 2 displays the XRD ϕ-scan patterns of STO (202), YBCO (108), and $LaAlO_3$ (202) reflections. All of these ϕ-scans exhibit modulation patterns, i.e., peaks occurring every 90°, indicating that the YBCO and STO films do not have in-plane misorientations. The crystallinity of the STO film was verified by the XRD ω

scan. The full width half maximum of (200) STO is about 1000 seconds. Figure 3 is the HRTEM micrograph showing the cross section of a STO/YBCO/LaAlO$_3$ sample. The interface sharpness between STO and YBCO is in the atomic layer range.

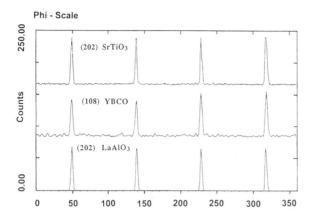

Figure 2. XRD φ scan of (202) STO, (108) YBCO and (202) LaAlO$_3$.

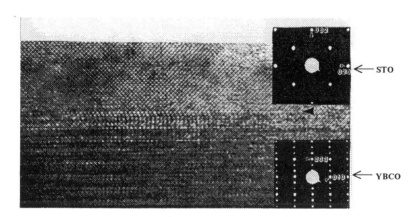

Figure 3. HRTEM image of (100) STO / (001) YBCO deposited by PE-MOCVD.

The electrical properties of the STO films were characterized by current - voltage, capacitance - voltage, capacitance and loss tanδ vs. frequency measurements. For the samples with a thickness of about 1500Å, the dielectric constant of the STO films is in the range of 250 - 320. The loss tanδ is about 10% at test frequency of 0.1 MHz. The leakage current density of the STO films in the range of $\mu A/cm^2$ at field of $10^5 V/cm$. It was found that the electrical properties depend on the YBCO polishing process and the growth conditions.

The AC magnetic susceptibility measurement of YBCO films before and after STO film growth is shown in Figure 4. There is a broad transition of the YBCO film before STO film growth on it. The transition became sharper after STO deposition. The PE-MOCVD growth of the STO film on top of YBCO film did not degrade the superconductor property of the bottom YBCO layer.

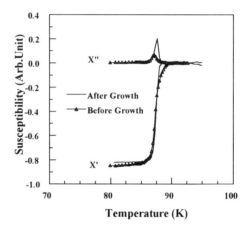

Figure 4. AC magnetic susceptibility measurement of YBCO films before and after STO film growth. The upper curves (with positive values) represent the imaginary part, and the lower curves (with negative values) represent the real part.

Summary

The epitaxial (100) $SrTiO_3$ thin film growth was successfully grown on YBCO substrates by a PE-MOCVD technique. There are no mixing phases and no misorientation in the deposited STO films. The interface between the STO films and the YBCO substrates is sharp (in atomic layer range). The electrical and structural properties of the grown $SrTiO_3$ films demonstrated the potential device applications.

References

1. T. Sakum et al., Appl. Phys. Lett. **57** (23), 2431 (1990).
2. H. Yamaguchi et al., Jap. J. Appl. Phys. **30** (9B), 2197 (1991).
3. H. Yamaguchi et al., Proceedings of the 8th ISAF, 1992, pp.285.
4. R.F. Pinizzotto et al., Mat. Res. Soc. Symp. Proc. **243**, 1992, pp.463.
5. E. Fujii et al., IEDM 1992, pp.267.
6. H. Yamaguchi et al., Jap. J. Appl. Phys. **30** (9B), 2193 (1991).
7. J. Zhao, et al., Appl. Phys. Lett. **56** (23), 2342 (1990).
8. S. Liang, et al., presented at the 1993 MRS Fall Meeting, Boston, MA, 1993 (unpublished).

DEPOSITION OF TITANIUM OXIDE FILMS FROM METAL-ORGANIC PRECURSOR BY ELECTRON CYCLOTRON RESONANCE PLASMA-ASSISTED CHEMICAL VAPOR DEPOSITION

ATSUSHI NAGAHORI AND RISHI RAJ
Department of Materials Science and Engineering, Cornell University, Ithaca, NY 14853

ABSTRACT

Titanium oxide thin films were deposited at room temperature by the Electron Cyclotron Resonance Plasma-Assisted Chemical Vapor Deposition (ECR-PACVD) method. Effects of deposition temperature, microwave power and plasma gas pressure were investigated. Both Ar and O_2 plasma were used to deposit films from titanium isopropoxide ($Ti(OC_3H_7)_4$) precursor. Transparent titanium oxide films were obtained with O_2 plasma. Refractive index and thickness were measured by ellipsometry and the films were characterized by x-ray diffraction.

INTRODUCTION

Electron cyclotron resonance (ECR) plasma has been widely used for deposition and etching in recent years[1,2]. Compared with other plasma techniques, e.g., DC and RF plasma, ECR plasma can be efficiently created with high plasma density at low pressure (10^{-3}-10^{-5} torr) and is useful at low temperature[3]. Therefore, low volatile metal-organic precursor for MOCVD can be applied with this technique and high yield and low temperature depositions are expected.

EXPERIMENTAL

The schematic diagram of the ECR-PACVD system is shown in Figure 1. The microwave power (2.45 GHz) was introduced into the plasma chamber through a rectangular wave guide and a quartz window. Two magnetic coils which were placed outside the plasma chamber maintained the plasma condition (875 G of magnetic flux density). High purity Ar (99.999%) or O_2 (99.995%) was chosen as a plasma gas.

A plasma stream was introduced into the deposition chamber and directed at a specimen placed at the center of the deposition chamber. Titanium isopropoxide (purity 97%) was used as

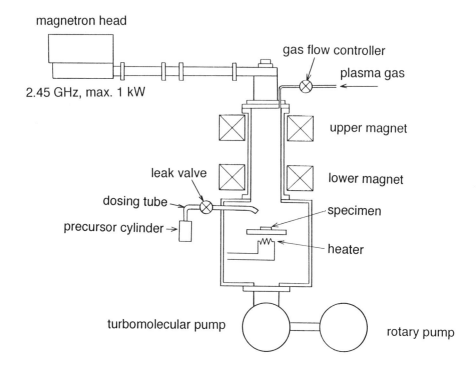

Fig. 1. Schematic diagram of ECR-PACVD system

the metal-organic precursor. The liquid precursor was reserved in a precursor cylinder and vapor was fed into the deposition chamber through a dosing tube. Both the precursor cylinder and the dosing tube were maintained at room temperature. The chamber base pressure was below 10^{-7} torr. A turbomolecular pump and a mechanical rotary pump were used to pump the system.

Titanium oxide films were deposited on sapphire (0001) and silicon (111) substrates. The substrate temperature was controlled from room temperature to 750°C during deposition with Ar plasma but the substrates were not heated during deposition with O_2 plasma. A gas flow controller was used for controlling the gas flow rate of plasma gas from 2.5 sccm. to 10.0 sccm and the chamber pressure was kept from 2.5×10^{-4} torr to 10^{-3} torr. The precursor was introduced into the deposition chamber using a leak valve and partial pressure of precursor in the chamber was 10^{-5} torr during deposition. Thicknesses and refractive indexes of the films were measured by ellipsometry. The profilometer was also used for measuring thickness.

RESULTS AND DISCUSSION

Depositions with Ar plasma

Figure 2 shows the deposition rate of titanium oxide films decreases with substrate temperature. The films deposited at lower temperature (below 500°C) were not transparent. The x-ray diffraction data shows Ti_2O was deposited instead of TiO_2 (Figure 3).

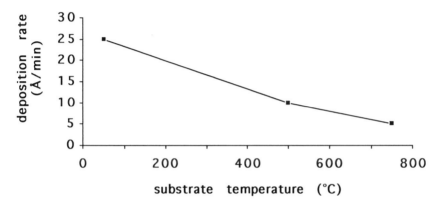

Fig. 2. Deposition rate of titanium oxide films with substrate temperature

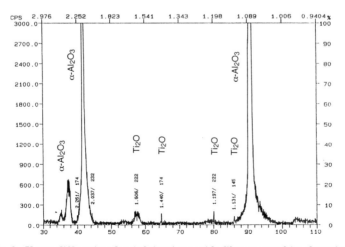

Fig. 3. X-ray diffraction data of titanium oxide film on sapphire deposited at room temperature with Ar plasma

Depositions with O_2 plasma

Transparent titanium oxide films were obtained with O_2 plasma. Effects of microwave power on deposition rate and refractive index of titanium oxide films are shown in Figures 4 and 5.

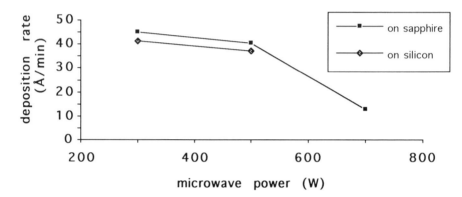

Fig. 4. Deposition rate of titanium oxide films with microwave power for the oxygen pressure of 10^{-3} torr at room temperature

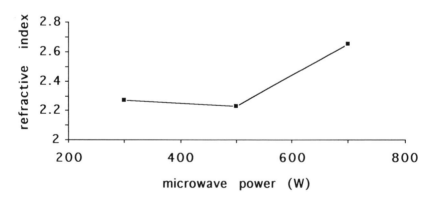

Fig. 5. Refractive index of titanium oxide films on sapphire with microwave power for the oxygen pressure of 10^{-3} torr at room temperature

Plasma gas pressure affects the deposition rate and refractive index of titanium oxide films (Figures 6 and 7). A high deposition rate was obtained at lower O_2 pressure because the precursor stream from the dosing tube does not spread out due to a long mean free path in high vacuum. On the other hand, since high O_2 pressure causes high plasma density, the deposition rate increases with high O_2 pressure.

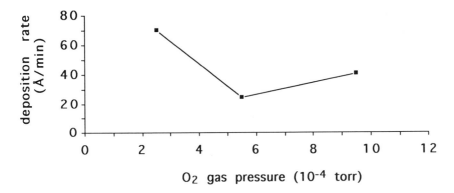

Fig. 6. Deposition rate of titanium oxide films with pressure in oxygen at room temperature, microwave power 500W, on sapphire

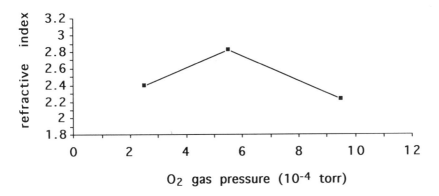

Fig. 7. Refractive index of titanium oxide films with pressure in oxygen at room temperature, microwave power 500W, on sapphire

Annealing effect

Films grown on silicon were annealed for one hour at 800°C in O_2 atmosphere. Refractive index of the film improved from n=1.99 to 2.53 (grown with O_2 plasma, O_2 pressure 10^{-3} torr, microwave power 500W). X-ray diffraction shows the as deposited films contain amorphous and anatase phases, and after annealing, titanium oxide was transformed into rutile phase.

CONCLUSION

Titanium oxide films were deposited at room temperature by ECR-PACVD with O_2 plasma. A maximum deposition rate of 70 Å/min was obtained with an O_2 pressure 2.5×10^{-4} torr and microwave power of 500W.

ACKNOWLEDGMENT

This work made use of the facilities of the Materials Science Center funded by the National Science Foundation under award No: DMR-9121654 and this work supported by a Grant from Nippon Mektron, Ltd., Tokyo, Japan.

REFERENCE

1. T. Ono, M. Oda, C. Takahashi and S. Matsuo, J. Vac. Sci. Technol. **B4**, 696 (1986).
2. J. Asmussen, J. Vac. Sci. Technol. **A7**, 883 (1989).
3. Y. Z. Hu, M. Li, J. Simko, J. Andrews and E. A. Irene, in <u>CVD-XI, proceedings of the 11th international conference on Chemical Vapor Deposition, 1990</u>, edited by K. E. Spear and G. W. Cullen (Electrochemical Society, Pennington, NJ, 1990),pp. 166-172.

HETEROEPITAXIAL Si-ZrO$_2$-Si BY MOCVD

ANTON C. GREENWALD,[*] NADER M. KALKHORAN,[*] FEREYDOON NAMAVAR,[*]
ALAIN E. KALOYEROS,[**] AND IOANNIS STATHAKOS[**]
[*]Spire Corporation, Bedford, MA 01730
[**]Department of Physics, State University of New York at Albany, Albany, NY 22222

ABSTRACT

The objective of this research was to demonstrate heteroepitaxial growth of yttria stabilized cubic zirconia on single crystal silicon substrates by chemical vapor deposition (CVD) using metalorganic source materials. We succeeded in depositing extremely smooth, well aligned films of zirconia on silicon substrates, both the <100> and <111> orientations, without an oxide interfacial layer. Experimental variables investigated included varying zirconia source materials, substrate temperatures, oxygen concentration, gas flow rates, yttria doping, substrate orientation, and cobalt-silicide as an oxygen diffusion barrier. ZrO$_2$ films were predominantly tetragonal when deposited in the absence of oxygen while cubic phase material could be put down at 750°C with oxygen background. Films deposited from TMHD zirconium contained no measurable carbon contamination. Deposits from trifluoro-acetylacetonate Zr contained small amounts of fluorine, even in the presence of water vapor, and some carbon when hydrogen was used as a diluent gas.

INTRODUCTION

Silicon-on-insulator (SOI) technologies are used to achieve high voltage isolation, radiation hardness, and higher speed circuits. Widespread application of these technologies has been limited by the quality of the silicon and/or cost. Chemical vapor deposition of epitaxial yttria stabilized zirconia (YSZ) on silicon with a silicon top layer offers the possibility of lower cost and improved quality material.

Thin film deposition of YSZ on silicon has been explored by many researchers for electronic properties. Only deposition under ultra-high vacuum conditions by electron beam evaporation[1] or pulsed laser deposition[2] has produced epitaxial growth. Ultra-high vacuum deposition techniques are not low cost nor high throughput, and therefore have low commercial potential. Very high quality films of zirconia have been deposited by CVD without concern for epitaxial growth.[3,4,5,6] The goal of this research is to obtain an epitaxial layer of zirconia on silicon by a low cost, high throughput process suitable for large scale commercialization.

The key technical problem faced is the initial growth of the YSZ film, the very first atomic layer. The substrate has a native oxide on the surface which inhibits epitaxial growth. Silicon, Si:Ge alloys, III-V compounds *etc.*, have been grown epitaxially on silicon by first etching this native oxide film *in situ* with HCl gas or reducing it with hydrogen at elevated temperatures, typically over 900°C. However, deposition of an oxide film on silicon tends to form the native oxide on the substrate instantaneously. Furthermore, ZrO$_2$ is an oxygen ion conductor at high temperatures and growth of an SiO$_2$ layer seems assured. Epitaxial growth has only been achieved when the ZrO$_2$ film can form an initial epitaxial mono-atomic layer before the substrate surface converts to SiO$_2$.

Our approach to this problem was to start the growth of the oxide ZrO_2 film using only the source material diluted in hydrogen or an inert (argon) gas. The source materials tested are beta-diketonates with twice as many oxygen atoms in this organic molecule as required to synthesize ZrO_2. In the absence of additional oxygen, this research showed that thermal decomposition of the betadiketonates yields near stoichiometric ZrO_2.

RESEARCH DESCRIPTION

Experiments were performed at Spire Corporation in a low pressure, hot wall CVD reactor (base pressure about 0.1 torr),[7] and at the State University of New York (SUNY) at Albany using a cold wall CVD reactor that can be pumped down to a base pressure under 10^{-7} torr. Source materials, either $Zr(TMHD)_2$ where TMHD is 2,2,6,6 tetramethyl -3,5-heptanedionate or $Zr(TFACAC)_2$ where TFACAC is trifluoro-acetylacetonate and $Y(TMHD)_2$, were placed in sealed containers maintained at constant temperature. All gas flow rates and temperatures were electronically monitored and controlled to better than one percent accuracy.

Single crystal silicon wafers of both <100> and <111> orientation, polished on one side were used as substrates for most of the work. One <111> wafer with a heteroepitaxial cobalt-silicide coating 40 nm thick was prepared by ion implantation and high temperature thermal annealing following the recipe of Namavar et al.[8] This coating was resistant to oxidation (at 800°C in pure oxygen for 30 minutes the $CoSi_2$ surface was unchanged and not oxidized) and this might be a preferred alternate substrate material as it does not form a native oxide.

The experimental variables which can affect the rate of film growth, composition, or the crystal structure are: substrate temperature, reactor pressure, total gas flow rate, the partial pressure (in the reactor) of Zr source material, the partial pressure of Y source material, the partial pressure of oxygen or other oxidant, and the chemical nature of the Zr source, the Y source, the carrier gas, and the oxidant. Note that the partial pressures of Zr and Y in the reactor are dependant variables; the corresponding independent variables are the temperature of the Zr source, the temperature of the Y source, and the carrier gas flow into the Zr and Y sources. This gives a total of twelve variables which had to be optimized.

After experimenting with the use of hydrogen for a carrier gas, which led to excess carbon the final film, argon was used as the carrier gas and oxygen as the oxidant. All deposition experiments were run at about 5 torr. Source temperatures were 140°C for Y(TMHD), 120°C for Zr(TFACAC) and either 180 or 235 °C for Zr(TMHD). Nominal optimal settings for other parameters were: 750°C substrate temperature, 500 sccm oxygen flow, 200 sccm argon for Zr carrier flow, and 40 sccm argon for Y carrier flow, and an additional 250 sccm argon for diluent flow into the reactor.

In Spire's reactor, the deposition rate of ZrO_2 increased proportionally with increasing oxygen concentration for the maximum flow (500 sccm) tested. Even though the chemical content of films deposited without the addition of oxygen gas were stoichiometric ZrO_2, the significant increase of reaction rate with addition of oxygen for both Zr metalorganic sources implies that strong chemical reactions are occurring on the surface. This may explain the difference in structure observed between films deposited with oxygen and films deposited without oxygen.

Zirconia films were analyzed for fluorine content by Auger Spectroscopy. Films deposited from Zr(TMHD) contained no fluorine contamination. Films deposited from Zr(TFACAC) in the presence of water vapor had a residual fluorine content 0.5 atomic %, but no fluorine was detected for a similar sample deposited in oxygen alone. This is the opposite result to what we expected - that water vapor would react with fluorine to form volatile HF and reduce the amount of fluorine in film compared to the dry oxygen case.

The structure of films deposited with high oxygen flows, as determined by X-ray diffraction, is shown for a sample deposited directly on <111> silicon in Figure 1. The sample is predominantly the desired cubic phase, opposite to the results obtained without oxygen.

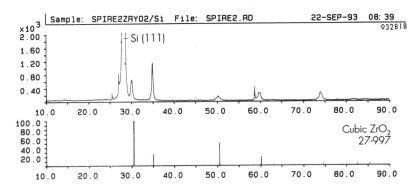

Figure 1   *X-ray diffraction spectra from ZrO_2 film deposited at 760°C with 200 sccm oxygen flow showing cubic phase.*

Depositions on $CoSi_2$ used both TFACAC and TMHD zirconium. With TFACAC, and without the addition of yttrium source material but other parameters as previously described, we obtained the results shown in the scanning electron microscope (SEM) picture in Figure 2 below. The large grains are obviously aligned to the cubic structure of substrate. The average film thickness was 100 nm. The height of the visible grains above this surface was about 500 nm. This demonstrated epitaxy; but the non-planar surface was unacceptable for electronic circuit fabrication. Initial experiments with Zr(TMHD) produced smooth films, aligned but not epitaxial.

At SUNY Albany ZrO_2 films were grown in a custom-made, cold-wall type, stainless steel CVD reactor with base pressures typically in the 10^{-7} to 10^{-6} torr range. A stainless steel parallel plate configuration was employed for the radio frequency plasma (13.56 MHz) generation, with the substrate placed on the bottom electrode. A specialized shower head, designed for uniform gas dispersion and reactant mixing, was employed for reactant delivery. Whenever needed, the plasma was used for *in-situ* substrate cleaning prior to deposition.

Structural, compositional, and electrical characteristics of the films were examined using X-ray diffraction (XRD) scanning electron microscope (SEM), Rutherford Backscattering(RBS) and X-ray photoelectron spectroscopy (XPS). XRD analyses of the CVD produced ZrO_2 films, as shown in Figure 3a, indicated a highly oriented polycrystalline cubic phase for films produced at 500°C. With an increase in substrate temperature, the films exhibited a phase change from cubic to monoclinic for films produced at 650 and 750°C, Figures 3b and 3c respectively.

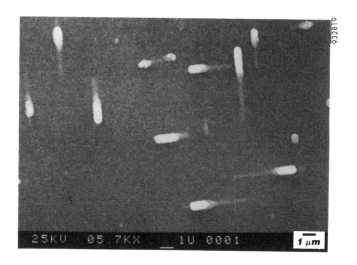

Figure 2 *SEM microphotograph of aligned (epitaxial) ZrO_2 on cobalt silicide on silicon.*

XPS showed no carbon nor fluorine contamination in the films deposited over a wide range of temperatures and confirmed that the films were indeed stoichiometric ZrO_2. An interface contamination layer was observed between the ZrO_2 film and the silicon substrate and was identified as SiO_2. This layer impeded epitaxial growth and a modified process was successfully identified to eliminate it.

YSZ films were grown at SUNY in the next step of this research program. Substrates were <100> and <111> silicon. Deposition temperatures wee in the range 550 to 750°C, and the source chemicals for elemental Y and Zr were respectively $Y(TMHD)_3$ and $Zr(TMHD)_2$. The sources were maintained at a temperature of 120 and 235°C respectively which would nominally yield sublimation pressure of about 10 torr for both chemicals. No oxygen was used during film growth. Instead, the precursors were carried into the reactor by 300 sccm of argon at a reactor pressure of 4 torr. After deposition the films were annealed *in-situ* for five minutes in a flow of 10 sccm of oxygen.

Structural, compositional, and electrical characteristics of these films were measured as before. XPS results showed that the films were YSZ with a composition of 12 atomic% yttria. There was no fluorine nor carbon contamination. Also, there was no SiO_2 interface contamination as shown in the three dimensional XPS plot in Figure 4 for the Si 2p state at the film-substrate interface. The film was a highly oriented cubic phase of YSZ, but not epitaxial.

This experiment was repeated using higher substrate temperatures during deposition, 750° to 950°C. All other parameters were exactly as described in this section. The results were analyzed as before. The deposited films were stoichiometric YSZ with a composition of $(ZrO_2)_{0.73}(Y_2O_3)_{0.27}$ without fluorine nor carbon contamination. As in the deposition at lower temperatures, there was no SiO_2 interface contamination, and the film structure was a highly oriented cubic phase of YSZ. The deposition rate for this experiment was in excess of 100 nm/min.

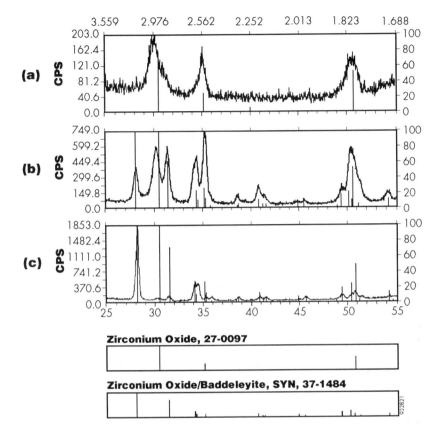

Figure 3   XPS from ZrO_2 films deposited by SUNY at: a) 500°C, b) 650°C, and c) 750°C showing change in phase from cubic to tetragonal with increasing temperature.

CONCLUSIONS

The MOCVD process deposited stoichiometric YSZ films on silicon. The material was highly oriented but not epitaxial except on $CoSi_2$. In the best case, the oxygen interface formation between the substrate and the YSZ film was completely suppressed, but epitaxial growth was still not observed.

Work by P. Legagneux et al.[9] and Holzschuh[10] has shown that oxygen partial pressure plays a key role in determining the composition of the YSZ phase and in formation of the appropriate epitaxial texture. Excess oxygen was found to have a detrimental effect similar to the case when no oxygen was employed. In addition, it is known that a higher deposition temperature facilitates the growth of larger grains and helps to form a denser film by coalescence of grains through a diffusion-controlled mechanism. In this mechanism, adatoms and sub-critical

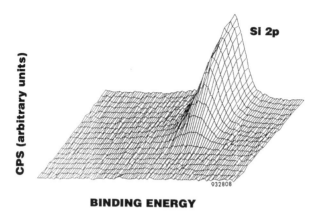

Figure 4 *Three dimensional XPS spectrum of the Si 2p state at the YSZ film-substrate interface indicating the lack of an interface layer of SiO_2 for initial growth without added oxygen.*

clusters diffuse over the substrate surface and are captures by stable islands. In contrast, lower deposition temperatures promote increased nucleation resulting in the formation of smaller grained films and, because of reduced surface diffusion, a higher density of grain boundaries. Clearly the growth of epitaxial YSZ required precise and simultaneous control of growth temperature and oxygen partial pressure. Spire believes very strongly that further process optimization will allow epitaxial growth of the film as has been observed for experiments in ultra-high vacuum conditions.

REFERENCES

1. H. Fukumoto et al., J. Appl. Phys. **66**, 616 (1989).
2. D.K. Fork et al., Appl. Phys. Lett. **57**, 1137 (1990).
3. Cheol Seong Huang and Myeong Joon Kim, J. Mater. Res. **8**, 1361 (1993).
4. S.B. Desu et al., in Chemical Vapor Deposition of Refractory Metals and Ceramics ed. T.M. Bresmann and B.M. Gallois *(Mater. Res. Soc. Proc.* **168**, Pittsburgh, PA, 1990) pp. 349-356; and in Chemical Vapor Deposition of Refractory Metals and Ceramics II ed. T.M. Besmann et al., *(Mater. Res. Soc. Proc.* **250**, Pittsburgh, PA, 1992) pp.323-329.
5. M. Balog et al., J. Electrochem. Soc. **126**, 1203 (1990).
6. M. Schieber et al., Appl. Phys. Lett. **58**, 301 (1991).
7. A. Greenwald et al., in Ferroelectric Thin Films II ed. A.I. Kingon et al., *(Mater. Res. Soc. Proc.* **243**, Pittsburgh, PA, 1992) pp. 457-462.
8. F. Namavar et al., in Phase Formation and Modification by Beam-Solid Interaction ed. G.S. Was et al., *(Mater. Res. Soc. Proc.* **235**, Pittsburgh, PA, 1993) pp. 285-292.
9. P. Legagneux et al., Appl. Phys. Lett., **53**, 1506 (1988).
10. H. Holzschuh and H. Shur, Appl. Phys. Lett., **59**, 470 (1990).

This research was supported in part by NSF Grant #III-9260636.

PART IV

Modeling

THERMOCHEMICAL DATA FOR CVD MODELING FROM AB INITIO CALCULATIONS

PAULINE HO* AND CARL F. MELIUS**
*Sandia National Laboratories, Albuquerque, NM 87185-0601
**Sandia National Laboratories, Livermore, CA 94551-0969

ABSTRACT

Ab initio electronic-structure calculations are combined with empirical bond-additivity corrections to yield thermochemical properties of gas-phase molecules. A self-consistent set of heats of formation for molecules in the Si-H, Si-H-Cl, Si-H-F, Si-N-H and Si-N-H-F systems is presented, along with preliminary values for some Si-O-C-H species.

INTRODUCTION

Developing a good model for a CVD process generally requires some description of the chemistry occurring in the process, either from an equilibrium or chemical kinetics point of view. Unfortunately, thermochemical data for many CVD systems of interest are often incomplete or inconsistent.

However, quantum chemistry has now reached the point where *ab initio* electronic-structure calculations yield usable thermochemical information for certain classes of compounds. For a number of years, we have been using the BAC-MP4 method, which combines calculations done at the MP4 level of theory (see next section) with empirical bond-additivity corrections (BACs), to obtain an extensive, self-consistent set of thermochemical data.

This method has been applied to over 2800 species, primarily molecules made up of first-row elements that are relevant to combustion,[1] but including a number of molecules relevant to the CVD of silicon-containing compounds. In this paper, we first review some of these earlier results before presenting some recent work on Si-O-C-H compounds. Detailed discussions of the Si-H, Si-H-Cl, Si-H-F, Si-N-H and Si-N-H-F systems are available elsewhere,[2-5] as well as similar work on Si-C-H and Si-C-Cl-H that is not included in this paper.[6,7]

THEORETICAL METHODS

The methods used in this work have been described in the literature,[1,4,5] so only a short description is included here.

Electronic structure calculations were done using the Gaussian codes.[8] Equilibrium geometries and harmonic vibrational frequencies were obtained at the HF/6-31G* level of theory (restricted Hartree-Fock theory,[9] RHF, for closed shell molecules, and unrestricted Hartree-Fock theory,[10] UHF, for open shell molecules, using the 6-31G* basis set[11,12]). This level of theory provides sufficiently accurate equilibrium geometries, but does not provide total energies suitable for determining reaction energies involving the breaking of covalent bonds. Vibrational frequencies calculated at this level of theory are known[13] to be systematically larger than experimental values, so each calculated frequency has been scaled by dividing it by 1.12 (determined by averaging the ratio of calculated to experimental vibrational frequencies over many known species). To determine atomization enthalpies, the effects of electron correlation were included in the calculation by performing single-point calculations at the HF/6-31G* geometries using Møller-Plesset (MP) perturbation theory[14-16] with single, double, triple and quadruple substitutions and the 6-31G** basis set.[11,12] This level of theory is generally a good compromise between the need for chemical accuracy in the calculated energies and limits on molecular size imposed by computational limitations. For most molecules, fourth-order perturbation calculations were done, but this was not possible for larger molecules like TEOS (tetraethoxysilane). In these cases, second-order perturbation calculations were done, which yields usable atomization enthalpies but with greater uncertainties than the MP4 calculations.

Empirical BACs are used to account for systematic errors in the *ab initio* calculations which result primarily from basis-set truncation. BACs are determined for each level of perturbation theory and depend mainly on the bond type and bond length, but there are small additional corrections for neighboring atoms. The BACs are obtained from reference compounds which have well-established experimental heats of formation or from analogous correction parameters for related classes of compounds. For Si–H, Si–Cl, Si–F and Si–C bonds, SiH_4, $SiCl_4$, SiF_4 and $Si(CH_3)_3$ were used as reference compounds, while Si_2H_6 was the reference compound for Si–Si bonds. For Si–N bonds, however, no good reference species was available, so we used parameters similar to those for Si–C bonds. For Si–O bonds, the BACs were based on a theoretical value obtained for H_3SiOH at a higher level of theory[17] (vide infra). There are additional BAC terms for spin, both for open-shell molecules and for closed-shell molecules which are UHF unstable. These terms are discussed in detail elsewhere.[1,4,5] The sum of the BAC terms are combined with the electronic energy from the perturbation theory calculation and the zero-point energy to obtain the heat of formation at 0K. Statistical-mechanics equations involving the calculated geometries and scaled frequencies are then used to determine entropies, heat capacities, enthalpies, and free energies at other temperatures.[1]

The error estimates given in the tables are determined via a systematic, but ad hoc, procedure using the results from electronic structure calculations at lower levels of theory.[1,4] These uncertainties only reflect the applicability of the calculational methods to the various species. As discussed in more detail elsewhere,[3,4] the actual uncertainties may be somewhat larger for species for which we do not have good reference species, for example molecules with Si–N or Si–O bonds or with short Si–Si (multiple) bonds. In the case of the larger molecules that were only studied at the MP2 level of theory, we have not developed a method for systematically estimating errors, but consider the uncertainties in the heats of formation to be in the range of 5–7 kcal/mole.

RESULTS AND DISCUSSION

Tables I and II list our calculated heats of formation and entropies at 298K for molecules in the Si-H, Si-H-Cl, Si-H-F, Si-N-H and Si-N-H-F systems. Detailed information on energies, vibrational frequencies and geometries from the *ab initio* calculations, can be found in the full papers.[2-5]

There are some general trends observable in the heats of formation. For example, the heats of formation of the $SiH_nCl_{(4-n)}$, $SiH_nF_{(4-n)}$, and $SiH_n(NH_2)_{(4-n)}$ ($0 \leq n \leq 4$) species show nearly linear dependences on n, indicating that the successive replacement of Si–H bonds with Si–Cl, Si–F or Si–NH_2 bonds is accompanied by monotonic changes in stabilization. This is also true for the replacement of Si–F bonds by Si–NH_2 bonds in the $SiF_n(NH_2)_{(4-n)}$ species. In contrast, the heats of

Table I: Calculated Enthalpies (kcal mol^{-1}) and Entropies (cal mol^{-1} K^{-1}) for Si-H, Si-H-Cl and Si-H-F Compounds

Species	ΔH_f° (298) [a]	S° (298)
SiH$_4$ [b]	8.19±1.00	48.9
SiH$_3$	47.43±1.01	51.8
SiH$_2$	64.80±2.20	49.5
SiH	91.01±1.14	46.0
SiCl$_4$ [b]	−158.39±1.00	79.7
SiCl$_3$	−76.00±1.58	75.6
SiCl$_2$	−36.15±3.73	67.5
SiCl ^{2}Π	37.76±2.34	55.8
SiH$_3$Cl	−32.03±1.06	60.0
SiH$_2$Cl$_2$	−74.42±1.09	68.8
SiHCl$_3$	−117.14±1.05	75.4
SiH$_2$Cl	7.91±1.13	62.7
SiHCl$_2$	−33.98±1.38	71.1
SiHCl	15.85±2.5	60.1
Si$_2$H$_6$ [b]	19.11±1.00	65.9
Si$_2$H$_5$	55.39±1.02	70.0
H$_3$SiSiH	74.91±2.73	68.0
H$_2$Si=SiH$_2$ (^{1}A$_g$)	62.90±2.30	65.6
HSiSiH$_2$	102.05±2.48	66.3
Si(H$_2$)Si	95.62±1.70	58.8
Si(H)Si	118.82±2.86	59.1
Si$_2$ ($^3\Sigma_g^-$)	145.79±1.31	54.8
Si$_3$H$_8$	28.36±1.03	83.9
TS1 [c]	87.94±2.25	64.6
TS2 [c]	104.77±2.71	69.0
TS3 [c]	70.91±2.19	69.1
TS4 [c]	66.48±1.58	68.3
SiF$_4$ [b]	−385.99±1.00	68.2
SiF$_3$	−237.42±1.87	67.8
SiF$_2$	−149.86±4.00	61.4
SiF	−12.42±2.89	52.7
SiHF$_3$	−288.63±1.34	66.6
SiH$_2$F$_2$	−186.38±1.47	62.8
SiH$_3$F	−85.50±1.21	57.0
SiHF$_2$	−139.57±1.89	65.1
SiH$_2$F	−42.16±1.37	59.7
SiHF	−35.70±2.85	57.1
Si$_2$F$_6$	−569.62±3.47	97.2

[a] See text for discussion of error estimates.
[b] Reference compound
[c] Transition states for the 1) H$_3$SiSiH → H$_2$Si=SiH$_2$ isomerization, 2) Si$_2$H$_6$ → H$_2$Si=SiH$_2$ + H$_2$ reaction, 3) Si$_2$H$_6$ → H$_3$Si=SiH + H$_2$ reaction, and 4) Si$_2$H$_6$ → SiH$_2$ + SiH$_4$ reaction.

formation for the SiH_n, $SiCl_n$, SiF_n, SiH_nNH_2 and SiF_nNH_2 compounds do not vary smoothly with n. In addition, the first and third successive Si–X (X = H, Cl, F) bond dissociation energies (BDEs) are significantly higher than the second and fourth successive BDEs. These observations are manifestations of the change in hybridization of the silicon from sp^3 to s^2p^2 at SiX_2.

Our calculated heats of formation compare favorably with the literature, especially for the saturated species. For example, for the $SiH_nCl_{(4-n)}$ and $SiH_nF_{(4-n)}$ species, our values agree quite well with those given in the JANAF tables[18] and by Walsh.[19] Our values for Si_3H_8 and Si_2F_6 also agree well with those given in the CATCH tables.[20] For the molecules with Si-N bonds, however, virtually no data are available in the literature for comparison with our heats of formation. For H_3SiNH_2, Jolly and Bakke[21] have estimated a heat of formation of -12 kcal mol^{-1}, which agrees well with our calculated value of -11.45 kcal mol^{-1}. The literature data for radical species are much sparser. The heat of formation of SiH_2 has been the subject of much recent study,[22] and our current calculated value of 64.8 kcal mol^{-1} is in good agreement with the value of 65.2 now recommended by Walsh.[23] For two isomers of Si_2H_4, our calculated values for H_3SiSiH and $H_2Si=SiH_2$ are 74.9 kcal mol^{-1} and 62.9 kcal mol^{-1}, respectively, which agree well with the experimental values of 74.6±2.0 and ≤62.3 kcal mol^{-1} obtained by Walsh.[23]

In some cases, substantial differences exist in the literature and our results support one value over another. For example, our values of -76.0 and -237.4 kcal mol^{-1} for $SiCl_3$ and SiF_3, respectively, support the values in the literature of -80 and -239 kcal mol^{-1} derived from bond dissociation energy measurements,[19] rather than the values of -93.3 and -259.4 kcal mol^{-1} derived from effusion/mass-spectrometric studies.[18] There are, of course, other cases in which disagreement between our values and literature values cannot be explained simply, some cases where disagreements may not be significant because the literature values are very uncertain, and many cases where no literature value exists.

We have also applied our calculational methods to a number of transition states that give insight into decomposition mechanisms. For example, several transition states relevant to disilane decomposition are included in Table I. Although disilane decomposes primarily to SiH_2 and SiH_4, it can also form $Si_2H_4+H_2$. Our results indicate that, although $H_2Si=SiH_2$ has a lower heat of formation than H_3SiSiH, disilane decomposition will initially produce H_3SiSiH because of the high transition state energy for the $Si_2H_6 \rightarrow H_2Si=SiH_2+H_2$ reaction. H_3SiSiH then has a low barrier for isomerization to $H_2Si=SiH_2$. Transition states for the decomposition of silylamine are also included in Table II. In contrast with disilane decomposition, the silylamine decomposition pathway for H_2 elimination is significantly lower in energy than the pathway for silylene elimination.

Table III lists calculated thermochemical parameters for species in the Si-O-C-H system. For the smaller species, calculated heats of formation are given for both the BAC-MP4 and BAC-MP2 levels of theory. For the larger molecules, only results from BAC-MP2 calculations are listed. In general, the BAC-MP2 numbers appear to be within a few kcal/mole of the BAC-MP4 numbers, with no systematic variations. In several cases, Table III lists heats of formation that were estimated using bond additivity methods because the MP2 calculations have not finished yet. We expect that these estimated values will be within 2-3 kcal/mole of the BAC-MP2 values.

The heats of formation listed in this table are somewhat preliminary; the BAC for Si-O bonds may need refinement. The present correction for Si-O bonds was developed

Table II. Calculated Enthalpies (kcal mol^{-1}) and Entropies (cal mol^{-1} K^{-1}) for Si-N-H and Si-N-H-F Compounds

Species	ΔH_f° (298) [a]	S° (298)
H_3SiNH_2	-11.45 ± 1.47	65.7
H_3SiNH	51.32 ± 1.38	65.5
H_2SiNH_2	28.03 ± 1.26	65.8
H_3SiN	96.56 ± 1.31	59.5
H_2SiNH	40.99 ± 3.26	59.9
$HSiNH_2$	26.33 ± 1.62	59.8
$H_2SiN\ ^2A_1$	141.40 ± 4.40	57.2
$HSiNH$	84.80 ± 4.96	60.6
$SiNH_2$	48.67 ± 1.55	58.6
$HSiN$	92.98 ± 24.3	54.8
$SiNH$	38.39 ± 5.41	51.7
$SiN\ ^2\Pi$	130.72 ± 5.67	52.3
$SiN\ ^2\Sigma^+$	115.55 ± 8.85	52.0
$H_3SiNHSiH_3$	-14.32 ± 2.03	83.5
$H_3SiNSiH_3$	48.90 ± 2.49	83.7
$H_2Si(NH_2)_2$	-37.30 ± 2.65	74.4
$HSi(NH_2)_3$	-64.86 ± 3.80	82.8
$HSi(NH_2)_2$	5.62 ± 2.23	75.3
$Si(NH_2)_4$	-92.88 ± 5.15	90.0
$Si(NH_2)_3$	-18.77 ± 3.21	84.3
TS5 [b]	62.67 ± 3.45	62.8
TS6 [b]	51.72 ± 1.18	64.3
F_3SiNH_2	-317.89 ± 2.93	80.0
F_3SiNH	-249.65 ± 2.21	82.2
F_2SiNH_2	-167.24 ± 1.48	77.2
F_3SiN	-200.02 ± 1.59	73.9
F_2SiNH	-146.94 ± 1.38	70.9
$FSiNH_2$	-80.01 ± 1.92	65.8
F_2SiN	-63.11 ± 3.47	70.5
$FSiNH$	-13.57 ± 2.63	68.9
$FSiN$	54.37 ± 8.57	63.7
$F_3SiNHSiH_3$	-320.19 ± 3.16	96.0
$F_3SiNSiH_3$	-252.84 ± 3.80	97.6
$F_2Si(NH_2)_2$	-247.30 ± 4.52	82.8
$FSi(NH_2)_3$	-170.70 ± 5.19	88.0
$FSi(NH_2)_2$	-95.47 ± 2.54	79.7

[a] See text for discussion of error estimates.
[b] Transition states for the 5) $H_3SiNH_2 \rightarrow SiH_2 + NH_3$, and 6) $H_3SiNH_2 \rightarrow HSiNH_2 + H_2$ reactions.

during work on Si-O-H compounds and is based on a higher level (G2) calculation for H_3SiOH.[17] There is, however, some literature data for Si-O-C-H species that may support the use of a different reference compound for the correction. Walsh[23] lists heats of formation of –119.5, –281.8 and –314.3 kcal/mole for $(CH_3)_3SiOH$, $Si(OCH_3)_4$, and $Si(OC_2H_5)_4$, respectively. These values are somewhat less negative than our values, and may warrant an adjustment after calculations are done for species such as TEOS and $(CH_3)_3SiOSi(CH_3)_3$.

The ΔH°_fs in Table III were used to analyze possible TEOS decomposition reactions, shown in Table IV. The first two reactions, involving simple breaking of Si–O and O–C bonds, are very endothermic (≥ 100 kcal/mole), and will not be significant under the conditions used for TEOS CVD. The next two reactions, involving the formation of $O=Si(OC_2H_5)_2$, are also quite endothermic (≥ 70 kcal/mole). These reactions are multicenter reactions, and are expected to have substantial kinetic barriers in addition to the endothermicities. These reactions will probably also not be significant under TEOS CVD conditions. The last two reactions, involving the formation of the $HSi(OC_2H_5)_3$ and $HOSi(OC_2H_5)_3$ species, have lower endothermicities of 46 and 11 kcal/mole, respectively. These are both 4-center reactions and are expected to have additional kinetic barriers, but the endothermicities indicate that the last reaction is the most reasonable pathway for TEOS decomposition. To investigate this further, transition states for the elimination of ethylene from $CH_3CH_2OSi(OH)_3$ and $CH_3CH_2OSiH_3$ were studied and the results included in Table III. These calculations confirmed the presence of substantial kinetic barriers. For the $CH_3CH_2OSi(OH)_3 \rightarrow C_2H_4 + Si(OH)_4$ and $CH_3CH_2OSiH_3 \rightarrow C_2H_4 + SiH_4$ reactions, the activation energies are 70 and 68 kcal/mole, respectively, while the endothermicities are 12 kcal/mole. These results are consistent with recent experiments by Lin and coworkers,[24] who report that $k = 5 \times 10^{13} \exp(-61500/RT)$ for the gas phase decomposition of TEOS.

CONCLUSION

The combination of high level *ab initio* electronic-structure calculations and empirical bond-additivity corrections has proven useful in obtaining thermochemical properties of gas-phase molecules for modeling CVD processes. In some cases, the calculated values help choose between conflicting measurements in the literature; in other cases, the calculated values are the only ones available. In addition to providing information on reaction endo/exothermicities, these calculational methods can also provide energetic and geometric information on transition states. Thus, likely reaction pathways can be identified and evaluated.

The calculational methods described in this paper can generally be applied to compounds comprised of elements in the first three rows of the periodic table and are limited to molecules with a relatively small number of heavy atoms. This is primarily caused by computational requirements that scale with higher powers of the number of atoms. The results for the smaller molecules, can, however, be used to develop bond-additivity methods for reliably estimating the thermochemistry of larger gas-phase species and possibly surface species. These types of calculations are also limited by the need for a reference species for each bond type. For many CVD systems, particularly MOCVD systems, good experimental data are sparse enough that we have to resort to estimates based on analogous species, leading to larger uncertainties in the calculated values.

Table III. Preliminary Values for ΔH_f° (298) at Two Levels of Theory (kcal mol^{-1}) and Entropies (cal mol^{-1} K^{-1}) for Some Si-O-C-H Compounds

Species	ΔH_f° (298) [a]		S° (298)
	BAC-MP4	BAC-MP2	
Si(OCH$_3$)$_4$		−285.5	123.2
Si(OH)(OCH$_3$)$_3$		−294.8	113.2
Si(OCH$_3$)$_2$(OH)$_2$	−304.8±6.5	−302.5	103.9
SiOH(OCH$_3$)$_2$	−172.3±2.2	−172.7	98.3
O=Si(OCH$_3$)$_2$	−174.9±8.4	−174.0	92.3
CH$_3$OSi(OH)$_3$	−316.0±6.0	−314.2	91.9
Si(OH)$_2$OCH$_3$	−180.9±1.8	−181.9	88.2
O=SiOHOCH$_3$	−184.1±8.0	−183.6	81.3
CH$_3$OSi(OH)$_2$O	−242.3±5.9	−239.1	92.8
CH$_2$OSi(OH)$_3$	−270.1±6.7	−268.5	92.1
Si(OC$_2$H$_5$)$_4$		−320.0 [c]	
Si(OH)(OC$_2$H$_5$)$_3$		−321.0	137.1
Si(OH)$_2$(OC$_2$H$_5$)$_2$		−321.9	117.5
CH$_3$CH$_2$OSi(OH)$_3$	−324.6±5.7	−322.8	99.0
CH$_3$CH$_2$OSiH$_3$	−67.5±1.1	−67.8	79.9
HSi(OC$_2$H$_5$)$_3$		−235.0	132.4
(CH$_3$)$_3$SiOH	−121.1±1.1	−121.2	90.9
TS7 [b]	−254.2±3.1	−252.8	101.5
TS8 [b]	1.0±7.1	1.4	84.5
Si(OC$_2$H$_5$)$_3$		−187.0 [c]	
OSi(OC$_2$H$_5$)$_3$		−248.8 [c]	
O=Si(OC$_2$H$_5$)$_2$		−189.1 [c]	
Si(OH)$_4$	−325.2±5.6	−323.9	81.9
SiH(OH)$_3$	−238.4±2.8	−238.7	76.4
SiH$_2$(OH)$_2$	−152.0±1.2	−153.0	68.7
H$_3$SiOH	−67.5±1.0	−68.4	61.8
Si(OH)$_3$	−189.9±1.6	−191.4	77.6
OSi(OH)$_3$	−251.6±5.4	−248.9	82.0
O=Si(OH)$_2$	−192.1±7.4	−192.0	70.0
SiO	−25.0±4.0	−23.3	50.5

[a] See text for discussion of error estimates.
[b] Transition states for the 7) CH$_3$CH$_2$OSi(OH)$_3$ → C$_2$H$_4$ + Si(OH)$_4$ and 8) CH$_3$CH$_2$OSiH$_3$ → C$_2$H$_4$ + SiH$_4$ reactions.
[c] Estimated using bond additivity

Table IV. Thermochemistry for various TEOS decomposition pathways

Reaction		ΔH_{rxn}(298)
Si(OC$_2$H$_5$)$_4$	→ Si(OC$_2$H$_5$)$_3$ + OC$_2$H$_5$	132
	→ OSi(OC$_2$H$_5$)$_3$ + C$_2$H$_5$	100
	→ O=Si(OC$_2$H$_5$)$_2$ + C$_2$H$_4$ + C$_2$H$_5$OH	86
	→ O=Si(OC$_2$H$_5$)$_2$ + O(C$_2$H$_5$)$_2$	71
	→ HSi(OC$_2$H$_5$)$_3$ + O=C(CH$_3$)H	46
	→ HOSi(OC$_2$H$_5$)$_3$ + C$_2$H$_4$	11

ACKNOWLEDGMENTS

We acknowledge the contributions of M. E. Coltrin and J. S. Binkley to the work on the Si-H and Si-H-Cl species, and those of M. C. Lin, M. R. Zachariah and M. D. Allendorf to the work on the Si-O-C-H species. This work was primarily supported by the U.S. Department of Energy under contract No. DE-AC04-94AL85000, with some support for the work on Si-N-H-F compounds from DARPA through WRDC/MLBC contract No. F33615-89-C-5628.

REFERENCES

1. C.F. Melius, "Thermochemistry of Hydrocarbon Intermediates in Combustion: Application of the BAC-MP4 Method", in *Springer-Verlag DFVLR Lecture Notes*, (Springer Verlag, 1990).
2. P. Ho, M.E. Coltrin, J.S. Binkley and C.F. Melius, J. Phys. Chem. **89**, 4647 (1985).
3. P. Ho, M.E. Coltrin, J.S. Binkley and C.F. Melius, J. Phys. Chem. **90**, 3399 (1986).
4. P. Ho and C.F. Melius, J. Phys. Chem. **94**, 5120 (1990).
5. C.F. Melius and P. Ho, J. Phys. Chem. **95**, 1410 (1991).
6. M.D. Allendorf and C.F. Melius, J. Phys. Chem. **96**, 428 (1992).
7. M.D. Allendorf and C.F. Melius, J. Phys. Chem. **97**, 720 (1993).
8. See, for example, M.J. Frisch, J.S. Binkley, H.B. Schlegel, K. Raghavachari, C.F. Melius, R.L. Martin, J.J.P. Stewart, F.W. Bobrowicz, C.M. Rohlfing, L.R. Kahn, D.J. DeFrees, R. Seeger, R.A. Whiteside, D.J. Fox, E.M. Fluder, S. Topiol, J.A. Pople, *Gaussian 86*, (Carnegie-Mellon Quantum Chemistry Publishing Unit, Carnegie-Mellon University, Pittsburgh, PA 15213, 1986).
9. C.C. Roothan, J. Rev. Mod. Phys. **23**, 69 (1951).
10. J.A. Pople, R.K. Nesbet, J. Chem. Phys. **22**, 571 (1954).
11. P.C. Hariharan, J.A. Pople, Theor. Chim. Acta **28**, 213 (1973).
12. M.M. Francl, W.J. Pietro, W.J. Hehre, J.S. Binkley, J.A. Pople, J. Chem. Phys. **77**, 3654 (1982).
13. J.A. Pople, H.B. Schlegel, R. Krishnan, D.J. DeFrees, J.S. Binkley, M.J. Frisch, R.A. Whiteside, Int. J. Quant. Chem. **S15** 269 (1981).
14. J.A. Pople, J.S. Binkley, R. Seeger, Int. J. Quantum Chem. **S10**, 1 (1976).
15. R. Krishnan, J.A. Pople, Int. J. Quantum Chem. **14**, 91 (1978).
16. R. Krishnan, M.J. Frisch, J.A. Pople, J. Chem. Phys. **72**, 4244 (1980).
17. Michael R. Zachariah, private communication.
18. *JANAF Thermochemical Tables*, J. Phys. and Chem. Ref. Data **14**, Supplement 1 (1985).
19. R. Walsh, J. Chem. Soc., Faraday Trans. I **79**, 2233 (1983).
20. J.B. Pedley, B.S. Iseard, *CATCH Tables* (University of Sussex, 1972) available from NTIS, Number AD-773468.
21. W.L. Jolly, A.A. Bakke, J. Amer. Chem. Soc. **98**, 6500 (1976).
22. See, for example, J.E. Baggott, H.M. Frey, K.D. King, P.D. Lightfoot, R. Walsh, I.M. Watts, J. Phys. Chem. **92**, 4025 (1988) and references therein.
23. R. Walsh, in *The Chemistry of Organic Silicon Compounds*, edited by S. Patai and Z. Rappoport (John Wiley and Sons Ltd., 1989), p. 371.
24. M.C. Lin, private communication; J.C.S. Chu, J. Breslin, N.S. Wang, and M.C. Lin, Mater. Lett. **12**, 179 (1991).

THERMODYNAMIC SIMULATION OF MOCVD YBa$_2$Cu$_3$O$_{7-x}$ THIN FILM DEPOSITION

C. Bernard[1], F. Weiss[2], A. Pisch[1,2] and R. Madar[2], [1]L.T.P.C.M. - ENSEEG, BP.75, 38402 St Martin d'Hères, France. [2]LMGP, BP.46, 38402 St Martin d'Hères, France.

ABSTRACT

Chemical Vapour Deposition of YBa$_2$Cu$_3$O$_{7-x}$ superconductors using organometallic precursor materials involves the formation of a great variety of condensed phases (oxides, carbonates, hydroxides, carbides, hydrides) and a complex gas phase. Therefore, the possibility of calculating the stability ranges of the different phases susceptible of being deposited and the influence of the main experimental parameters (precursor partial pressure, oxygen partial pressure, total pressure, temperature) is very attractive. Recently, some initial attempts have been published, but the authors did not have sufficiently complete and reliable thermodynamic data on the species involved in the process. These results will be discussed. Furthermore, a homogeneous and complete set of data for the whole system will be presented, and the reactions of the deposition process will be simulated on the basis of this data set. The most important results will be compared to the previous calculations and to recent experimental work.

INTRODUCTION

Of the thin film materials that are superconducting at high Tc, YBa$_2$Cu$_3$O$_{7-x}$ is the one that has been studied the most. It has been known since 1987 [1], it can be synthesised without much problem and it has an advantageous critical temperature. For its preparation, most of the conventional thin film methods have been used. Good quality low temperature deposits have been obtained by vapour phase physical deposition. However, these methods require the use of a high vacuum and thus apply only to small deposition areas. To overcome this drawback, MOCVD is a particularly good method. Studies already performed with MOCVD have been described in several publications [2]-[6] with, in particular, a recent review by Leskela et col. [6].

Film preparation by the powder method gives a single stoichiometric composition, at least as far as the Y, Ba and Cu elements are concerned. On the other hand, using CVD, the results appear to be much more uncertain and a secondary phase, albeit often in very small proportions, is nonetheless practically always present in the deposited layers. The role of this phase has still not been defined, and it is not necessarily easy to understand the influence of the Ba/Y and Cu/Y ratios on the critical temperature and flux, Tc and Jc, [7]. It is clear, however, that an understanding of the nature of this phase (CuO, Y$_2$O$_3$, etc.) and its proportions may help the experimental worker to optimise his deposits. The thermodynamic simulation of what happens on the reactor substrate, during the deposition process, in relation to experimental parameters, may therefore prove to be extremely beneficial. Studies have already been carried out in this direction [8] [9]. However, in view of the complex nature and number of phases present, the disagreements on phase diagrams involved and the absence or wide dispersion of thermodynamic data, it has not been possible to perform a very detailed study. Thanks to a European contract BREU 0203C (superdata), performed within the context of Brite Euram, experimenters and optimisation specialists have made a number of advances in the Y-Ba-Cu-O system. As a result, a more comprehensive simulation of the process has been made possible.

PRINCIPLE OF THE THERMODYNAMIC APPROACH

The thermodynamic simulation of the deposition process, as will be described in the case of superconductors, is in fact only the first stage of our normal approach [10]. The complex nature of this particular case justifies this presentation. The stage in question consists in minimising the total Gibbs free energy of the chemical system involved. Assuming that thermodynamic equilibrium is achieved in the reactor, the composition of the gas phase and of the deposited condensed phases is deduced for a given set of experimental conditions: total pressure, temperature, partial pressures of carrier gases.

To minimise the Gibbs energy, a complete and coherent set of thermodynamic data on all species likely to be represented at equilibrium must be available. The first step will therefore be to review all these species, excluding initially all those which include C and H. The Y-Ba-Cu-O tetrahedron in the vicinity of the Y_2O_3-BaO-CuO plane will be described for temperatures approaching 800°C and for oxygen pressures of just a few torrs, conditions which correspond to conventional MOCVD deposits.

PHASES PRESENT - THERMODYNAMIC DATA

On the occasion of the Brite superdata contract, data relating to the Y-Ba-Cu metallic system and to the corresponding complex diagrams were optimised, but the oxygen pressures used are such that these phases do not appear in the reactors. The metal-oxygen binary phases were also defined with particular emphasis being given to the Cu-O system, but although it was possible to produce better thermodynamic data, nothing new was found regarding the phases present. On the other hand, the M-M'-O ternary phases and the quaternay phase merit a detailed description which will be compared to values given in the literature and to the choices made by the authors of previous simulations [8] [9].

For the sake of coherency, the phases of this system were synthesised before characterising them (RX, ATD, EDX) and before measuring their thermodynamic properties; two preparation methods were considered: sintering of the oxides Y_2O_3, BaO/BaO_2 and CuO under oxygen or artificial air, or dilution of these same oxides in nitric acid with drying and sintering under oxygen.

Y-Ba-O ternary phase

Four phases are known in the literature: Y_2BaO_4, $Y_4Ba_3O_9$, $Y_2Ba_2O_5$ and $Y_2Ba_4O_7$. It has been shown that only the first two compounds are real oxides and that the other two are CO_2-stabilised oxycarbonates which cannot be obtained in a carefully controlled atmosphere. $Y_4Ba_3O_9$ is very sensitive to the processing conditions, so that it can only be prepared using Y_2O_3 and BaO/BaO_2 without dissolution and by sintering in pure flowing oxygen. Otherwise, impurity phases such as Y_2BaO_4 and Y_2O_3 appear, as confirmed by X-ray diffraction. Both real oxides are very sensitive to air humidity and, even at room temperature, a relatively fast evolution of the prepared samples can be observed.

Calorimetry measurements were carried out by A. Rais et al. [11] and A. Watson [12]. They are shown in table 1 with data from the literature and with the values that the present authors have selected and which fit the $BaO-Y_2O_3$ phase diagram [16]. For their thermodynamic simulation, Harsta et col. [8] used the results of Lee and Lee [17] whose data stabilise the four compounds, while Vahlas et col. [9] are in agreement with the results of Frase and Clarke [18] which stabilise $Y_2Ba_4O_7$ and Y_2BaO_4 at 950° C.

Table I
Thermodynamic data for the Y-Ba-O compounds.

	$\Delta H^o_{f,ox}$ kJ mol^{-1}	$\Delta S^o_{f,ox}$ J mol^{-1} k^{-1}	Temperature range K	References
Y_2BaO_4	-120.13	4.08	774-1220	[13]
	-128.31	-5.211	850-1220	[14]
	-30.138			[12]
	0.4 ± 14.4			[15]
	-61.31 ± 7.07			[11]
	-33.84*	-1.57*		
$Y_4Ba_3O_9$	-120.13	4.7		[14]
	-50.119			[12]
	-61.35*	+19.38*		

* Values used in the simulation.

Y-Cu-O ternary phases

Two ternary phases are known: $Y_2Cu_2O_5$ and $YCuO_2$. They can both be rapidly synthesised at temperatures higher than 900° C even in natural air. No CO_2 solubility is known. The phase with the lower oxygen content was produced while sintering the basic oxide mixture in pure argon.

Table II
Thermodynamic data for the Y-Cu-O compounds.

	$\Delta H^o_{f,ox}$ kJ mol^{-1}	$\Delta S^o_{f,ox}$ J mol^{-1} k^{-1}	Temperature range K	References
$Y_2Cu_2O_5$	-19.45	-3.56	973-1223	[21]
	20.7 ± 2.5	22.8 ± 1.7	1173-1340	[22]
	10.91	13.41	1097-1292	[23]
	-72.579	-4.24	923-1273	[24]
	11.21	15.07	873-1323	[14]
	18.47	21.9	860-980	[13]
	14.3	17.0	1097-1292	[25]
	12.796	18.4	1050-1250	[26]
	-7.45	-16.37	1058-1329	[27]
	56.36	71.8	953-1110	[28]
	9.1	14.0	1025-1220	[29]
	12.1 ± 2.7			[30]
	8.8 ± 7.1			[15]
	9.2 ± 3.2			[31]
	14.4 ± 4.36			[11]
	6.95*	15.06*		[20]
$YCuO_2$	32.977	-4.95	923-1273	[24]
	-5.82	-44.8	1000-1300	[32]
	5.573	5.57	1058-1329	[27]
	-5.8	-3.9	873-1023	[14]
	22.89	24.39	1115-1273	[28]
	-2.46*	5.12*		[20]

*Values used in the simulation they correspond to 298° K values, and they are used with $\Delta Cp \neq 0$.

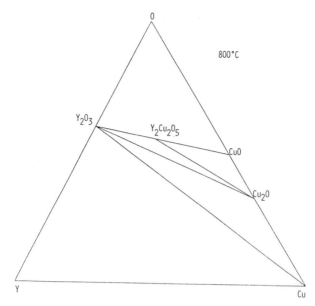

Figure 1: Y-Cu-O isothermal section at 800°C.

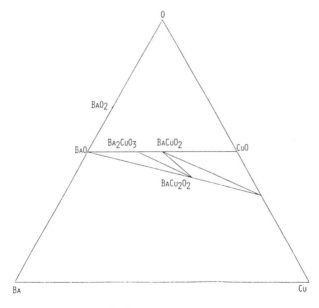

Figure 2: Ba-Cu-O isothermal section at 800°C.

The conventional representation of the $YO_{1.5}$-CuO_x system [19], which places CuO and Cu_2O on the same temperature—copper mole fraction diagram, is not sufficiently clear, especially for the case studied here. In this particular case, it will be necessary to work under a reduced oxygen pressure in order to deposit, in the first stage, the $YBa_2Cu_3O_6$ phase which is under the Y_2O_3-BaO-CuO plane. It is thus necessary to visualise the O-Y-Cu face, cf. figure 1 [20]. As regards the thermodynamic data of these two compounds, numerous measurements have been taken and are given in table 2 with the values selected for this study [20].

From the simulation standpoint, Harsta et col. [8] again use the results of Lee and Lee [17]. As a result, they only take into account $Y_2Cu_2O_5$, while Vahlas et col. [9] take the data of Kale and Jacob [14].

Ba-Cu-O ternary phase

Five compounds are reported in the literature:

- $BaCuO_2$ which can be produced without any problem, but a certain CO_2-stability content is known from earlier work. This contamination is very difficult to determine due to the complicated crystallographic structure of the compound.

- $BaCu_2O_2$ and Ba_3CuO_4 which exist only at low or very low oxygen partial pressures and are normally prepared in pure argon atmosphere.

- $Ba_2CuO_{3+\delta}$: this compound has two different crystallographic structures: the high temperature phase with $\delta = 0.1$ and a tetragonal cell. Lowering the temperature (T<810°C) results in the formation of an orthorhombic phase with $\delta = 0.3$. These compounds are very unstable when stocked in humid or CO_2-containing atmospheres.

- $Ba_2Cu_3O_{5+\delta}$: this compound on the copper-rich side of the diagram is non-stoichiometric and can only be prepared at higher oxygen pressures.

In the experimental conditions of the study described here, only three compounds were studied: $BaCuO_2$, $BaCu_2O_2$ and Ba_2CuO_{3+x}. The experimental phase diagram of BaO-CuO_x is not well known [19]. Like in the Y-Cu-O ternary, it is the Ba-Cu-O face towards 800°C which is of most interest, and more especially the equilibria below the BaO-CuO line. The solution indicated on figure 2 has been retained which is in agreement with the experimental results obtained at 850°C by Beyers and Ahn [33] and by K. Borowiec et al. [36]. On the other hand, it differs from the thermodynamic model made by Voronin and Degterov [34] who propose the CuO-$BaCu_2O_2$ equilibrium at 827°C instead of the $BaCuO_2$-Cu_2O equilibrium.

The measured or estimated thermodynamic data are given in table III.

For the simulation, Harsta et col. [8] take into account only $BaCuO_2$ and Ba_2CuO_3 with values that have been estimated by Lee and Lee [17] by comparison with the $CaO-CuO_x$ system. As for Vahlas et col. [9] they have estimated the free energies of these two compounds by adjustment on an isothermal section at 950°C $YO_{1.5}$-BaO-CuO.

Table III
Thermodynamic for the compounds Ba-Cu-O

	$\Delta H^o_{f,ox}$ kJ mol^{-1}	$\Delta S^o_{f,ox}$ J mol^{-1} k^{-1}	T. range K	$\Delta G^o_{f,ox}$ J mol^{-1}	References
$BaCuO_2$	-63.4	5.15	736-1102		[35]
	-42.9	-7.8	1023-1223		[36]
	-40.102	-5.6	1100-1180		[26]
	-62.9 ± 4.4				[15]
	-86.0				[37]
	-83.8 ± 3.1				[31]
	-85.2 ± 2.7				[30]
	-9.368+	23.82+			[17]
	-41.278+	-4.125+			[9]
	-108.41 ± 2.8				[11]
	-44.2 ± 14.0				[12]
	-38.485*	-5.5*			
				-38819.934-3.268·T +1.079·T Ln T+	[34]
$BaCu_2O_2$	-30.358	-0.9	1023-1223		[36]
	-48.798	-6.7	980-1120		[26]
	-39.09*	-6.7*			
				-46080.068 + 3.518·T +0.399·T Ln T+	[34]
Ba_2CuO_3	-10.353+	28.8+			[17]
	-5.451+	24.117+			[9]
	-63.74*	-24.86*			
				-76243 + 35.201·T+	[34]

+ Estimated values
* Values used in the simulation

Y-Ba-Cu-O quaternary phase

A lot of work on the phase diagram of the quaternary system has been recently published [38] [41]. The diagram can be divided into three different zones according to the metal composition:

- Y-rich: only one compound exists in the Y-rich part of the phase diagram: Y_2BaCuO_5, which is stoichiometric in oxygen and stable in natural air, so it can easily be synthesised and stored without further precautions.

- Ba-rich: in the Ba-rich part of the diagram, the situation is very complicated because of the low stability of the compounds against moisture and a very high solubility range of CO_2 [42]. Under carefully controlled conditions, only one real oxide is confirmed with the nominal composition $YBa_4Cu_3O_9$ (slightly non-stoichiometric in oxygen) and a second one is presumed to exist. There are two different compositions given in the literature: $YBa_6Cu_3O_{11}$ and $YBa_5Cu_2O_9$. In this work, the $YBa_5Cu_2O_9$ compound has been prepared, but because of the very high reactivity of this phase (after a few minutes in natural air its colour changed from dark-grey to brown-yellow) a final conclusion is difficult to draw.
Two other phases with compositions $Y_2Ba_8Cu_5O_{18}$ and $YBa_3Cu_2O_x$ can only be obtained using $BaCO_3$ as starting material and are therefore clearly CO_2-stabilised.

- Cu-rich: three different compounds are known, having the same basic layered structure but varying in the copper content:

    $YBa_2Cu_3O_{6+x}$ $0 < x < 1$ superconductor with $T_c = 91$ K
    $Y_2Ba_4Cu_7O_{14+x}$ unstable at low P_{O_2}, Tc ~ 90 K
    $YBa_2Cu_4O_8$ unstable at low P_{O_2}, stable at very high P_{O_2}, Tc ~ 80 K

$YBa_2Cu_3O_{6+x}$ can easily be prepared at temperatures higher than 900°C without any further control of environmental parameters. Low-temperature oxygen treatment leads to the superconducting O_7 phase with good superconducting properties. The formation of the other two compounds depends strongly on the oxygen partial pressure and the particle size distribution of the starting material. Use of the standard powder technique always resulted in the formation of $YBa_2Cu_3O_{6+x}$ and CuO due to faster growth kinetics. Dissolving the oxide with a following drying step before sintering decreases the grain size and therefore the wanted compounds can rapidly be obtained.

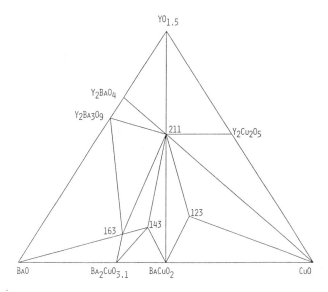

Figure 3: $YO_{1.5}$-BaO-CuO isothermal pseudoternary section at 800°C.

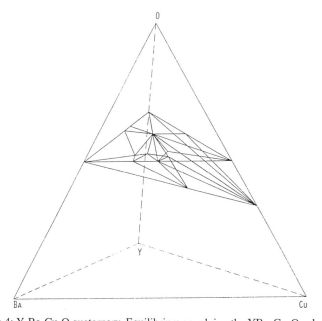

Figure 4: Y-Ba-Cu-O quaternary. Equilibrium envolving the $YBa_2Cu_3O_7$ phase.

To describe the isothermal section $YO_{1.5}$-BaO-CuO around 800°C, it is necessary to take into account three compounds $YBa_2Cu_3O_{6.5}$, Y_2BaCuO_5, $YBa_4Cu_3O_{8.5}$ plus, on the barium-rich side, a compound which may be $YBa_5Cu_2O_9$ or $YBa_6Cu_3O_{11}$. This gives the plot shown on figure 3 which is in agreement with the isothermal section at 900°C of Osamura and Zhang [41]. Given the compounds and tielines taken into consideration, this section is very different from those used by Harsta et col. [8] or Vahlas et col. [9] in their simulation of the MOCVD process.

Under the deposition conditions described in this paper, namely 800°C and a few torrs of oxygen pressure, it is the $YBa_2Cu_3O_6$ phase which is most likely to be deposited. This places the equilibria involved below the plane of the section in figure 3 with occurrence of the phases represented in figures 1 and 2. Although difficult to work out, the tetrahedric representation in figure 4 illustrates the equilibria concerned.

As in the case of the ternary phases, there is little thermodynamic data concerning the compounds. Such data that are available are given in table 4 together with the data selected by the present authors by optimising all available information: thermodynamic measurements - phase diagram.

Table IV
Thermodynamic data for the compounds Y-Ba-Cu-O.

	$\Delta H^o_{f,ox}$ kJ mol^{-1}	$\Delta S^o_{f,ox}$ J mol^{-1} k^{-1}	References
Y_2BaCuO_5	48.6 ± 3.0		[30]
	-92.0		[37]
	-76.17 ± 22.0		[43]
	-24.4	38.0	[25]
	-72.5	79.3	[13]
	-152.47	-25.76	[44]
	-61.7 ± 2.5	-6.5 ± 2.2	[26]
	-53.92 ± 5.42		[11]
	-53.40*	-3.16*	
$YBa_4Cu_3O_{8.5}$	-264.9 ± 9.4	-125.0 ± 8.8	[26]
	-241.96 ± 41		[12]
	-210.03*	-74.51*	
$YBa_2Cu_3O_6$	-94.1 ± 7.6		[30]
		-4.44 ± 3.0	[45]
	-131.43 ± 5.93		[11]
	-89.80*	-7.4*	

* Values used in the simulation.

MODELING OF THE $YBA2CU3O_{7-X}$ PHASE

The state of order in the basal oxygen plane of the $YBa_2Cu_3O_{7-x}$ is strongly correlated to the superconducting properties of this material [46,47]. The tetragonal-to-orthorhombic structural phase transitions associated with the oxygen ordering have been studied by neutron powder diffraction [48,50] and X-ray diffraction [51]. At low temperature, the observed stable or orthorhombic structures (OrthoI, OrthoII) [52-53] consist of alternating full and empty Cu-O chains. The 2D Ising model with asymmetric next-nearest-neighbour interactions (ASYNNNI), proposed originally by de Fontaine et al. [54,55], successfully explained the ground state of the system. A set of such pair interactions, as derived from first principles, LMTO total energy calculations and the Connolly-Williams method, was provided by Sterne and Wille [56]. Using these pair interactions, statistical thermodynamic calculations according to the Cluster Variation Method (CVM) have been performed, giving good quantitative agreement with experimental phase diagrams, in particular, for the higher temperature regions [57-59]. An improved numerical procedure of the CVM [60,61] has been used to calculate the (c, T) phase diagram with the Sterne and Wille parameters for lower temperatures as well [62].

In the Brite superdata contract, A. Pasturel and C. Colinet [63] present a O-O pair interaction calculation in the basal plane according to the embedded cluster method (ECM). On the basis of this calculation, it was possible to determine values in remarkable agreement with the results of the other methods, but the phase diagram shows a quite considerable deviation from the orthoI-orthoII transition line, specifically due to the high sensitivity of renormalised effective interaction pairs (Vi). In view of this great sensitivity of the phase diagram to the Vi value, it was decided to use a less physical model but which allows these pair energies to be fitted to the experimental information available: the sublattice compound energy model [64,65].

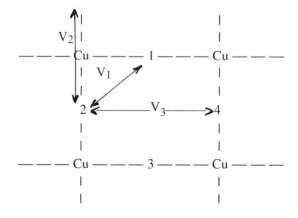

Figure 5: Sketch of the basal C-O plane.

The schematic representation of the Cu-O basal plane indicated on figure 5 corresponds to De Fontaine's preliminary description [54], but instead of taking two interpenetrating square sublattices, W. Huang and B. Sundman [66] chose four sublattices to distribute the oxygen atoms, two-by-two parallel (1 and 3 or 2 and 4) and two-by-two perpendicular (1 and 2 or 3 and 4). Three interaction energies were used: V_1 between two perpendicular sublattices, V_2 which is an interaction energy at the second-nearest neighbours above a copper atom in each lattice, V_3 which is the interaction between two parallel sublattices.

The Gibbs free energy of the $YBa_2Cu_3O_{7-x}$ phase can be written for two molecule units,

$YBa_2Cu_6O_{12}(O, Va)_1 (O, Va)_1 (O,Va)_1 (O, Va)_1$:

$$\begin{aligned}
G_m = \; & y_{Va}^1 \; y_{Va}^2 \; y_{Va}^3 \; y_{Va}^4 \; {}^0G_{Y_2Ba_4Cu_6O_{12}} \\
+ & y_{Va}^1 \; y_{Va}^2 \; y_{Va}^3 \; y_O^4 \; {}^0G_{Y_2Ba_4Cu_6O_{13}} \\
+ & y_{Va}^1 \; y_{Va}^2 \; y_O^3 \; y_{Va}^4 \; {}^0G_{Y_2Ba_4Cu_6O_{13}} \\
+ & y_{Va}^1 \; y_O^2 \; y_{Va}^3 \; y_{Va}^4 \; {}^0G_{Y_2Ba_4Cu_6O_{13}} \\
+ & y_O^1 \; y_{Va}^2 \; y_{Va}^3 \; y_{Va}^4 \; {}^0G_{Y_2Ba_4Cu_6O_{13}} \\
+ & y_{Va}^1 \; y_{Va}^2 \; y_O^3 \; y_O^4 \; {}^0G_{Y_2Ba_4Cu_6O_{14}}{}^{(A)} \\
+ & y_{Va}^1 \; y_O^2 \; y_O^3 \; y_{Va}^4 \; {}^0G_{Y_2Ba_4Cu_6O_{14}}{}^{(A)} \\
+ & y_O^1 \; y_{Va}^2 \; y_{Va}^3 \; y_O^4 \; {}^0G_{Y_2Ba_4Cu_6O_{14}}{}^{(A)} \\
+ & y_O^1 \; y_O^2 \; y_{Va}^3 \; y_{Va}^4 \; {}^0G_{Y_2Ba_4Cu_6O_{14}}{}^{(A)} \\
+ & y_{Va}^1 \; y_O^2 \; y_{Va}^3 \; y_O^4 \; {}^0G_{Y_2Ba_4Cu_6O_{14}}{}^{(B)} \\
+ & y_O^1 \; y_{Va}^2 \; y_O^3 \; y_{Va}^4 \; {}^0G_{Y_2Ba_4Cu_6O_{14}}{}^{(B)} \\
+ & y_{Va}^1 \; y_O^2 \; y_O^3 \; y_O^4 \; {}^0G_{Y_2Ba_4Cu_6O_{15}} \\
+ & y_O^1 \; y_{Va}^2 \; y_O^3 \; y_O^4 \; {}^0G_{Y_2Ba_4Cu_6O_{15}} \\
+ & y_O^1 \; y_O^2 \; y_{Va}^3 \; y_O^4 \; {}^0G_{Y_2Ba_4Cu_6O_{15}} \\
+ & y_O^1 \; y_O^2 \; y_O^3 \; y_{Va}^4 \; {}^0G_{Y_2Ba_4Cu_6O_{15}} \\
+ & y_O^1 \; y_O^2 \; y_O^3 \; y_O^4 \; {}^0G_{Y_2Ba_4Cu_6O_{16}}
\end{aligned}$$

$$+ RT \sum_{4}^{i=1} (Y_O^1 \ln Y_O^1 + Y_{Va}^1 \ln Y_{Va}^1)$$

$$+ y_{Va}^1 \; y_{Va}^3 \; y_O^2 \; y_{Va}^2 \; y_O^4 \; y_{Va}^4 \; L_{Va:\,O,\,Va:Va:O,\,Va}$$
$$+ y_{Va}^2 \; y_{Va}^4 \; y_O^1 \; y_{Va}^1 \; y_O^3 \; y_{Va}^3 \; L_{O,\,Va:Va:O,\,Va}$$

where the y_s correspond to the fractions of sites on each of the four sublattices: for example, y_{Va}^2 is the fraction of sites occupied by vacancies (Va) on the sublattice 2. For each sublattice, $y_{Va}^i + y_O^i = 1$, R is the ideal gas constant and L_i are the interaction parameters. The Gibbs energies of the pure end-member components, which are obtained if each sublattice is filled with species of one type only: vacancy or oxygen atom, correspond to:

$^0G_{Y_2Ba_4Cu_6O_{12}}$ = U_0 = GP 1230$_6$
$^0G_{Y_2Ba_4Cu_6O_{13}}$ = U_1 = GP 1230$_6$ + 0.25GO$_2$ + 2V$_2$
$^0G_{Y_2Ba_4Cu_6O_{14}}^{(A)}$ = U_{2A} = GP 1230$_6$ + 0.5GO$_2$ + 2V$_1$ + 4V$_2$
$^0G_{Y_2Ba_4Cu_6O_{14}}^{(B)}$ = U_{2B} = GP 1230$_6$ + 0.5GO$_2$ + 4V$_2$ + 2V$_3$
$^0G_{Y_2Ba_4Cu_6O_{15}}$ = U_3 = GP 1230$_6$ + 0.75GO$_2$ + 4V$_1$ + 6V$_2$ + 2V$_3$
$^0G_{Y_2Ba_4Cu_6O_{16}}$ = U_4 = GP 1230$_6$ + GO$_2$ + 8V$_1$ + 8V$_2$ + 4V$_3$

where GP1230$_6$ = -89802.1 + 7.4 · T + 2G$_y^o$ + 4G$_{Ba}^o$ + 6G$_{Cu}^o$ + 6G$_{O_2}^o$

twice the Gibbs energy of formation of YBa$_2$Cu$_3$O$_6$ with reference to the elements. The following values were used:

GO$_2$ = -259847 - 169.682T + 2.GH$_{SER}^{O_2}$

$L_{Va:O, Va:Va:O,Va}$ = $L_{O,Va:VA:O, Va:Va}$ = 30000
V_1 = 16000
V_2 = -8000
V_3 = 8000

which correspond to a preliminary approach by Huang and Sundman [66] where L_i and V_1, V_3 values have been optimised from information on the experimental phase diagram and where V_2 was taken to be -0.5V_1 according to Kikuchi and Choi [60]. The expression of GO$_2$ was optimised using the vapour pressure measurements of Lindemer et col. [67]. The phase diagram calculated from these data is close to that proposed by Kikuchi and Choi [60] figure 5.

In the Brite superdata contract, Huang and Sundman subsequently developed more complicated versions in order to use the results of the physicists Pasturel and Colinet [63]. However, for the optimisation of the MOCVD deposit, which involves a large number of species, it proved more convenient to use a simplified version.

Two similar approaches based on the compound energy model have been published recently [68,69]. These involve different assumptions with regard to the possible sublattices. As far as the authors of the MOCVD simulations are concerned, Harsta et col. [8] ignored the stoichiometric variation by considering only the compound YBA$_2$Cu$_3$O$_{6.5}$ while Vahlas et col. [9] used Lindemer's model [67]. This model accounts for the variation in solid solution

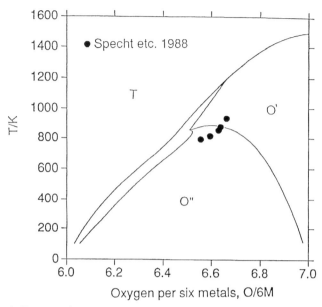

Figure 6: Concentration versus temperature (C, T) order-disorder phase diagram [66] and Specht and col. measurements [51].

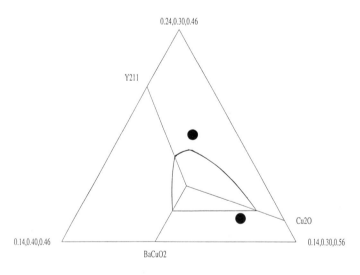

Figure 7: Ternary CVD diagram for a total pressure of 5 torr an oxygen partial pressure of 2 torr and 825°C. The principal phase deposited is $YBa_2Cu_3O_{6.08}$. The contours represent an isoyield of 90% of 123 phase. The points correspond to the experimental deposits. ●

stoichiometry as a function of temperature and oxygen partial pressure. Unfortunately, it does not give the order-disorder transition at the correct temperature. Vahlas et col. [9] were therefore obliged to remove the orthorhombic phase whenever it appeared in their calculations instead of the tetragonal phase.

OPTIMISATION OF MOCVD $YBa_2Cu_3O_{7-x}$ THIN FILM DEPOSITION

A simulation of the on-going reactions during thin film deposition of the superconducting phase will be presented. The ß-dicetonates of Y, Ba and Cu were evaporated in various mole fractions, evaporated, transported using argon gas, mixed with oxygen to react at a heated reaction zone. To avoid computing minimization problems, all phases with almost no probability to appear like carbides, hydrides, have not been taken into account, as already mentioned in the case of the pure metals. This leads to the following list:

Y, Ba, Cu, C, Y_2O_3, $BaCO_3$, BaH_2O_2, BaO, BaO_2, $CuCO_3$, CuH_2O_2, CuO, Cu_2O, $Y_2Cu_2O_5$, $YCuO_2$, $BaCuO_2$, $BaCu_2O_2$, Ba_2CuO_3, Y_2BaCuO_5, $YBa_4Cu_3O_{8.5}$, Y_2BaO_4, $Y_4Ba_3O_9$.

$O_2(G)$, $H_2(G)$, Ar(G), Y(G), OY(G), $OY_2(G)$, $O_2Y(G)$, $O_2Y_2(G)$, Ba(G), BaHO(G), $BaH_2O_2(G)$, BaO(G), Cu(G), CuO(G), C(G), CHO(G), $CH_2O(G)$, $CH_4O(G)$, CO(G), $CO_2(G)$, $C_2O(G)$, $C_3O_2(G)$, $C_2H_4O(G)$, H(G), HO(G), $HO_2(G)$, $H_2O(G)$, $H_2O_2(G)$, O(G), $O_3(G)$.

There is no thermodynamic data available for the precursor materials in the literature. They were discribed in order to obtain complete decomposition of these compounds at the deposition temperature.

Our standard deposition conditions are the following: The total gas pressures is a few torr with an oxygen partial pressure of about 2 torr. The vapor pressures of the organometallic compounds are a few millitorrs, which means that they were highly diluted in the gas phase. The substrate temperature is around 800°C. The simulation of the deposition process using these experimental conditions leads to the following diagram which represents the deposited phases as a function of the initial gas phase. In figure 7 an isoyield of 90 mole% of the superconducting 123 phase has been drawn like already proposed by Vahlas et al. [9] in their simulation. One can find the well known result that it is almost impossible to obtain the superconducting phase alone and therefore it is interesting to choose the secondary phase simultanously deposited as a function of process parameters. The morphology of thin films with different compositions indicated in figure 7 are represented as SEM photos in figure 8.

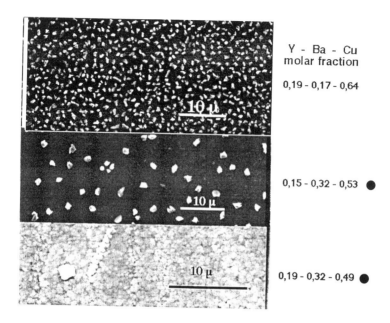

Figure 8: SEM photos of MOCVD thin films with various compositions.

CONCLUSION

To establish a correct and reliable ensemble of the thermodynamic functions to describe the chemical reactions taking place in an MOCVD process of $YBa_2Cu_3O_{7-x}$, a critical evaluation of the experimental data, which is rare and often contradictory, is essential. The data presented here has been homogenized to satisfy most of the existing informations, but is still not perfect. Nevertheless a simulation gives useful information concerning the process, like:

- optimization of nature and quantity of secondary phases, co-deposited in classical MOCVD,

- choice of solvants in an aerosol CVD-process, where the orgonometallic precursors are dissolved. The formation of carbonates is often reported and the simulation gives useful limits of the phase stability of the superconductor versus CO_2,

- in the case of fluorine-based precursors, a deposition zone without any BaF_2 contamination in an in-situ process is desired.

REFERENCES

[1] M.K. Wu, J.R. Ashburn, C.J. Torng, P.H. Hor, R. Meng, L. Gao, Z.J. Huang, Y.Q Wang and C.W. Chu, Phys. Rev. Lett. **58**, 908 (1987).

[2] A. Erbil, K. Zhang, B.S. Kwak and E.P. Boyd SPIE Procd. **1187**, 104 (1989).

[3] M.L. Hitchman, D.D. Gilliland, D.J. Cole-Hamilton and S.C. Thompson, in New Materials and their Applications 1990. Ed. (D. Holland, Institute of Physics, Bristol, 1990), p.305.

[4] L.M. Tonge, D.S. Richeson, T.J. Marks, J. Zhao, J. Wang, B.W. Wessels, H.O. Marcy and C.R. Kannewurf, Adv. Chem. Ser. **226**, 351 (1989).

[5] J. Zhao and P. Norris in Synthesis and Characterisation of High-Temperature Superconductors Ed. (J.J. Pouch et al., Trans. Tech. Publications, Aedermannsdorf 1993), p.130.

[6] M. Leskela, H. Mösla and L. Niinisto, Supercond. Sci. Technol. **6**, 627 (1993).

[7] J. Zhao and P. Norris, Thin Solid Films **206**, 122 (1991).,

[8] A. Harsta and J.O. Carlsson, J. Crystal Growth **110**, 631 (1991).

[9] C. Vahlas and T.M. Besmann, J. Am. Ceram. Soc. **10**, 2679 (1992).

[10] C. Bernard and R. Madar in Chemical Vapour Deposition of Refractory Metals and Ceramics, Ed. By T.M. Besmann and B.M. Gallois (Mat. Res. Soc. Procd. 168, Pittsburg, PA, 1990) pp.3-17.

[11] A. Rais, M. Ganteaume, J.C. Mathieu and J. Rogez, CNRS, Marseille, Brite Euram project BREU 0203C.

[12] A. Watson, Shefield, Brite Euram project BREU 0203C.

[13] A.M. Azad and O.M. Sreedharan, J. Mat. Sci. Lett. **8**, 67 (1989).

[14] G.M. Kale and K.T. Jacob, Solid State Ionics **34**, 247 (1989).

[15] Y. Idemoto, J. Takahaschi and K. Fueki, Phys. C **194**, 177 (1992).

[16] L.M. Lopato, I.M. Maister and A.V. Shevchenko, Izv. Akad. Nauk SSSR, Neorg. Mat. **8**, 5, 861 (1972).

[17] B.J. Lee and D.N. Lee, J. Am. Ceram. Soc. **72**, 314 (1989).

[18] K.G. Frase and D.R. Clarke Adv. Ceram. Mat. **2**, 295 (1987).

[19] R.S. Roth, K.L. Davis and J.R. Dennis, Adv. Ceram. Mat. **2**, 303 (1987).

[20] K. Hack, Aachen, Brite Euram project BREU 0203C.

[21] V. Wiesner, G. Krabbes, M. Riestschel, Mat. Res. Bull. **24**, 1261 (1989).

[22] Yu.D. Tretyakov, A.R. Kaul and N.V. Makukhin, J. Sol. St. Chem. **17**, 183 (1976).

[23] R. Pankajavalli and O.M. Sreedharan, J. Mater. Sci. Lett. **7**, 714 (1988).

[24] K. Borowiec and K. Kolbrecka J. Less-Common Met. **613**, 143 (1990).

[25] C.Ji. Fan and Z. Zhao, J. Less-Common Met. **161**, 49 (1990).

[26] G.F. Voronin, S.A. Degterov and Yu.Ya. Skolis Procd. 3rd German Soviet Bilat. Seminar on High Tc Supercond. Karlsruhe 1990.
[27] F. Glathe, H. Oppermann and W. Reichelt, Z. Anorg. Allg. Chem. **606**, 41 (1991).
[28] R.M. Suzuki, S. Okada, T. Oishi and K. Ono, Mat. Trans., JIM 31, **10**, 1078 (1990).
[29] R. Shimpo and Y. Nakamura, J. Jap. Inst. Met. **54**, 4, 549 (1990).
[30] Z. Zhou and A. Navrotsky, J. Mat. Res. **7**, 11, 2920 (1992).
[31] A. Navrotsky in Chemistry of Electronic Ceramic Materials (NIST special publication 804, Gaithersburg, MD, 1990), p. 370.
[32] S.N. Mudretsova, I.A. Vasilyeva and Zh. V. Filippova, Zh. Fiz. Khim. **63**, 3108 (1989).
[33] R. Beyers and B.T. Ahn, IBM Research Report RJ 7797 (71914), 1990.
[34] G.F. Voronin and S.A. Degterov submitted to J. Solid State Chem. (1993).
[35] A.M. Azad, O.M. Sreedharan and K.T. Jacob, J. Mater. Sci. **26**, 3374 (1991).
[36] K. Borowiec and K. Kolbrecka J. Alloys and Compounds **176**, 225 (1991).
[37] L.R. Morss, S.E. Dorris, T.B. Lindemer and N. Naito, Eur. J. Solid. State Inorg. Chem. **27**, 327 (1990).
[38] D.M. de Leeuw, C.A.H.A. Mutsaers, C. Langereis, H.C.A. Smoorenburg and R.J. Rommers, Physica C, **152**,39 (1988).
[39] F. Abbattista, M. Vallino, D. Mazza, M. Luco-Borlera and C. Brisi, Mater. Chem. Phys., **20**, 191 (1988).
[40] R. Beyers and B.T. Ahn, Ann. Rev. Mat. Sci. **21**, 335 (1991).
[41] K. Osamura and W. Zhang, Z. Metallkde. **84**, 522 (1993).
[42] P. Karen and A. Kjekshus, J. of sol. State Chem. **94**, 298 (1991).
[43] F.G. Garzon, I.D. Raistrick, D.S. Ginley and J.W. Halloran, J. Mater. Res. **6**, 5, 885 (1991).
[44] R. Pankajavalli and O.M. Sreedharan, J. Mater. Sci. Lett. **8**, 225(1989).
[45] G.F. Voronin and S.A. Degterov, Physica C, **176**, 387 (1991).
[46] J.D. Jorgensen, B.W. Veal, W.K. Kwok, G.W. Crabtree, A. Umezawa, L.J. Nowicki and A.P. Paulikas, Phys. Rev. B **36**, 5731 (1987).
[47] J.D. Jorgensen, D.G. Hinks, H. Shaked, B. Dabrowski, B.W. Veal, A.P. Paulikas, L.J. Nowicki, G.W. Crabtree, W.K. Kwok, A. Umezawa, L.H. Nunez, B.D. Dunlap, C.U. Segre and C.W. Kimball, Physica B **156-157**, 877 (1989).
[48] J.D. Jorgensen, M.A. Beno, D.G. Hinks, L. Soderholm, K.J. Volin, R.L. Hitterman, J.D. Grace, I.K. Schuller, C.U. Segre, K. Zhang and M.S. Kleefisch, Phys. Rev. B 36, 3608 (1987).
[49] J.D. Jorgensen, B.W. Veal, A.P. Paulikas, L.J. Nowicki, G.W. Crabtree, H. Claus and W.K. Kwok, Phys. Rev. B **41**, 1863 (1990).

[50] N.H. Andersen, B. Lebech and H.F. Poulsen, J. Less-Common Metals **164-165**, 124 (1990).
[51] E.D. Specht, C.J. Sparks, A.G. Dhere, J. Brynestad, O.B. Cavin, D.M. Kroeger and H.A. Oye, Phys. Rev. B **37**, 7426 (1988).
[52] M.A. Alario-Franco, C. Chaillout, J.J. Capponi, J. Chenavas and M. Marezio, Physica C **156**, 455 (1988).
[53] J. Reyes-Gasca, T. Krekels, G. Van Tendeloo, J. Van Landuyt, W.H.M. Bruggink, M. Verweij and S. Amelinckx, Solid State Communications **70**, 269 (1989).
[54] D. de Fontaine, L.T. Wille and S.C. Moss, Phys. Rev. B **36**, 5709 (1987).
[55] L.T. Wille and D. de Fontaine, Phys. Rev. B **37**, 2227 (1988).
[56] P.A. Sterne and L.T. Wille, Physica C 162-164, 223 (1989).
[57] G. Ceder, M. Asta, W.C. Carter, M. Kraitchman, D. de Fontaine, M.E. Mann and M. Sluiter, Phys. Rev. B **41**, 8698 (1990).
[58] D. de Fontaine, G. Ceder and M. Asta, J. Less-Common Metals **164-165**, 108 (1990).
[59] M. Asta, D. de Fontaine, G. Ceder, E. Salomons and M. Kraitchman, J. Less-Common Metals **168**, 39 (1991).
[60] R. Kikuchi and J. Choi, Physica C **160**, 347 (1989).
[61] V.E. Zubkus, S. Lapinskas and E.E. Tornau, Phys. Stat. Sol. B **156**, 93 (1989).
[62] V.M. Matic, Physica A **184**, 571 (1992).
[63] A. Pasturel and C. Colinet, Grenoble, Brite Euram project BREU 0203C.
[64] M. Hillert and L.I. Staffansson, Acta Chem. Scand. **24**, 3618 (1970).
[65] J.O. Andersson, A.F. Guillermet, M. Hillert, B. Jansson and B. Sundman, Acta Metall., **34**, 437 (1986).
[66] W. Huang and B. Sundman, Stockholm, Brite Euram project BREU 0203C.
[67] T.B. Lindemer, J.F. Hunley, J.E. Gates, A.L.Sutton Jr., J. Brynestad, C.R. Hubbard and P.K. Gallagher, J. Am. Ceram. Soc., 10, 1775 (1989).

LOW TEMPERATURE CVD OF TiN FROM Ti(NR$_2$)$_4$ AND NH$_3$: FTIR STUDIES OF THE GAS-PHASE CHEMICAL REACTIONS

BRUCE H. WEILLER
The Aerospace Corporation, Mechanics and Materials Technology Center,
PO Box 92957/M5-753, Los Angeles, CA 90009-2957

ABSTRACT

The gas-phase chemical reaction between Ti(NMe$_2$)$_4$ and NH$_3$ is a critical step in the Metallorganic Chemical Vapor Deposition (MOCVD) of TiN at low temperatures. We have examined this reaction using a flow-tube reactor coupled to an FTIR spectrometer. A sliding injector provides control over the reaction time and the kinetics of reactive species can be measured as a function of the partial pressure of an added reagent. The disappearance of Ti(NMe$_2$)$_4$ was measured as a function of reaction time and NH$_3$ pressure at 26 °C. The resulting bimolecular rate constant is $(1.1 \pm 0.1) \times 10^{-16}$ cm^{3}molecules^{-1}s^{-1}. Dimethylamine is observed as a direct product from this reaction consistent with other studies. We have also measured the rate constant using ND$_3$ and find a substantial isotope effect, $k_h/k_d = 2.4 \pm 0.4$. This indicates that H-atom transfer is involved in the rate limiting step. We show that these results can be explained by a mechanism comprised of transamination reactions with NH$_3$.

INTRODUCTION

TiN is a material with many properties that make it useful for a range of applications. It has high hardness (almost as hard as diamond), high electrical conductivity (greater than titanium metal), a high melting point and it is chemically inert. The potential applications of this material include wear resistant coatings, diffusion barriers for metallization in integrated circuits, and optical or thermal control coatings for spacecraft components that would be less susceptible to erosion by small particles.

The reaction of Ti(NMe$_2$)$_4$ with NH$_3$ is one of the most promising methods to deposit TiN since it proceeds at low temperatures and gives high quality films with good conformal coverage.[1] An interesting feature of this system is that surface reactions between Ti(NMe$_2$)$_4$ and NH$_3$ could not be observed under UHV conditions.[2] Gas-phase pre-reaction between Ti(NMe$_2$)$_4$ and NH$_3$ was required for the formation of low-carbon TiN films. In order to optimize this process and design improved CVD reactors, quantitative kinetics data on this gas-phase reaction is needed. Earlier we published a preliminary report on the use of a flow-tube reactor coupled to an FTIR spectrometer to study this reaction.[3] Now we have obtained the bimolecular rate constants for the reaction of Ti(NMe$_2$)$_4$ with NH$_3$ and ND$_3$ by measuring the decay of Ti(NMe$_2$)$_4$ as a function of time and NH$_3$ and ND$_3$ pressures. We also demonstrate that the mixing time is not significant in these measurements by the independence of the results on the use of Ar or He as the buffer gas. Finally we show our kinetics results can be successfully modeled using a simple mechanism involving a series of transamination reactions.

EXPERIMENTAL

The experimental apparatus is described in detail elsewhere.[4] Briefly, the flow-tube reactor is a 1-m long, 1.37" i.d. teflon-coated stainless-steel tube equipped with a sliding injector port that allows the distance from the focus of the IR beam to be varied. The observation region is a standard cross (NW-40) equipped with purged windows, purged capacitance manometers, and a throttle valve controller to maintain constant pressure. Both the flow tube and the observation region are wrapped with heat tape for temperature regulation when desired. The IR beam from the FTIR spectrometer (Nicolet 800) is focused at the center of the flow tube perpendicular to the flow. The focusing optics and the IR detector are enclosed in plexiglass boxes and the entire beam path is purged with dehumidified, CO$_2$-free air. Mass flow meters measure the separate flows of buffer (He or Ar), bubbler, purge, and the flow of a dilute mixture of NH$_3$ in buffer (6.0%). The buffer gas and NH$_3$ flows are mixed and fed into the side arm of the flow tube while the Ti(NMe$_2$)$_4$

mixture is flowed into the sliding injector. All chemicals were used as received from the following suppliers: Ti(NMe$_2$)$_4$ (Strem); Ar (Matheson, UHP grade); NH$_3$ (Matheson, electronic grade); He (Spectra Gases, UHP grade). The spectrometer was operated at 8 cm^{-1} resolution and 256 scans were averaged. For kinetics measurements integrated intensities were used.

RESULTS AND DISCUSSION

Figure 1 shows the IR spectra obtained when Ti(NMe$_2$)$_4$ is reacted with NH$_3$ in the flow tube reactor. The top spectrum (a) is in the absence of NH$_3$. The relatively intense NC$_2$ symmetric stretch at 949 cm^{-1} is a good signature for Ti(NMe$_2$)$_4$, and we use it to monitor the number density of Ti(NMe$_2$)$_4$. The middle four spectra result from the addition of 0.245 torr of NH$_3$ and increasing the distance between the injector and observation region from 10 cm to 80 cm at constant NH$_3$ pressure. The flow rate was 170 sccm giving the reaction times listed in the caption. In spectra (b) through (e), we see NH$_3$ bands at 968 and 932 cm^{-1} (v_2) and at 1628 cm^{-1} (v_4).[5]

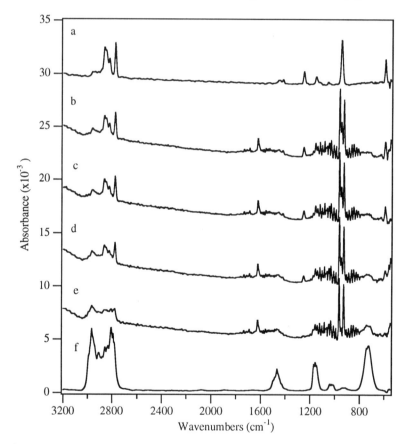

Figure 1. IR spectra for the reaction of Ti(NMe$_2$)$_4$ with NH$_3$. The top spectrum (a) is prior to NH$_3$ addition and the bottom spectrum (f) is for a sample of HNMe$_2$. Spectra (b) through (e) result from the addition of 0.245 torr NH$_3$ to the sample in (a) and the following reaction times: b) 0.69, c) 1.37, d) 2.75, e) 5.49 s.

These spectra clearly show the disappearance of Ti(NMe$_2$)$_4$ with increasing reaction time. The bottom spectrum is for a sample of HNMe$_2$ (Aldrich) at 8 cm^{-1} resolution. By comparison we can clearly see the formation of HNMe$_2$ as reported by Dubois et al. and expected for a transamination reaction.[2] It should be noted that our results are not limited by mixing rates (see below) and HNMe$_2$ is a direct product from the reaction of NH$_3$ with Ti(NMe$_2$)$_4$. In our earlier work we did not observe the formation of HNMe$_2$ due to poor signal-to-noise ratio and insufficient reaction time.[3]

When the integrated intensity of the 949 cm^{-1} band is plotted vs. time on a semi-log plot, a linear dependence is observed as expected for a pseudo first order reaction. This is consistent with our estimate of the Ti(NMe$_2$)$_4$ partial pressure of ~0.01 torr using the available vapor pressure data.[6] This is much less than the NH$_3$ pressure (0.1 to 0.4 torr) and therefore we expect an exponential decay for [Ti(NMe$_2$)$_4$]: [Ti(NMe$_2$)$_4$] = [Ti(NMe$_2$)$_4$]$_0$exp(-k$_{bi}$[NH$_3$]t), where k$_{bi}$ is the bimolecular rate constant. Furthermore, we expect a linear relationship between the logarithm of the IR absorbance and the reaction time: $\ln(A/A_0) = -k_{obs}t$, $k_{obs} = k_{bi}[NH_3]$. Here A is the integrated absorbance of the Ti(NMe$_2$)$_4$ band, A_0 is the average integrated absorbance of the Ti(NMe$_2$)$_4$ band in the absence of NH$_3$ and k_{obs} is the observed decay constant. The experiment was repeated for a series of NH$_3$ pressures. The data were fit to straight lines using a weighted least squares routine and the slopes of these lines give the observed decay constants (k_{obs}) at each NH$_3$ pressure. Similar data for ND$_3$ is shown in Figure 2.

When the observed decay constants (k_{obs}) are plotted against NH$_3$ pressure (Figure 3), we find a linear dependence with a zero intercept as expected for pseudo-first order conditions. The slope of this plot gives the bimolecular rate constant $k_{bi} = (1.1 \pm 0.1) \times 10^{-16}$ cm^3(molec.-s)$^{-1}$. Apparently, this is the first measurement of the rate constant for a transamination reaction. Although the rate constant represents only about a 10^{-6} reaction probability, it is faster than expected considering that the temperature is only 26 °C and that both Ti(NMe$_2$)$_4$ and NH$_3$ are closed-shell, stable molecules.

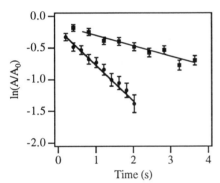 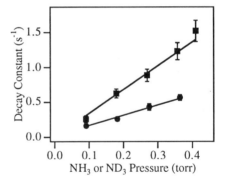

Figure 2. Plot of ln(A/A$_0$) vs. time for the reaction of Ti(NMe$_2$)$_4$ with ND$_3$ using He buffer gas. The ND$_3$ pressures are 0.0915 (squares) and 0.364 torr (circles).

Figure 3. Plot of the observed decay constants (k_{obs}) vs. NH$_3$ (squares) and ND$_3$ (circles) pressure. He buffer gas was used.

For the ND$_3$ data, we find the rate constant to be substantially reduced from the NH$_3$ value. Figure 3 shows that a plot of decay constant vs. ND$_3$ density gives a rate constant $k_d = (4.5 \pm 0.5) \times 10^{-17}$ cm^3(molec.-s)$^{-1}$ or $k_h/k_d = 2.4 \pm 0.4$. This indicates a primary isotope effect and that H-atom transfer is involved in the rate limiting step of this reaction. This is consistent with labeling studies that showed the amine proton in the product HNMe$_2$ originates from NH$_3$.[7] Below we discuss the mechanistic implications of this result.

In order to test for the possibility that the observed rates are affected by the mixing time, we have repeated these measurements in Ar buffer gas. The calculated diffusional mixing time in Ar at 25 °C is about three times longer than in He. If the mixing time was significant, this would be reflected in a smaller observed rate constant in Ar than in He. Figure 4 shows a plot of $\ln(A/A_0)$ vs. time for three NH_3 pressures in Ar. As described above for He buffer gas, the slopes of these lines are plotted vs. NH_3 pressure in Figure 5. The resulting bimolecular rate constant is $(8.4 \pm 0.8) \times 10^{-17}$ $cm^3(molec.-s)^{-1}$ and is within experimental error of the value in He. Therefore we conclude that mixing is not important under these experimental conditions. This is supported by the calculated diffusional mixing times of 0.07 and 0.2 s in He and Ar, respectively. With the enhanced mixing due to turbulence, the true mixing times will be considerably shorter. Our shortest observation time is 0.25 s and therefore the mixing time should not be significant for our measurements. This conclusion is also supported by our ability to reproduce the literature rate constant for the reaction of O_3 with isobutene.[4]

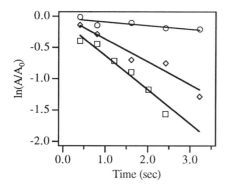 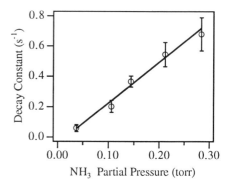

Figure 4. Plot of $\ln(A/A_0)$ vs. time for the reaction of $Ti(NMe_2)_4$ with NH_3 using Ar buffer gas. The NH_3 pressures are 0.0373 (circles), 0.106 (diamonds) and 0.212 torr (squares).

Figure 5. Plot of the observed decay constants (k_{obs}) vs. NH_3 pressure using Ar buffer gas.

The mechanism of the gas-phase reaction is of considerable interest since it appears to determine what species reach the surface of the growing TiN film. Our kinetic data is based on the disappearance of the NC_2 stretch of the dimethyl amido group. A reasonable supposition is that the frequency of this band is relatively independent of the other ligands attached to the Ti center. Given the low resolution of our measurements, we would not expect to resolve the various species with different numbers of NMe_2 groups. Therefore, in order to account for the complete disappearance of the 949 cm^{-1} band, all Ti species with NMe_2 groups would have to be removed. One simple mechanism that could account for this is a series of transamination reactions that convert $Ti(NMe_2)_4$ into $Ti(NH_2)_4$, a species that has not been characterized:

$Ti(NMe_2)_4 + NH_3$ → $Ti(NMe_2)_4 \cdot (NH_3)$ (1)

$Ti(NMe_2)_4 \cdot (NH_3)$ → $Ti(NMe_2)_4$ (2)

$Ti(NMe_2)_4 \cdot (NH_3)$ → $Ti(NMe_2)_3(NH_2) + HNMe_2$ (3)

$Ti(NMe_2)_3(NH_2) + 3 NH_3$ → $Ti(NH_2)_4 + 3 HNMe_2$ (4)

Here we propose that NH_3 reversibly forms a weak adduct with the electron deficient Ti center in $Ti(NMe_2)_4$ and that this adduct is the precursor to the rate limiting transamination reaction. We have also assumed that all reactions subsequent to (3) are fast and kinetically insignificant. Therefore we have grouped them in (4).

By integrating the resulting differential equations, it possible to show that this mechanism can quantitatively reproduce our results. The important observations we need to account for are: 1) a first order decay of Ti(NMe$_2$)$_4$, 2) a first order dependence of the rate on NH$_3$ pressure, and 3) the kinetic isotope effect. For the simulation we have used the program Acuchem[8] with some reasonable assumptions and estimates of the rate constants: 1) the addition of the second NH$_3$ is fast, k_1 = 1.00 x 10^{-11} cm^3(molec.-s)$^{-1}$, and reversible, k_2 = 9.80 x 10^6 s^{-1}, 2) the rate constant for H-atom transfer is k_{3h} = 104 s^{-1} and is responsible for the isotope effect, k_{3d} = 43.3 s^{-1}, and 3) all subsequent reactions are fast and not kinetically significant.

Figure 6a shows the results of such a simulation for [NH$_3$] = 0.44 torr. Ti(NMe$_2$)$_4$ decays exponentially via a small concentration of Ti(NMe$_2$)$_4$•(NH$_3$) with concomitant formation of Ti(NH$_2$)$_4$ and four HNMe$_2$ molecules. Figure 6b shows that when the densities of all Ti species containing NMe$_2$ groups are summed and the log is taken, a linear temporal dependence is observed in agreement our results.

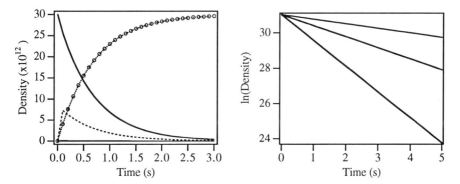

Figure 6. Simulation of the reaction of Ti(NMe$_2$)$_4$ with NH$_3$ using the mechanism in equations (1) through (4). a) left, the density of various species as a function of time. Ti(NMe$_2$)$_4$ (—), Ti(NMe$_2$)$_4$•(NH$_3$) (x 20, ---), Ti(NH$_2$)$_4$ (•••), HNMe$_2$ (x 0.25, circles), the sum of Ti species containing NMe$_2$ groups (diamonds). The NH$_3$ pressure is 0.438 torr. b) right, the log of the sum of all species with NMe$_2$ groups as a function of time. The NH$_3$ pressures are 0.078, 0.188, and 0.438 torr.

Figure 7 shows a comparison between the simulated data and the experimental results for both NH$_3$ and ND$_3$. This mechanism gives good quantitative agreement with results. The purpose of this exercise was to determine if this simple mechanism can account for our data. It should be noted that we do not imply that reactions (1) trough (4) account for all of the gas-phase chemistry. Clearly other reactions do occur since we observe the formation of yellow powder in the flow tube that is consistent with polymeric compounds as discussed by Dubois et al.[2] These products may form from the polymerization of imido species (L$_2$Ti=NR) or even bis-imido species (RN=Ti=NR). In any event, reactions (1) through (4) may account for the initial gas phase chemistry between Ti(NMe$_2$)$_4$ and NH$_3$.

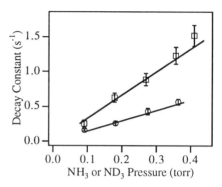

Figure 7. Comparison of simulated data with experimental results. The markers are the experimental results from Figure 3 and the solid lines are the simulated data. The squares are for NH_3 and the circles are for ND_3.

Work in progress is aimed at the direct spectroscopic identification of the intermediates in this reaction and correlating them with the quality of the TiN deposited. Temperature dependent kinetics measurements are planned as well as similar measurements with related precursors such as $Ti(NEt_2)_4$.

ACKNOWLEDGMENTS

This work was supported by The Aerospace Sponsored Research Program. The laboratory assistance of B. V. Partido is gratefully acknowledged. We thank Professor R. G. Gordon and Dr. L. H. Dubois for helpful discussions and communication of results prior to publication. We also thank Dr. N. Cohen for helpful discussions regarding kinetics issues and Dr. R. L. Martin for the loan of the O_3 generator.

REFERENCES

1. a) R. M. Fix, R. G. Gordon, and D. M. Hoffman, Mat. Res. Soc. Symp. Proc. **168**, 357 - 362 (1990). b) R. M. Fix, R. G. Gordon, and D. M. Hoffman, J. Am. Chem. Soc. **112**, 7833 - 7835 (1990), c) K. Ishihara, K. Yamazaki, H. Hamada, K Kamisako, and Y. Tarui, Jpn. J. Appl. Phys. **29**, 2103 - 2105 (1990). d) R. M. Fix, R. G. Gordon, and D. M. Hoffman, Chem. Mater. **3**, 1138 - 1148 (1991).
2. L. H. Dubois, B. R. Zegarski, and G. S. Girolami J. Electrochem. Soc. **139**, 3603 (1992).
3. B. H. Weiller, MRS Symp. Proc. **282**, 605 (1993).
4. B. H. Weiller, MRS Symp. Proc., in press
5. G. Herzberg, Infrared and Raman Spectra of Polyatomic Molecules, 1st ed. (Van Nostrand Reinhold Company, New York, 1945), p. 295.
6. The vapor pressure at 25 °C is 0.11 torr, D. Roberts, The Schumacher Corporation, personal communication.
7. J. A. Prybyla, C.-M. Chiang, and L. H. Dubois J. Electrochem. Soc. **140**, 2695 (1993).
8. W. Braun, J. T. Herron, D. K. Kahaner, Int. J. Chem. Kinet. **20**, 51 (1988).

THERMODYNAMIC CONSTRAINTS FOR THE *IN SITU* MOCVD GROWTH OF SUPERCONDUCTING Tl-Ba-Ca-Cu-O THIN FILMS

WILLIAM L. HOLSTEIN
DuPont Central Research and Development, Experimental Station, P. O. Box 80304, Wilmington, DE 19808-0304

ABSTRACT

In spite of several attempts, superconducting Tl-Ba-Ca-Cu-O thin films have not been successfully prepared *in situ* by metal organic chemical vapor deposition (MOCVD). Preparation of a phase by MOCVD requires that it be thermodynamically stable with respect to its decomposition into volatile species and other condensed phases. For MOCVD growth of Tl-Ba-Ca-Cu-O compounds in the presence of oxygen from reagents containing only C-H or C-H-O ligands, $Tl_2O(g)$ and $TlOH(g)$ exhibit appreciable volatility. If reagents with ligands containing fluorine are used, the formation of volatile $TlF(g)$ must also be considered. Thermodynamic data for these materials are compiled, and thermodynamic relationships between these gases, $H_2O(g)$ and $HF(g)$ are established. The thermodynamic stability of $TlOH(g)$ and $TlF(g)$ makes the *in situ* growth of Tl-Ba-Ca-Cu-O compounds by MOCVD more difficult than their *in situ* growth by physical vapor deposition processes, for which $Tl_2O(g)$ is the only volatile Tl-containing species present.

INTRODUCTION

Metal organic chemical vapor deposition (MOCVD) is a versatile technique for the *in situ* growth of a wide range of semiconductor and oxide compounds. The application of MOCVD to the preparation of high temperature superconducting phases has been extensively investigated [1], and both *in situ* $YBa_2Cu_3O_7$ [2,3] and *in situ* $Bi_2Sr_2CaCu_2O_8$ [4,5] thin films have been successfully prepared. With the development of single-source MOCVD [6], in which the reagents are mixed and injected into the gas stream, many of the limitations of MOCVD attributable to the low vapor pressures and thermal stabilities of the reagents have been overcome.

While considerable progress has been made with $YBa_2Cu_3O_7$ and $Bi_2Sr_2CaCu_2O_8$ thin films, the *in situ* preparation of Tl-Ba-Ca-Cu-O phases by MOCVD has not yet been demonstrated. Interest in these phases arises from their high superconducting transition temperatures and low microwave surface resistance [7]. In several works, precursor films containing the elements Ba, Ca, and Cu have been deposited by MOCVD, which were then converted into Tl-Ba-Ca-Cu-O phases during a separate high temperature thallination step [8-13], but such post-deposition thallination processes are not suitable for the preparation of multilayer structures.

By contrast, *in situ* $TlBa_2CaCu_2O_7$ and $TlBa_2(Ca,Y)Cu_2O_7$ thin films have been successfully prepared at substrate temperatures of around 825 K by rf magnetron sputtering of BaCaCu and BaCaYCu oxide targets at O_2 and Ar partial pressures of 100 mtorr while simultaneously evaporating thallium oxide [14,15]. The need for a significant overpressure of volatile $Tl_2O(g)$ during the growth process was demonstrated. The successful *in situ* growth of these phases by physical vapor deposition suggests that their preparation by MOCVD should also be possible. After all, MOCVD allows for greater flexibility in the selection of oxygen partial pressure. However, as shown below, an additional complication for the growth of Tl-containing materials by MOCVD results from the volatility of Tl-containing compounds other than $Tl_2O(g)$ generated by oxidation of the reagents' organic ligands. When reagents containing C-H and C-H-O ligands are used, thermodynamically stable $TlOH(g)$ is generated. When reagents with fluorinated ligands are used, thermodynamically stable $TlF(g)$ is also generated.

THERMODYNAMIC RELATIONSHIPS

The phase diagram for the Tl-O system has only recently been fully established [16]. Three thermodynamically stable condensed thallium oxide phases exist: $Tl_2O(c,l)$, $Tl_4O_3(c,l)$ and $Tl_2O_3(c)$. Their stability regimes depend on temperature and oxygen partial pressure. $Tl_2O(g)$ is the only known gaseous Tl-O compound [17]. Its thermodynamic properties are compiled in Table 1 for the temperature range 700-1200 K ($P° = 1$ atm) [16-18].

In MOCVD processes, volatile thallium hydroxide, $TlOH(g)$, forms through the reaction of $Tl_2O(g)$ and water vapor, which is generated within the reactor by the oxidation of organic ligands

$$Tl_2O(g) + H_2O(g) \stackrel{K_1}{=} 2\, TlOH(g) \ . \tag{1}$$

Inorganic hydroxides usually exhibit low volatility, but $TlOH(g)$ is an exception. Its thermodynamic properties have recently been measured from 940 K to 1020 K [19]. From this data, thermodynamic properties for $TlOH(g)$ were estimated for the temperature range 700-1200 K by assuming that $\Delta H_1°$ is independent of temperature (Table 2).

The organometallic compounds of some elements exhibit low volatility. This is notably the case for Ba compounds, and it makes the reproducible preparation of Ba-containing films by MOCVD difficult. Fluorinated β-diketonate compounds of barium (and sometimes other elements as well) are often used because of their higher vapor pressures [10-13]. Water vapor must also be present, so as to allow fluorine to be removed as hydrogen fluoride, $HF(g)$. Otherwise, the barium in the films is converted to thermodynamically stable $BaF_2(c)$, preventing the incorporation of Ba into the oxide crystal structure.

For the *in situ* growth of Tl-Ba-Ca-Cu-O compounds from fluorinated reagents, the reaction of $HF(g)$ with $TlOH(g)$ produces volatile $TlF(g)$

$$TlOH(g) + HF(g) \stackrel{K_2}{=} TlF(g) + H_2O(g) \ . \tag{2}$$

The thermodynamic properties of $TlF(g)$ have been the subject of several studies [20-22]. Its thermodynamic properties are compiled in Table 3 for the temperature range 700-1200 K [23].

At equilibrium, the vapor pressures of $Tl_2O(g)$, $TlOH(g)$, $TlF(g)$, $HF(g)$, and $H_2O(g)$ are related by the equilibrium constants for Reactions (1) and (2)

$$K_1 = \frac{P_{TlOH}^2}{P_{Tl_2O}\, P_{H_2O}} \ , \tag{3}$$

$$K_2 = \frac{P_{TlF}\, P_{H_2O}}{P_{TlOH}\, P_{HF}} \ . \tag{4}$$

The equilibrium constants K_1 and K_2 calculated from the thermodynamic data in Tables 1-3 and additional tabulated data for $H_2O(g)$ [18] and $HF(g)$ [23] are compiled in Table 4.

We consider the *in situ* growth of $TlBa_2CaCu_2O_7$. Growth conditions must be selected so that the vapor pressure of $Tl_2O(g)$ is high enough to prevent its decomposition

$$2\, TlBa_2CaCu_2O_7(c) \stackrel{K_3}{=} 4\, BaCuO_2(c) + 2\, CaO(c) + Tl_2O(g) + \frac{3}{2} O_2(g) \ . \tag{5}$$

Thus for the *in situ* growth of $TlBa_2CaCu_2O_7$, it is necessary that

$$P_{Tl_2O} \geq K_3/P_{O_2}^{3/2} \ . \tag{6}$$

Table 1. Thermodynamic Properties of $Tl_2O(g)$.

T (K)	J/mol-K			kJ/mol			log K_f
	$C_p°$	$S°$	$-(G°-H_{298}°)/T$	$H°-H_{298}°$	$\Delta H_f°$	$\Delta G_f°$	
700	56.741	364.172	332.510	22.163	-10.296	-48.866	3.647
800	57.192	371.779	336.953	27.861	-12.194	-54.247	3.542
900	57.587	378.538	341.205	33.600	-14.074	-59.390	3.447
1000	57.947	384.624	345.248	39.377	-15.940	-64.325	3.360
1100	58.284	390.163	349.083	45.189	-17.795	-69.074	3.280
1200	58.605	395.249	352.721	51.033	-19.639	-73.654	3.206

Table 2. Thermodynamic properties of $TlOH(g)$.

T (K)	J/mol-K			kJ/mol			log K_f
	$C_p°$	$S°$	$-(G°-H_{298}°)/T$	$H°-H_{298}°$	$\Delta H_f°$	$\Delta G_f°$	
700	47.102	318.479	292.516	18.174	-100.530	-120.490	8.991
800	47.938	324.792	296.134	22.926	-101.888	-123.250	8.048
900	48.766	330.637	299.790	27.762	-103.200	-125.840	7.304
1000	49.582	335.846	303.168	32.678	-104.472	-128.290	6.701
1100	50.380	340.629	306.382	37.672	-105.702	-130.610	6.202
1200	51.152	344.936	309.308	42.754	-106.896	-132.660	5.775

Table 3. Thermodynamic Properties of $TlF(g)$.

T (K)	J/mol-K			kJ/mol			log K_f
	$C_p°$	$S°$	$-(G°-H_{298}°)/T$	$H°-H_{298}°$	$\Delta H_f°$	$\Delta G_f°$	
700	37.138	275.315	254.477	14.587	-194.304	-238.542	17.787
800	37.338	280.288	257.399	18.311	-195.354	-244.774	15.972
900	37.496	284.653	260.192	22.053	-196.406	-250.875	14.550
1000	37.627	288.653	262.843	25.811	-197.461	-256.672	13.407
1100	37.741	292.245	265.355	29.578	-198.516	-262.737	12.446
1200	37.843	295.533	267.735	33.357	-199.571	-268.518	11.681

Values for K_3 are not accurately known. From high temperature equilibrium studies, we estimate K_3 to be about 4×10^{-5} atm$^{5/2}$ at 1150 K [24], while from *in situ* physical vapor deposition growth of $TlBa_2CaCu_2O_7$ at 800 K, we estimate K_3 to be on the order of 10^{-11} atm$^{5/2}$ at 800 K [14,15]. From these values, estimated values of K_3 for temperatures of 700-1200 K are listed in Table 4. (These values may be in error by as much as an order of magnitude.) From Eq. (6), estimated minimum required values of P_{Tl_2O} for the *in situ* growth of $TlBa_2CaCu_2O_7$ when $P_{O_2} = 0.01$ atm are 1×10^{-8} atm at 800 K, 2×10^{-6} atm at 900 K, 3×10^{-4} atm at 1000 K and 8×10^{-3} atm at 1100 K. For $P_{O_2} = 0.1$ atm, the required P_{Tl_2O} is about 30× lower. *In situ* growth of Tl-Ba-Ca-Cu-O phases requires a combination of low temperature and high P_{O_2}.

ANALYSIS OF *IN SITU* Tl-Ba-Ca-Cu-O MOCVD PROCESSES

We examine the *in situ* MOCVD growth of $TlBa_2CaCu_2O_7$, first considering the use of reagents containing only C-H or C-H-O ligands. The oxidation of Ba, Ca, and Cu reagents within the MOCVD reactor leads to H_2O, CO_2, and nonvolatile Ba, Ca, and Cu species, which diffuse to the wafer surface and deposit on it. The oxidation of the Tl reagent, however, in addition to forming H_2O and CO_2, leads to volatile $Tl_2O(g)$ and TlOH(g). Incorporation of Tl_2O into the film occurs only when growth conditions are thermodynamically favorable [Eq. (6)].

The equilibrium vapor pressures of thallium-containing species (i) normalized to the feed stream partial pressure of the Tl reagent (abbreviated as Tl*), P_i/P_{Tl^*}, at 1000 K are plotted in Figure 1 as a function of the elemental H/Tl ratio in the feed stream. As H/Tl increases, P_{TlOH} increases and P_{Tl_2O}

Table 4. Values for the equilibrium constants K_1, K_2 and K_3. (Note: estimated values for K_3 contain considerable uncertainty.)

T (K)	log K_1	log K_2	log K_3
700	-1.250	3.706	-14.1
800	-0.738	3.091	-11.0
900	-0.338	2.611	-8.6
1000	-0.020	2.217	-6.7
1100	0.241	1.912	-5.1
1200	0.458	1.642	-3.8

decreases, and in order to meet the thermodynamic criteria of Eq. (6), a higher feed stream partial pressure of Tl reagent must be used. As an example, we consider a feed stream containing a 20× surplus of Tl reagent and a stoichiometric ratio of Ba, Ca, and Cu. The most commonly used ligand is the β-diketonate compound 2,2,6,6-tetramethyl-3,5-heptanedione, abbreviated as thd, which contains 19 H atoms. For a feed stream containing 20:2:1:2 Tl(thd):Ba(thd)$_2$:Ca(thd)$_2$:Cu(thd)$_2$, H/Tl = 28.5 and $P_{Tl_2O}/P_{Tl^*} \cong 0.059$. The feed stream H/Tl elemental ratio can be reduced by use of reagents containing ligands with reduced hydrogen content. Thallium cylopentadiene, Tl(C$_5$H$_5$), is volatile, as is Cu(acac)$_2$ [acac = 2,4-pentanedione, with 7 H atoms]. For a feed stream containing 20:2:1:2 Tl(C$_5$H$_5$):Ba(thd)$_2$:Ca(thd)$_2$:Cu(acac)$_2$, H/Tl = 12.1 and $P_{Tl_2O}/P_{Tl^*} \cong 0.111$. The H/Tl ratio can be further minimized by direct volatilization of thallium oxide [16] rather than use of a thallium organometallic compound, but such a process requires considerable redesign of the reagent delivery system.

Figure 2 shows the normalized equilibrium partial pressure of $Tl_2O(g)$, P_{Tl_2O}/P_{Tl^*}, as a function of the H/Tl elemental ratio in the feed stream at temperatures of 800, 900, 1000, and 1100 K. The fraction of Tl reagent converted to $Tl_2O(g)$ decreases with increasing temperature. In addition, K_3 also increases with increasing temperature. These factors combine to make the *in situ* MOCVD growth of Tl-Ba-Ca-Cu-O compounds at elevated temperatures very difficult. Since the required thallous oxide overpressure decreases with increasing oxygen partial pressure [Eq. (6)], MOCVD growth of these materials is favored at high P_{O_2}. MOCVD growth of oxide compounds is usually carried out at total pressures less than 0.02 atm [1]. The onset of deleterious buoyancy forces must be considered for $P_{O_2} > 0.01$ atm in vertical rotating disk reactors and for $P_{O_2} > 0.1$ atm in traditional cold-wall horizontal reactors. Therefore, MOCVD reactors must be carefully designed when high oxygen partial pressures are present [25].

If fluorinated reagents are used, TlF(g) is readily formed. The normalized equilibrium partial pressures of Tl species, P_i/P_{Tl^*}, are plotted in Figure 3 as a function of the F/Tl elemental ratio in the feed stream at 1000 K for a fixed Tl/H elemental ratio of 5. For F/Tl < 1, almost all of the F is tied up as TlF(g), decreasing P_{Tl_2O} accordingly. For F/Tl > 1, virtually all of the Tl is present as TlF(g), and P_{Tl_2O} is very small. The use of fluorinated reagents during the *in situ* MOCVD growth of Tl-Ba-Ca-Cu-O thin films should be avoided.

CONCLUSIONS

In situ growth of Tl-Ba-Ca-Cu-O thin films by physical vapor deposition can be achieved in the presence of a sufficient overpressure of $Tl_2O(g)$. *In situ* growth by MOCVD is complicated by the fact that oxidation of organometallic reagents results primarily in TlOH(g) (and TlF(g) if fluorinated ligands are used) rather than $Tl_2O(g)$, lowering the overpressure of

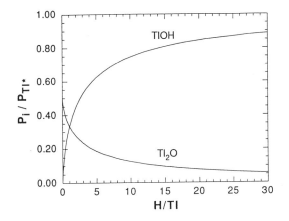

Figure 1. Equilibrium partial pressures of volatile Tl species (Tl_2O and TlOH) normalized to the partial pressure of Tl reagent in the feed stream, P_i/P_{Tl^*}, at 1000 K as a function of the H/Tl elemental ratio in the feed stream for the case where there is no fluorine in the feed stream.

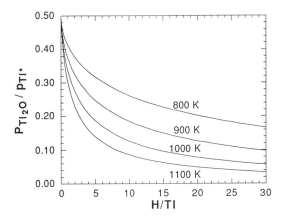

Figure 2. Equilibrium partial pressure of Tl_2O normalized to the partial pressure of Tl reagent in the feed stream, P_{Tl_2O}/P_{Tl^*}, as a function of the H/Tl elemental ratio in the feed stream for temperatures of 800, 900, 1000, and 1100 K.

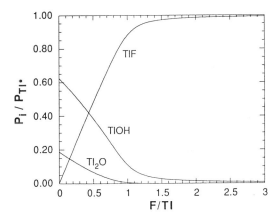

Figure 3. Equilibrium partial pressures, P_i, of volatile Tl species (Tl_2O, TlOH, and TlF) normalized to the partial pressure of Tl reagent in the feed stream P_{Tl^*} at 1000 K as a function of the F/Tl elemental ratio in the feed stream for the case where the H/Tl elemental ratio in the feed stream is 5.

$Tl_2O(g)$ and requiring much higher feed stream partial pressures of the Tl reagent than would otherwise be expected. The thermodynamic stability of $TlF(g)$ is so high that growth of Tl-Ba-Ca-Cu-O phases for feed streams containing F/Tl ratios greater than unity will not be possible. Indeed, F-containing reagents should be avoided entirely. The amount of Tl-containing reagent in the feed stream required to generate a sufficient P_{Tl_2O} to allow for preparation of these phases can be minimized by growth at low temperatures and high P_{O_2}, as well as by minimizing the H/Tl elemental ratio in the feed stream. The latter is achieved by selection of reagents with low hydrogen content and possibly by direct evaporation of thallium oxide.

ACKNOWLEDGEMENTS

This work was supported in part by the U. S. Air Force under Contract F33615-90-C-5924. The author expresses his gratitude to T. L. Aselage, D. W. Face, D. J. Kountz. L. A. Parisi, J. P. Nestlerode and F. M. Pellicone for helpful discussions.

REFERENCES

1. M. Leskelä, H. Mölsä and L. Niinistö, Supercond. Sci. Technol. **6**, 627 (1993).
2. P. A. Zawadski, G. S. Tompa, P. E. Norris, C. S. Chern, R. Caracciolo, and B. H. Kear, J. Electron. Mater. **19**, 357 (1990).
3. S. Matsuno, F. Uchikawa, and K. Yoshizaki, Japn. J. Appl. Phys. **29**, L947 (1990).
4. J. Zhang, J. Zhao, H. O. Marcy, L. M. Tonge, B. W. Wessels, T. J. Marks, and C. R. Kannewurf, Appl. Phys. Lett. **54**, 1166 (1989).
5. K. Endo, S. Hayashida, J. Ishiai, Y. Matsuki, Y. Ikedo, S. Misawa, and S. Yoshida, Japn. J. Appl. Phys. **29**, L294 (1990).
6. R. Hiskes, S. A. DiCarolis, R. D. Jacowitz, Z. Lu, R. S. Feigelson, R. K. Route, and J. L. Young, J. Crystal Growth **128**, 781 (1993).
7. W. L. Holstein, L. A. Parisi, Z.-Y. Shen, C. Wilker, M. S. Brenner, and J. S. Martens, J. Supercond. **6**, 191 (1993).
8. K. Zhang, E. P. Boyd, B. S. Kwak, A. C. Wright, and A. Erbil, Appl. Phys. Lett. **55**, 1258 (1989).
9. D. S. Richeson, L. M. Tonge, J. Zhao, J. Zhang, H. O. Marcy, T. J. Marks, B. W. Wessels, and C. R. Kannewurf, Appl. Phys. Lett. **54**, 2154 (1989).
10. G. Malandrino, D. S. Richeson, T. J. Marks, D. C. DeGroot, J. L. Schindler, and C. R. Kannewurf, Appl. Phys. Lett. **58**, 182 (1991).
11. N. Hamaguchi, R. Gardiner, P. S. Kirlin, R. Dye, K. M. Hubbard, and R. E. Muenchausen, Appl. Phys. Lett. **57**, 2136 (1990).
12. N. Hamaguchi, R. Gardiner, and P. S. Kirlin, Appl. Surf. Science **48/49**, 441 (1991).
13. N. Hamaguchi, R. Boerstler, R. Gardiner, and P. Kirlin, Physica C **185-189**, 2023 (1991).
14. D. W. Face and J. P. Nestlerode, Appl. Phys. Lett. **61**, 1838 (1992).
15. D. W. Face and J. P. Nestlerode, IEEE Trans. Appl. Supercond. **3**, 1516 (1993).
16. W. L. Holstein, J. Phys. Chem. **97**, 4224 (1993).
17. D. Cubicciotti, and F. J. Keneshea, J. Phys. Chem. **71**, 808 (1967).
18. L. B. Pankratz, Thermodynamic Properties of Elements and Oxides, Bulletin No. 672, U. S. Bureau of Mines (1982).
19. W. L. Holstein, J. Chem. Thermodynamics **25**, 1287 (1993).
20. A. S. Kana'an, J. Chem. Thermodynamics **17**, 233 (1985).
21. F. Keneshea and D. Cubicciotti, J. Phys. Chem. **69**, 3910 (1965); **71**, 1958 (1967).
22. D. Cubicciotti and H. Eding, J. Chem. Eng. Data **10**, 343 (1965); **12**, 284 (1967).
23. L. B. Pankratz, Thermodynamic Properties of the Halides, Bulletin No. 674, U. S. Bureau of Mines (1984).
24. T. L. Aselage, E. L. Venturini, and S. B. van Deusen, unpublished.
25. W. L. Holstein, Prog. Crystal Growth Character. **24**, 111 (1992).

NUMERICAL ANALYSIS OF AN IMPINGING JET REACTOR FOR THE CVD AND GAS-PHASE NUCLEATION OF TITANIA

SULEYMAN A. GOKOGLU[1], G.D. STEWART[2], J. COLLINS[3] AND D.E. ROSNER[3]
[1]NASA Lewis Research Center, Cleveland, OH 44135
[2]Ohio Aerospace Institute, Brook Park, OH 44142
[3]Chemical Eng. Dept., Yale University, New Haven, CT 06520-8286

ABSTRACT

We model a cold-wall atmospheric pressure impinging jet reactor to study the CVD and gas-phase nucleation of TiO_2 from a titanium tetra-iso-propoxide (TTIP)/oxygen dilute source gas mixture in nitrogen. The mathematical model uses the computational code FIDAP and complements our recent asymptotic theory for high activation energy gas-phase reactions in thin chemically reacting sublayers. The numerical predictions highlight deviations from ideality in various regions inside the experimental reactor. Model predictions of deposition rates and the onset of gas-phase nucleation compare favorably with experiments. Although variable property effects on deposition rates are not significant (~11% at 1000K), the reduction of rates due to Soret transport is substantial (~75% at 1000K).

1. INTRODUCTION

Production of sophisticated materials with superior properties by CVD requires an understanding of coupled transport and chemical processes. This goal can be realized in research programs by combining both experiments and modeling.
Recent Yale University research focused on TiO_2 CVD with simultaneous gas-phase reaction leading to particle nucleation [1-2]. A simple asymptotic theory treated the onset of reactions leading to particle nucleation and reduced CVD rates at high surface temperatures. The theory assumed that all gas-phase reactions are restricted to a thin chemical sublayer adjacent to a hot CVD surface due to their large activation energies. They also developed a lab-scale, cold-wall, impinging jet CVD reactor [3]. Using experiments and appropriate asymptotic and numerical models, we expect to guide future reactor designs and scale-up, resulting in maximum CVD rates while avoiding harmful particle nucleation.
The present axisymmetric, numerical model points to the role of bouyancy and recirculation in the experimental reactor and clarifies the effect of Soret transport in modifying TTIP mass transfer rates. It provides the ability to capture the transition from surface kinetics to gas-phase transport control. Furthermore, it extends the earlier asymptotic theory [1] by removing the restriction that the activation energies of gas-phase reactions should be "large" (i.e. the chemical sublayer should be "thin").
For the case studied at Yale [1-3] for the CVD of TiO_2 from dilute TTIP/O_2 in N_2, the ratio of chemical sublayer to thermal boundary layer thickness is estimated to be 1/5 at the onset of particle nucleation (T≈1050K), which may be too large for the accuracy of the asymptotic theory. Hence, our numerical model may ultimately

be better able to capture the effects of homogeneous chemistry leading to an observed deposition rate fall-off above ~1050K.

2. EXPERIMENT

We use a cold-wall, atmospheric pressure, axisymmetric, impinging-jet reactor (Fig. 1). The liquid TTIP source is a constant temperature bubbler. After mixing TTIP vapor with excess oxygen (to "burn away" any co-deposited carbon) and diluting with nitrogen in a short mixing chamber, the gas jet emerges from a converging nozzle and impinges on a polished quartz substrate (diam. 1.3cm). The substrate and the alumina substrate holder are supported from below by an RF-heated graphite susceptor. The susceptor temperature is measured using a Pt-30%Rh/Pt-6%Rh thermocouple just below the substrate. This thermocouple reading is correlated with direct substrate surface temperature measurements made in situ using known melting point lacquers (with ~3-4% accuracy). Gas and surface temperatures at several other locations are also measured by thermocouples. The concentration of water vapor in the reactor is less than 3ppm. Deposition rates are measured by in situ interferometry and confirmed by ex situ weight gain. Further details of the experimental system and operating procedure are given in [3].

3. PHYSICO-CHEMICAL/NUMERICAL MODEL

Our modeling study adopts the finite-element-based computational fluid dynamics code FIDAP. The new version of FIDAP incorporates many phenomena relevant to CVD processes, such as temperature dependent fluid density, transport and thermodynamic properties, Soret diffusion, and gas-phase and surface chemical reactions.

We approximate the reactor by using a 2-D axisymmetric geometry (Fig. 2). The inlet and exit are placed sufficiently far from the deposition zone to eliminate numerical uncertainties on rate predictions. The numerical mesh above the substrate is fine enough to resolve the chemical sublayer. Temperature boundary conditions are interpolations of measured surface temperatures in the Yale axisymmetric impinging jet reactor.

We assume that the gas-phase decomposition of TTIP is first order with a rate constant of $3.96 \times 10^5 \exp(-8480/T)$ s^{-1} (T is in Kelvins) [4]. We also assume that the gas-phase products form nondepositing particles. We fit our low substrate temperature rate data to a Soret-corrected Arrhenius expression, $1.21 \times 10^9 \exp(-16480/T)$ m/s, to obtain a pseudo-first order surface reaction rate constant. This expression corresponds to a TTIP reactive sticking probability of unity at 987K. For simplicity, the surface reaction rate constant is kept at its 987K value for higher substrate temperatures so that the sticking probability does not exceed unity.

The Soret diffusion factor for dilute TTIP in N_2 is calculated from kinetic theory (the estimated Lennard-Jones parameters for TTIP are $\sigma=8.13$Å and $\epsilon/k=589$K) and fitted to the expression $\alpha_T=1.971[1-(223.6)/T]$. Unfortunately, FIDAP is currently restricted to constant Soret diffusivities, $D_T=\rho D \omega \alpha_T$, hence we evaluate the gas (N_2) density ρ, TTIP Fickian diffusivity D, and α_T at $T_{film}=(T_{nozzle}+T_{surface})/2$, and use the TTIP mass fraction ω at the T_{film} location for the corresponding diffusion limit calculation.

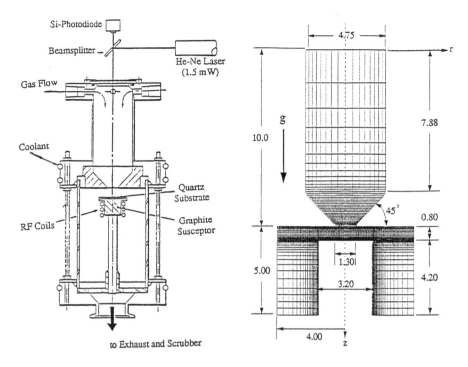

FIGURE 1: Schematic of the experimental reactor; after [1].

FIGURE 2: The computational domain. All dimensions in centimeters, cm.

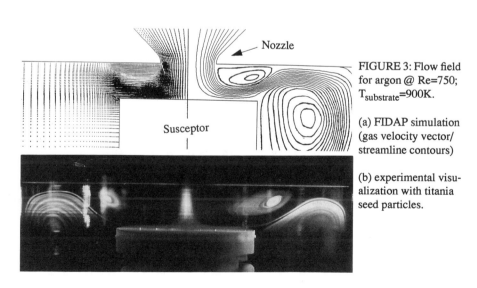

FIGURE 3: Flow field for argon @ Re=750; $T_{substrate}$=900K.

(a) FIDAP simulation (gas velocity vector/ streamline contours)

(b) experimental visualization with titania seed particles.

4. RESULTS AND DISCUSSION

First, we compare FIDAP predictions to the observed flow field inside the reactor. The experiments use argon seeded with fine titania particles for flow visualization. Figure 3 shows a typical comparison at a Reynolds number of 750 based on nozzle diameter at STP and 900K substrate temperature. Indeed, FIDAP can demonstrate recirculations a) as the jet emerges from the nozzle due to flow separation, and b) on the sides of the hot susceptor due to bouyancy-induced convection. These recirculations are not detrimental because from the photograph one can see that particles trapped in the recirculation regions do not diffuse into the reagent jet, and we estimate that reaction product vapor species are also unable to penetrate appreciably into the jet. The Richardson number based on nozzle to substrate distance is ~0.07 for this case. Lower flow rate (Re<100) measurements at higher substrate temperatures are hampered by buoyancy effects in the jet above the substrate. We discuss elsewhere [5] the operating conditions needed to avoid bouyancy-induced convection in such reactors. Our numerical analysis is capable of handling such non-ideal behavior.

For the results discussed below we fix the gas flow rate at 5160 sccm (Re≈500 for N_2) and the substrate-temperature-based TTIP concentration at $C_s=2.5 \times 10^{-6}$ Kgmol/m^3. Deposition rates are reported as **effective** reaction probabilities ϵ defined as $\epsilon = \dot{n}''/[(1/4)\bar{v}C_s]$, where $\dot{n}''$ and $\bar{v}$ are the molar flux and mean thermal speed of TTIP, respectively, evaluated at the surface.

Figure 4 depicts Soret diffusion and temperature-dependent gas transport-property effects on predicted rates at the stagnation point for transport controlled conditions. The rate reduction due to Soret transport is >75%, whereas the effect of variable properties is <15%, for T>1000K.

The model predicts reasonable deposition rates over the entire regime from surface kinetics to transport control by incorporating the surface kinetics extracted from our experiments (Fig. 5). With the gas-phase kinetics of [4], the model agreement with experiment is less satisfactory, though with similar trends. As the depletion of TTIP by the gas-phase reaction increases (i.e. chemical sublayer gets thicker) at higher substrate temperatures, thereby decreasing the TTIP surface flux (CVD rate), FIDAP's handling of Soret transport becomes more inconsistent. This is because the constant D_T, used by FIDAP as defined above for the calculation of Soret transport, does not exhibit the expected diminishing Soret contribution as TTIP concentration diminishes. Hence, our rate predictions with Soret effect above substrate temperatures of 1200K are not shown. The cause(s) of our underprediction of the steepness of the high temperature rate fall-off is currently under investigation.

The onset of homogeneous nucleation is inferred from a sudden drop in deposition rate at a certain substrate temperature. The reliability of the asymptotic theory prediction of this temperature depends on whether the chemical sublayer is indeed "much" thinner than the thermal boundary layer at the prevailing conditions. Figure 6 depicts the TTIP mass fraction profiles calculated by FIDAP at the stagnation point at a substrate temperature of 1200K with and without gas-phase reaction. We also plot the difference between the two mass fraction curves to demonstrate the extent of gas-phase reaction, as well as the corresponding temperature profile. The predicted rate at this temperature with gas-phase che-

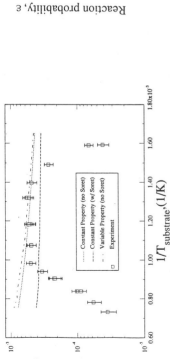

FIGURE 4: Predicted deposition rates with diffusion limitations showing variable property and Soret effects. Experimental data shown for reference.

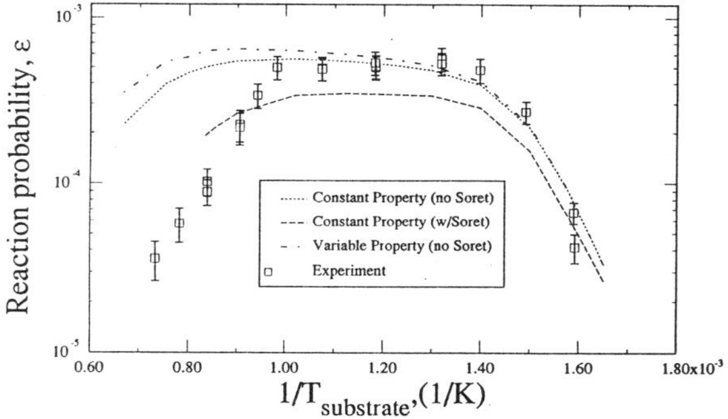

FIGURE 5: Comparison of predicted deposition rates (including both gas and surface kinetics) with experimental data.

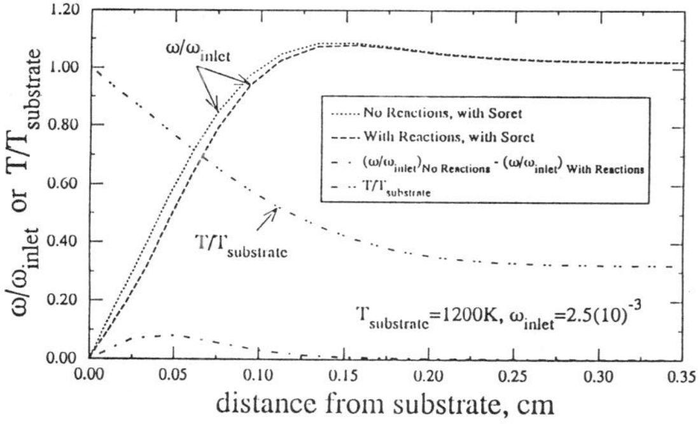

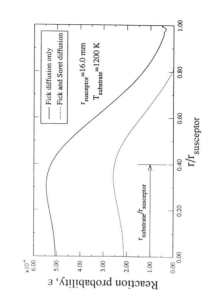

FIGURE 7: Radial variation of predicted deposition rates across susceptor surface.

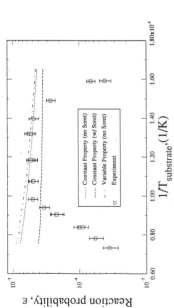

FIGURE 6: The effect of gas phase reaction on mass fraction profile at 1200K. Soret effect is included.

mistry is observably lower than the one without gas-phase chemistry. If the peak of this curve is taken as a measure of the thickness of the chemical sublayer, it is substantial (30%) compared with the thermal boundary layer thickness. The amplitude and width of the difference curve would be larger at even higher substrate temperatures. Therefore, under such conditions predictions of the full numerical approach are expected to be more reliable than predictions of the simple asymptotic theory. Also noteworthy is the slight "enrichment" in the mass fraction profiles away from the surface due to the Soret effect.

Figure 7 shows that the predicted rates on the susceptor are not uniform in the radial direction beyond the substrate radius (6.35mm vs. 16mm for the susceptor). However, because the deposition rate is roughly uniform across the substrate, we have compared rate measurements to stagnation point rate predictions (Figs. 4 & 5).

5. CONCLUSIONS

Our numerical model can correctly describe the flow field in our experimental atmospheric pressure, impinging jet reactor used for studying the CVD and gas-phase nucleation of titania. The model points to some buoyancy and flow recirculation effects inside the hot substrate reactor affecting the "ideal" stagnation point flow behavior. It can guide the interpretation of deposition rate measurements over a substrate temperature range (600-1400K) covering surface-kinetics-, gas-phase transport-, and gas-phase reaction-governed regimes. Soret transport effects on predicted rates [6] are significant (~75% reduction at 1000K), but variable property effects are modest (<15%). Our numerical model extends the capability of the earlier asymptotic theory to predict the onset temperature of nucleation and the associated reduced deposition rates to include conditions where the chemical sublayer is not "much" thinner than the thermal boundary layer. Future work will include effects of flow rate and TTIP concentration [7], and a more careful examination of the high temperature rate fall-off regime.

6. REFERENCES

1. D.E. Rosner, J. Collins and J.L. Castillo, Proc. Twelfth Int. Symp. on Chemical Vapor Deposition, edited by K.F. Jensen and G.W. Cullen, Proc. Vol. 93-2 (The Electrochem. Soc., Pennington, NJ, 1993), p. 41; also J.L. Castillo and D.E. Rosner, Yale Univ., HTCRE Lab. Publ. #196, June 1993, to be submitted.
2. J. Collins, D.E. Rosner, and J. Castillo, in Chemical Vapor Deposition of Refractory Metals and Ceramics II, edited by T.M. Besmann, B.M. Gallois and J.W. Warren (Mater. Res. Soc. Symp. Proc., **250**, 1992) pp. 53.
3. J. Collins, PhD Dissertation, Yale University, Department of Chemical Engineering, in preparation, 1993.
4. K. Okuyama, et al., AIChE J. **36** (3), 409 (1990).
5. G.D. Stewart, et al., in preparation, 1993.
6. Rosner, D.E.,Transport Processes in Chemically Reacting Flow Systems, Butterworth-Heinemann, Stoneham, MA, 3rd Print, 1990.
7. K.L. Siefering and G.L. Griffin, J. Electrochem. Soc. **137** (3), 814 (1990); ibid., **137** (4), 1206 (1990).

THEORETICAL MODELING OF CHEMICAL VAPOR DEPOSITION OF SILICON CARBIDE IN A HOT WALL REACTOR

FENG GAO AND RAY Y. LIN
University of Cincinnati, Department of Materials Science and Engineering, Mail Box #12, Cincinnati, OH 45221

ABSTRACT

A theoretical model, which describes the coupled hydrodynamics, mass transport and chemical reaction, has been developed to simulate chemical vapor deposition (CVD) of silicon carbide (SiC) from gas mixture of methyltrichlorosilane (MTS), hydrogen and argon in a hot wall reactor. In the model analysis, the governing equations were developed in the cylindrical coordinate, and solved numerically by using a finite difference method. A kinetic rate expression of CVD-SiC deposition from the gas mixture was obtained from this study. The deposition rate has an Arrhenius-type dependence on the deposition temperature and is first order with respect to the MTS concentration. Estimated activation energy is 254 kJ/mol. Predicted deposition rate profiles by the model analysis incorporated with the obtained kinetic rate expression showed excellent agreement with experimental data over a variety of applied deposition conditions.

INTRODUCTION

Chemical vapor deposition is an advanced materials processing technique to provide coatings over substrates for the purpose of improving properties of the substrates. Recently, SiC prepared by CVD processes has received more and more attentions, due to CVD-SiC's many attractive properties, such as oxidation resistance and high mechanical strength. As a result, many researches have been devoted to investigating CVD-SiC theoretically and experimentally. However, most of these studies has focused on either discovering the morphologies and properties of SiC deposits or relating the deposition rate with deposition conditions qualitatively [1-6]. Only a few of them have tried to establish a theoretical analysis for the SiC deposition process [7,8]. Since the CVD process is very complex and involves many variables, a theoretical modeling analysis would be valuable to predict kinetics of SiC deposition as well as the effect of changes in deposition conditions on the deposition rate.

The hot wall reactor used in this study is a conventional and widely applied reactor in the CVD processes. Large temperature gradients typically exist along the longitudinal axis of hot wall reactors. However, a detailed mathematical analysis for a CVD process in a hot wall reactor has not been developed up to now. In addition, the kinetic mechanism involved in SiC deposition from gas mixtures of methyltrichlorosilane (MTS), H_2 and Ar has not been well documented.

In the present works, the CVD-SiC from gas mixtures of MTS, H_2 and Ar, prepared in a hot wall tubular reactor at one atmosphere, is investigated experimentally and theoretically. The main purpose of this study is to develop a detailed mathematical model in which hydrodynamics, mass transport and chemical reactions are taken into consideration. With this model, the deposition behavior of the CVD process used in this research can be predicted. A kinetic rate expression for the CVD-SiC from the gas mixture of MTS, H_2 and Ar will be established based on the combination of model calculations and experimental data. The validity and ability of the model in predicting the deposition rate profile will be verified.

EXPERIMENTAL

The CVD-SiC depositions from the gas mixture of MTS, H_2 and Ar were conducted in a horizontal hot wall reactor. Substrates were small pieces of CVD-SiC pre-coated alumina. For each deposition run, substrates were placed on the inner surface of the reactor wall at various locations within the reactor as shown in Fig.1. Deposition rate of SiC deposit was obtained by measuring the weight gain of the substrate before and after deposition. And the deposition rate profiles were used for extracting the kinetic data of CVD-SiC deposition and for comparing with the model predictions. Temperature profile in the reactor was determined with a R-type thermocouple at various locations along the axis of the reactor tube. Three reactor temperature profiles, for which the maximum temperature are 1000C, 1050C and 1100C respectively, were selected in this investigation. Table 1 lists the deposition parameters selected in the experiments.

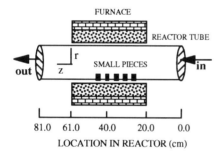

Fig.1 Arrangement of samples in the reactor.

Table1 Deposition conditions in experiments

Deposition Conditions	Selected in the experiments
Pressure	one atmosphere
Location of Sample	31.2, 34.2, 37.2, 40.2, 43.2
Temperature Profile	1000C 1050C and 1100C
MTS Mole Fraction	7.10% 4.78% and 3.24%
Inlet Flow Rate	100-110 and 200-220 sccm

MODEL DEVELOPMENT

A theoretical model, which describes the physical and chemical phenomena occurred in the reactor (shown in Fig.1) during the deposition, was developed as follows.

Assumptions

(1). Steady state flow has been established.
(2). Flow of the reaction gas mixture is considered to be laminar. Under the SiC deposition process of this study with Reynolds number < 100, this assumption is suitable.
(3). The gravitational effect is neglected due to the negligible temperature gradient in the radial direction within the hot wall reactor.
(4). In the flow direction, i.e. axis of the tube, convective transport is dominant.

(5). Because the deposition rate is slow and the reactant gas (MTS) is sufficiently diluted by nonreactive gases (Ar and H_2), the transport properties (ρ, μ, D) of the gas mixture depend only on the temperature and not on the concentrations of reaction gases.

(6). No gas phase reactions occur in the bulk during the deposition (this assumption allows that only initial reactants are considered in the model). Because the kinetic process of chemical reactions involved in CVD-SiC from an MTS precursor has not been well understood up to now, an overall heterogeneous reaction, which takes place on the substrate surface, was considered in the present model analysis.

(7). Since chemical vapor deposition is slow, the reaction heat associated with the deposition can be neglected, comparing the heat provided by the furnace. The radial temperature is considered as uniform based on the temperature measurements.

Governing Equations

Since the tubular reactor was used in this study, the governing equations were developed in cylindrical coordinates. It is assumed that the reaction mixture flows in the direction of coordinate z, where z = 0 at the inlet. The coordinate r is in the radial direction. r = 0 at the center line of the reactor, and r = r_0 at the surface of the reactor wall. Based on the theory of fluid dynamics under the above-mentioned assumptions, the continuity equations which govern the behavior of laminar flow within the reactor were derived, and their final forms are presented as follows.

(i). Overall mass continuity equation:

$$\rho(\frac{\partial V_r}{\partial r} + \frac{\partial V_z}{\partial z}) + V_r(\frac{\rho}{r}) + V_z\frac{\partial \rho}{\partial z} = 0 \qquad (1)$$

where V_z, V_r are axial and radial velocities of the gas mixture, and ρ is the density.

(ii). Conservation of momentum:

$$\rho(V_z\frac{\partial V_z}{\partial z} + V_r\frac{\partial V_z}{\partial r}) + \frac{\partial P}{\partial z} = \mu(\frac{\partial^2 V_z}{\partial r^2} + \frac{1}{r}\frac{\partial V_z}{\partial r}) \qquad (2)$$

where P is pressure, and μ is viscosity of the gas mixture.

(iii). Species mass balance equation:

$$V_z\frac{\partial X_A}{\partial z} + V_r\frac{\partial X_A}{\partial r} = D_{A\text{-mix}}(\frac{\partial^2 X_A}{\partial r^2} + \frac{1}{r}\frac{\partial X_A}{\partial r}) + (\frac{1}{\rho}\frac{\partial \rho D_{A\text{-mix}}}{\partial z})\frac{\partial X_A}{\partial z} \qquad (3)$$

where $D_{A\text{-mix}}$ is diffusivity of MTS in the gas mixture, and X_A is the MTS mole fraction.

(iv). Equation of state:

$$\rho = \frac{P\overline{M}}{RT} \qquad (4)$$

where $\overline{M}$ is the average molecular weight of the gas mixture, R is the gas constant.

(v). Boundary Conditions:

$$z = 0 \text{ (inlet)}: X_A = X_{Ao}; \quad V_r = 0; \quad V_Z = u_0; \quad \rho = \rho_0 \qquad (5)$$

$$r = r_0 \text{ (wall)}: V_r = V_Z = 0; \quad -\rho D_{A\text{-mix}}(\frac{\partial X_A}{\partial r}) = R_S \overline{M} \qquad (6)$$

$$r = 0 \text{ (center)}: \frac{\partial V_Z}{\partial r} = 0; \quad \frac{\partial X_A}{\partial r} = 0. \qquad (7)$$

where r_0 is radius of the reactor, u_0 is inlet velocity of the gas mixture, ρ_0 is initial density of the gas mixture, X_{A0} is inlet MTS mole fraction, and R_S --- consumption rate of MTS due to the deposition (mol./cm² sec).

Chemical kinetics

The kinetics of CVD-SiC from the gas mixture of MTS, H_2 and Ar has not received considerable attention. Up to now, detailed kinetics of the reactions are still poorly understood. Usually, an overall heterogeneous reaction, which takes place on the substrate surface, has been applied [9-12]. The kinetics is believed to be represented by

$$R_S = k_0 \exp(-\frac{\Delta G}{RT}) C_{MTS}^{\gamma} \qquad (8)$$

where ΔG is the activation energy of the deposition, k_0 is the preexponential, γ is the reaction order, and C_{MTS} is the concentration of MTS (mol/cm³).

The governing equations, which are a set of second-order partial differential equations, with the boundary conditions were solved numerically by using the finite difference method.

RESULTS AND DISCUSSION

To determine the kinetic expression of CVD-SiC from the gas mixture of MTS, H_2 and Ar, four sets of experiment with different deposition conditions were performed. X-ray diffraction analysis of deposition products on the substrates from these experiments shows that all deposits are pure β-SiC coatings. The deposition rates at the wall surfaces from each experiment were modeled as a polynomial function of location within the reactor by a curve fitting method. The values of deposition rate were then used as the boundary conditions for the governing equations. Therefore, MTS concentrations at the wall surfaces were obtained from the model calculations. After knowing the deposition rate and the concentration of MTS as well as the temperature (T) at the wall surfaces, the kinetic constants in Eq.(8) were obtained from the linear Arrhenius plot. All data from four experiments (at both 1000C and 1050C reactor temperature profiles) were used in making the Arrhenius plots. The deposition rates from each experiment and the fitted curves are shown in Fig.2. The calculated MTS concentrations from the model simulations are presented in Fig.3. A linear Arrhenius plot was obtained with the reaction order γ being equal to one. Thus, the kinetic expression for CVD-SiC from the gas mixture may be expressed as

$$R_S = 2.12 \times 10^8 \exp(-254 \times 10^3 / 8.134\, T)\, C_{MTS} \quad (g/m^2 sec.) \quad (9)$$

From the literature [12,13], Fitzer et al. have proposed that the activation energy for the CVD-SiC from gas mixture of MTS and H_2 are 225 and 206 kJ/mol. Considering the negative effect of argon carrier gas on the deposition rate, the activation energy ($\Delta G = 254$ kJ/mol) for the CVD-SiC from the gas mixture of MTS, H_2 and Ar obtained from this investigation is acceptable.

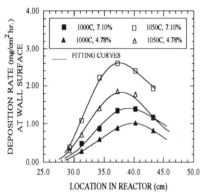

Fig.2 Deposition rates at wall surface from experiments and their fitting curves.

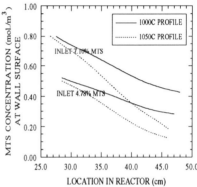

Fig.3 MTS concentration profiles predicted by the model analysis.

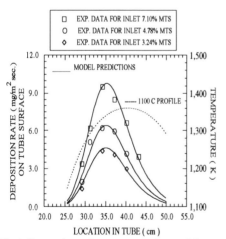

Fig.4 Comparisons between experimental data and model predictions as the temperature profile being 1100C and three inlet MTS mole fractions used.

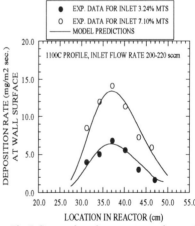

Fig.5 Comparison between experimental data and the model predictions as the inlet flow rate changed.

Using the kinetic data obtained from the present work, the model analysis was performed to predict the deposition rate profile for deposition at 1100C profile. The deposition rates obtained from both the experiments and model predictions are presented in Fig.4. It is obvious that the model predictions agree well with the experimental data for all three gas mixtures with different inlet MTS mole fractions.

To test the validity of the model analysis and the kinetic data obtained as inlet flow rate changed, two experiments, with the total flow rate being set at twice as much as those in the previous experiments were accomplished. Fig.5 shows the comparison between the experimental data and predicted deposition rate profiles. It is seen from this figure that the predicted values by the model analysis also agree well with the experimental results as the inlet flow rate changed.

CONCLUSIONS

CVD-SiC from gas mixture of MTS, H_2 and Ar, in a hot wall tubular reactor at one atmosphere, has been investigated experimentally and theoretically. A detailed mathematical model, which describes the interactions among gas-phase hydrodynamics, mass transport and chemical reaction, has been developed to predict the behavior of this CVD process. In the model analysis, the governing equations were developed in cylindrical coordinates and solved numerically along with appropriate boundary conditions by a finite difference method. The deposition rate of CVD-SiC from the gas mixture of MTS, H_2 and Ar can be presented in term of deposition temperature (T) and MTS concentration (C_{MTS}) as $R_S = 2.12 \times 10^8 \exp(-254 \times 10^3 / 8.314\, T)\, C_{MTS}$. These results are comparable with that obtained previously by other investigators. Validity of the model analysis and the kinetic rate expression obtained from this work were verified by the fact that the predicted deposition rate profiles from the model calculations show excellent agreement with experimental data over a variety of selected deposition conditions including reactor temperature profile, inlet MTS mole fraction and flow rate of the gas mixture.

REFERENCES

1. D. J. Cheng, et al., J. Electrochem. Soc. **134** (12), 3145 (1987).
2. S. Motojima, H. Yagi and N. Iwamori, J. Mater. Sci. Let. **5,** 13 (1986).
3. J. Chin, P. K. Gantzel and R. G. Hudson, Thin Solid Films, **40,** 57 (1977).
4. D. E. Cagliostro and S. R. Riccitiello, J. Amer. Ceram. Soc. **76** (1), 39 (1993).
5. F. Kobayashi, K. I. Kawa and K. Iwamoto, J. Crys. Growth, **28,** 395 (1976).
6. S. S. Shinozaki and H. Sato, J. Amer. Ceram. Soc. **61** (9/10), 425 (1978).
7. J. H. Koh and S. I. Woo, J. Electrochem. Soc. **137** (7), 2215 (1990).
8. M. D. Allendorf and R. J. Kee, J. Electrochem. Soc. **138** (3), 841 (1991).
9. S. M. Gupte and J. A. Tsamopoulos, J. Electrochem. Soc. **136** (2), 555 (1989); **137** (5) 1626 (1990); **137** (11), 3675 (1990).
10. R. R. Melkote and K. F. Jensen, Proc. 11th. Int. Conf. on CVD, 506 (1990).
11. K. Brennfleck et al., Proc. 9th. Int. Conf. on CVD, 649 (1984).
12. R. Fedou, F. Langlais and R. Naslain, Proc. 11th. Int. Conf. on CVD, 513 (1990).
13. E. Fitzer, D. Hegen and H. Strohmeier, Proc. 9th. Int. Conf. on CVD, 525 (1984).

GROWTH OF EPITAXIAL β–SiC FILMS ON POROUS Si(100) SUBSTRATES FROM MTS IN A HOT WALL LPCVD REACTOR

CHIEN C. CHIU, SESHU B. DESU[†] AND GANG CHEN
Department of Materials Science and Engineering, Virginia Polytechnic Institute and State University, Blacksburg, VA 24061–0237
†: To whom all correspondence should be addressed.

ABSTRACT:

β–SiC thin films were epitaxially grown on porous Si(100) substrates at 1150°C from methyltrichlorosilane (CH_3SiCl_3 or MTS) in a hot wall reactor by using low pressure chemical vapor deposition (LPCVD). The growth rate was 120 Å/min. Epitaxial β–SiC(100) thin films were obtained after a deposition time of 12.5 min. The crystallinity of the β–SiC films was affected by the deposition time. For instance, rotational β–SiC(100) crystals and polycrystalline β–SiC with a highly preferred orientation of (100) planes were observed for a deposition time of 50 min. The results from XRD and TEM indicated that polycrystalline β–SiC films with a preferred orientation of β–SiC(111) appeared after further increasing the deposition times (time ≥ 75 min). The change in crystallinity is attributed to increasing roughness with increasing deposition time.

INTRODUCTION:

Because of its large band gap, high saturated electron velocity, and high breakdown electric field, β–SiC is a promising candidate for high temperature, high frequency, and high power electronic devices. It is also a desirable protective coating for the devices operating at high temperatures due to the excellent resistance to oxidation, corrosion, and creep at elevated temperatures.

Chemical vapor deposition (CVD) is an attractive process to grow β–SiC on Si substrates. Although there has been significant progress in depositing single crystal β–SiC films on Si substrates, all of these efforts have been conducted by using a cold wall reactor, and at higher temperatures (T>1300°C). Stoichiometric β–SiC films were deposited on smooth Si substrates in a hot wall LPCVD reactor by using MTS precursor below 1300°C; however, the deposited β–SiC films were polycrystalline with a preferred orientation of (111) of β–SiC [1]. Furthermore, a large number of defects were produced by depositing β–SiC on smooth Si(100) substrates because of a large lattice constant mismatch (20%) between β–SiC deposit and the underlying Si substrate. Thus, by using porous Si substrates, it is expected that (i) the number of defects will be reduced due to the release of strain between the β–SiC deposit and Si substrates, and (ii) the formation of the preferred orientation of β–SiC(111) will be depressed and the β–SiC(100) will be promoted since the sites to nucleate β–SiC(111) might be eliminated [2]. Until now, no attempt has been made to deposit single crystal β–SiC on Si(100) in a hot wall LPCVD reactor. The advantages in using hot wall LPCVD reactors are (i) the ability to deposit single crystal β–SiC thin films simultaneously upon many Si wafers, and (ii) elimination of contamination from the susceptor by using very low deposition pressure [3].

In this study, a novel procedure to grow single crystal β–SiC films with the orientation of (100) on porous Si(100) substrates is presented by using a hot wall LPCVD reactor and the MTS–H_2 mixture. The effect of deposition time on the orientation of the films is also reported.

EXPERIMENTAL PROCEDURE:

The hot wall LPCVD reactor was a quartz tube with inner and outer diameters of 5.08 cm and 5.43 cm, respectively. A uniform temperature (± 5°C) zone of 6 cm was at

the center of the reactor. To obtain a constant and sufficient equilibrium vaporizing pressure of MTS, the bubbler was maintained at a constant temperature of 33°C. Furthermore, the needle valves and the tubing between bubbler and reactor were held at 70°C to prevent the condensation of MTS inside the valves and the tubing [1].
Single crystal Si(100) was used in this study. The reaction chamber was evacuated and purged with H_2. Then, the Si substrates were heated in flowing H_2 in order to remove any residue of the native oxide on the surface. Before the deposition of SiC films from the mixture of MTS and H_2, a porous surface with a thin layer of converted SiC on the Si substrate was obtained by reacting the Si(100) substrate with purified acetylene (99.6%) without flowing H_2 with a total pressure of 0.1 Torr. This process was performed at 1000°C for 5 min. The details of the process for obtaining porous Si substrates has been described elsewhere [4,5].
After a porous Si substrate was obtained at 1000°C, the temperature of the furnace was increased to 1150°C to carry out the CVD process at a total pressure of 1.8 Torr. During the CVD process, the flow rate of H_2 was kept at 300 sccm, and the H_2 to MTS ratio was calibrated to be 100. The samples were furnace cooled in flowing H_2 after the depositions were completed. The phase–analysis of the deposited β–SiC films was examined by XRD with a CuKα radiation and transmission electron microscopy (TEM). Scanning electron microscopy (SEM) was used to observe the film thickness and morphology.

RESULTS AND DISCUSSION:

The β–SiC films obtained at this temperature (1150°C) were shiny and mirror–like. A typical cross section SEM micrograph showing the surface morphology and thickness of the film is shown in Fig. 1. The film was deposited for 12.5 min with a growth rate of 120Å/min. It is interesting to note that no SiC deposit was observed on the facets inside the pore (Fig. 1) although the surface reaction is expected to be the controlling reaction under these deposition conditions [1]. No explanation for this is provided at this point. The XRD pattern of the sample is given in Fig. 2. The peaks for single crystal Si(100) substrates underneath the β–SiC film are at 69° and 62° for Si Kα(400) and Si Kβ(400), respectively. The (200) and (400) planes were the only diffraction planes from the deposited β–SiC thin films from the XRD pattern. This suggested that these films have a highly preferred orientation of β–SiC(100). In order to further examine the preferred orientation of the grown films by using TEM, the electron beam was normal to the surfaces of the samples under TEM. For the sample shown in Fig. 1, since the thickness was thin enough to be penetrated by the electron beam in the TEM, the specimens were prepared by back–etching the Si substrates with an etch rate of 34.8 μm/min from a solution of HF+HNO$_3$+CH$_3$COOH (3:5:3) [6]. The bright field plane–view TEM micrograph with the selected area diffraction (SAD) pattern is depicted in Fig. 3. In Fig. 3, the SAD pattern, which shows a transmission electron diffraction pattern with a zone axis of <001> for β–SiC, also confirmed the preferred orientation from the XRD pattern (Fig. 2). The dot pattern and the diffused ring pattern (the SAD pattern in Fig. 3) indicated that β–SiC single crystal coexisted with a trace amount of polycrystalline β–SiC during the epitaxial growth of the thin film. Thus the film is believed to be a mixture of single crystal β–SiC with the polycrystalline β–SiC embedded inside. The TEM micrograph in Fig. 3 shows a rectangular feature structure. This structure was probably caused by the nucleation process at the beginning of the CVD process. During this period, the β–SiC film was predominantly nucleated at the edges of the porous defects, which were perpendicular to each other and lying parallel to the silicon <110> direction [7]. Thus, the formation of a rectangular feature was favored. In addition to the small portion of polycrystalline β–SiC, planar defects such as stacking faults were also observed in matrix of single crystal β–SiC thin films from a dark field TEM image with a higher magnification, as indicated in Fig. 4.
Figure 5 shows the SAD pattern from the top layers and the XRD pattern of the β–SiC thin films with the deposition time of 50 min. In order to observe the crystal

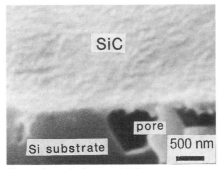

Fig. 1: A typical cross section scanning electronic micrograph of β–SiC thin films deposited at 1150°C for 12.5 min.

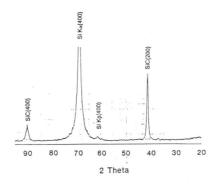

Fig. 2: XRD pattern of β–SiC thin films deposited at 1150°C for 12.5 min.

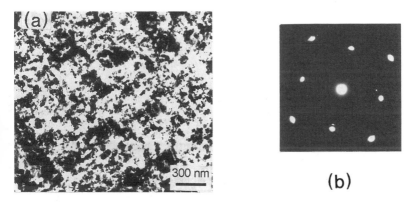

Fig. 3: (a) Bright field TEM micrograph of β–SiC thin film, and (b) the corresponding electron diffraction pattern of β–SiC with <001> zone axis. The sample was obtained at 1150°C for 12.5 min.

structure at the top surfaces of these samples, the samples were prepared by ion milling from the Si side at an angle of 15° to the surface. Again the electron beam was normal to the surfaces to examine the preferred orientation obtained in these samples. From both the SAD pattern and the XRD pattern given in Fig. 5, it is clear that the β–SiC films show partial epitaxy with β–SiC(100) ∥ Si(100) at the top surfaces of the samples. When compared to the sample shown in Fig. 3, the polycrystalline ring pattern is more prominent in the SAD pattern (Fig. 5). However, more than one dot patterns was observed for single crystal β–SiC from electron diffraction with the <001> zone axis. This indicated that nonepitaxial deposition was occurring in the matrix. The matrix on the top layers of the deposited β–SiC film consists of the crystals which are rotated perpendicularly to the <001> zone axis of β–SiC. The reason for this is likely due to the increasing surface roughness with deposition time, as shown later.

The same sample preparation method was employed to make the TEM samples for the films deposited for 100 min. The SAD and the XRD patterns of these samples are shown in Fig. 6. In contrast to the films deposited for 12.5 min (Fig. 3) and 50 min (Fig. 5), only the polycrystalline ring pattern of β–SiC was observed. Furthermore, the (111) plane from β–SiC also appeared in the XRD pattern (Fig. 6(a)). This indicats that the preferred orientation was (111) for the polycrystalline β–SiC on the top surfaces of these samples.

From the XRD and TEM results shown above, it is obvious that the crystallinity of the deposited β–SiC from the mixture of MTS and H_2 was affected by the deposition time. For short deposition time (*i.e.* 12.5 min), the matrix of the films was single crystalline in nature (β–SiC) with a preferred orientation of (100) planes. With increasing deposition time (50 min), the rotational crystals of β–SiC with the preferred orientation of (100) grew on the top surface of single crystal β–SiC(100) matrix. However, the preferred orientation of β–SiC(100) was preserved (Fig. 5). Further increasing the deposition time (100 min), polycrystalline β–SiC with the preferred orientation of (111) was observed on the top surfaces of the previously deposited polycrystalline β–SiC(100) grains, as shown in Fig. 6. In fact, the polycrystalline β–SiC with a preferred orientation of (111) was started to be deposited at the deposition time of 75 min. Jacobson reported a similar phenomenon by using $(CH_3)_2SiCl_2$ as the precursor for the β–SiC films on Si(100) substrates [8]. In contrast to our results, the preferred orientations are two of the {110} planes from β–SiC after the polycrystalline β–SiC(100) grains were observed. Therefore, the precursors used may play a role in obtaining the preferred orientation on the top surfaces of the films because of the differences in kinetic processes in the CVD procedure, such as surface adsorption, surface diffusion and various surface chemical reactions.

In addition, according to Sheldon et al. [2] and Cheng et al [9], it is believed that the surface roughness, which became prominent with increasing time, can create a large number of nucleation sites by using the MTS–H_2 gas system to deposit β–SiC films. As shown in Fig. 7, the topography of the SiC films was observed to be rougher with increasing time, a typical phenomenon for CVD processes. Therefore, after a certain deposition time, the growth of the preferred orientation of β–SiC(111) was favored and thus controlled the nucleation and the microstructure of the deposited materials. Additionally, with increase in film thickness, the influence from the modified Si(100) substrates, which was believed to cause the formation of β–SiC(100) at the early stage of deposition, will be tenuous in the growing front of the deposited β–SiC films. Thus, the (111) preferred orientation, which is the most commonly observed preferred orientation from MTS/H_2 gas system, occurred after a certain deposited thickness [1,2,9,10].

When comparing the results of this study to the β–SiC thin films deposited on smooth Si(100) substrates from MTS, it is interesting to note that, at the early stage of the CVD process, (100) direction is the only preferred orientation from β–SiC on porous Si(100) substrates instead of the preferred orientation from the (111) planes of β–SiC observed on smooth Si(100) substrates [1,2]. According to Sheldon et al. [2], the β–SiC(111) surfaces occurred because of the orientated microtwins between the β–SiC thin film and the underlying Si(100) substrate. These microtwins occurred in a {111}

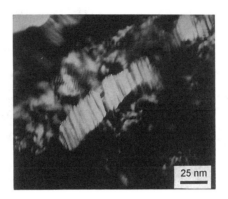

Fig. 4: Dark field TEM micrograph showing the stacking fault in β–SiC thin films.

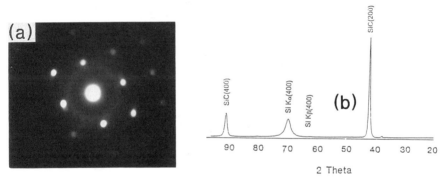

Fig. 5: (a) The SAD pattern from the top surface of the β–SiC film and (b) the XRD pattern of the β–SiC films deposited at 1150°C for 50 min.

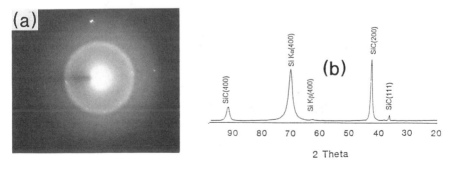

Fig. 6: (a) The SAD pattern from the top surface of the β–SiC film and (b) the XRD pattern of the β–SiC films deposited at 1150°C for 100 min.

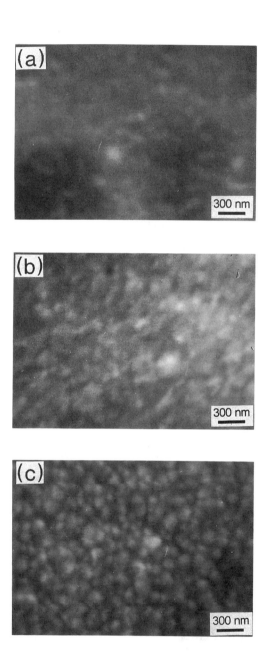

Fig. 7: The topography of the β–SiC thin films deposited at 1150°C for (a) 12.5 min, (b) 50 min, and (c) 100 min.

plane on the surface of smooth Si(100) substrate at the early stage of the CVD process. Thus the subsequent new {111} surfaces of the deposited β–SiC thin films were nucleated at the intersection between the SiC buffer layer and a {111} twin plane. Also the surface roughness on the converted Si surface, which is caused by the high density of defects due to the lattice mismatch between the Si(100) substrate and the β–SiC thin film, was believed to play a role in determining the orientation of the SiC deposit at the early stage of the CVD procedure [11,12,13]. For porous Si(100) substrates, it was believed that the growth of SiC film from MTS at the early stage of nucleation and the subsequent microstructure of the deposited materials were not controlled by the surface roughness of the modified Si(100) substrates. This is because, in the present study, the surfaces of the Si(100) substrates, after reacting with C_2H_2, were expected to be much rougher than those showing a smooth SiC buffer layer [1]. Similar types of defects also existed on the top surfaces of our samples, since the surfaces were also converted to SiC by using a hydrocarbon gas (C_2H_2). Thus the absence of the preferred orientation of (111) from the β–SiC films deposited on modified Si(100) substrates could be due to the elimination of the microtwins occurring in a {111} plane on the top surfaces. In fact, according to Schigeta et al [15], the porous nature of Si substrate can significantly reduce the defect density (i.e. stacking fault) between SiC and Si interface and favor the formation of single crystal β–SiC(100). Furthermore, the porous nature on the top surfaces of Si(100) substrates could provide the surface–energy anisotropy in the early stage of CVD process [14]. That is, films with the orientation that minimize the interfacial free energy should grow preferentially (e.g. β–SiC(100) in this study). The amount of strain due to lattice constant mismatch in the early stage of CVD process were also believed to be released by the porous nature of the Si(100) substrates. Therefore the formation of β–SiC(100) films with single crystal matrix and preferred orientation of (100) from β–SiC were favored. The same phenomenon was also described for the CVD β–SiC films on Si(100) substrates by using cold wall reactors [16,17].

SUMMARY:

Epitaxial β–SiC(100) thin films with single crystal matrix having smooth surfaces were grown at 1150°C in a hot wall LPCVD reactor by using MTS as the precursor for short deposition time (12.5 min). As far as we know, this is the lowest temperature used to grow single crystal β–SiC in a hot wall reactor and by using MTS.

With increasing deposition time, the rotational β–SiC(100) crystals and polycrystalline β–SiC were observed on the top surface of the previously deposited matrix of single crystal β–SiC(100). The (100) orientation from polycrystalline β–SiC can only be grown with deposition times shorter than 75 min in this study. Further increased the deposition time (time > 75 min), the appearance of the polycrystalline β–SiC(111) layers on the top surface of the polycrystalline β–SiC(100) film was observed.

REFERENCES:

1. C.C. Chiu, S.B. Desu, and C.Y. Tsai, J. Mat. Res. **8**, 2617 (1993).
2. B.W. Sheldon, T.M. Besmann, K.L. More, and T.S. Moss, J. Mat. Res. **8**, 1086 (1993).
3. F. Langlais, F. Hottier, and R. Cadoret, J. Cryst. Growth, **56**, 659 (1982).
4. C.C. Chiu and S.B. Desu, J. Mat. Res. **8**, 535–544 (1993).
5. C.C. Chiu, C.K. Kwok, and S.B. Desu, in _Chemical Vapor Deposition of Refractory Metals and Ceramics II_, edited by T.M. Besmann, B. M. Gallois, and J.W. Warren (Mat. Res. Soc. Symp. Proc. **250**, Pittsburgh, PA, 1992) pp. 179–185.
6. D.G. Schimmel, in _Quick Reference Manual for Silicon Integrated Circuit Technology_, edited by W.E. Beadle, J.C.C. Tsai, and R.D. Plummer (John Wiley & Sons, Inc., New York, 1985), p. 5–9.
7. R.C. Newman and J. Wakefield, in _Solid State Physics in Electronics and Communication_, edited by M. Desirant and J.L. Michels, p 319–328 (1960).

8. K.A. Jacobson, J. Electrochem. Soc. **118**, 1001 (1971).
9. D.J. Cheng, W.J. Shyy, D.H. Kuo, and M.H. Hon, J. Electrochem. Soc. **134**, 3145 (1987).
10. M.G. So and J.S. Chun, J. Vac. Sci. Technol. **A6**, 5 (1988).
11. C.H. Carter, Jr., R.F. Davis, and S.R. Nutt, J. Mat. Res. **1**, 811 (1986).
12. H.J. Kim, R.F. Davis, X.B. Cox, and R.W. Linton, J. Electrochem. Soc. **134**, 2269 (1987).
13. N. Bécourt, J.L. Ponthenier, A.M. Papon, and C. Jaussaud, Physica B, **185**, 79 (1993).
14. T. Yonehara, H.I. Smith, C.V. Thompson, and J.E. Palmer, Appl. Phys. Lett. **45**, 631 (1984).
15. M. Shigeta, Y. Fujii, K. Furukawa, A. Suzuki, and S. Nakajima, Appl. Phys. Lett. **55**, 1522 (1989).
16. P. Liaw and R.F. Davis, J. Electrochem. Soc. **132**, 642 (1985).
17. K. Ikoma, M. Yamanaka, H. Yamaguchi, and Y. Shichi, J. Electrochem. Soc. **138**, 3028 (1991).

PART V

Precursor Design and Delivery: Reactors

SYNTHETIC STRATEGIES FOR MOCVD PRECURSORS FOR HTcS THIN FILMS

HARRY A. MEINEMA*, KLAAS TIMMER, HANS L. LINDEN AND CAREL I.M.A. SPEE
TNO-Institute of Applied Physics, Department of Inorganic Materials Chemistry, P.O. Box 595, 5600 AN Eindhoven, The Netherlands

ABSTRACT

In recent years much attention has been given world-wide to the development of suitable MOCVD precursors for the deposition of HTcS thin films. Synthetic research has been and is concentrated on the development of superior Ba-, Sr-, Ca- and Y-precursors. Most emphasis is given to the synthesis of thermally stable volatile barium compounds. Synthetic strategies are based on encapsulating the central metal atom, by use of multidentate ligand systems and/or bulky substituents. Most attention is given to the development of thermally stable volatile ß-diketonate complexes, fluorine-free and fluorine-substituted, with auxiliary ligands. Thermally stable monomeric complexes of fluorine-substituted ß-diketonates with polyethers are by far the most volatile Ba-, Sr-,and Ca-precursors presently available. Low melting Y(thd)$_3$.L complexes where L is 4-Et- or 4-t-Bu-pyridine-N-oxide can be used as liquid yttrium MOCVD precursors at temperatures above 100°C. This paper gives a survey of the trends in recent research activity and developments in these areas.

INTRODUCTION

Since 1988, when the first results on MOCVD of high critical temperature superconducting (HTcS) thin films were reported, it has become apparent that the development of an effective MOCVD process for such layers depends crucially on the availability of suitable precursors, having sufficient volatility and thermal stability. Various types of metalorganic and organometallic compounds have been applied for the deposition of yttrium-, bismuth-, lead- or thallium-based superconducting thin layers. For Y-, Cu-, Bi-, Pb- and Tl-precursors commercially available compounds such as Y(thd)$_3$.xH$_2$O (X = 0 to 1), Cu(thd)$_2$ or Cu(hfac)$_2$, Ph$_3$Bi, Me$_4$Pb or Et$_4$Pb and CpTl or Tl(thd) were found to be suitable. The availability of suitable Ba-, Sr- and Ca- precursors, however, appeared to be much more of a problem. Therefore, in recent years much attention has been given world-wide to the development of thermally stable volatile Ba-, Sr- and Ca-compounds. Synthetic strategies have been based on encapsulating the central metal atom by the use of multidentate systems and/or bulky substituents. Most attention has been directed at the development of thermally stable fluorine-free and fluorine-substituted ß-diketonate complexes with auxiliary ligands. Based on this concept a series of novel Y-precursors have also been developed. This paper presents a survey of the trends in recent research activities and developments in these areas.

MOCVD-PRECURSORS

Barium-, Strontium-, Calcium-precursors

The development of thermally stable volatile Ba-, Sr- and Ca- compounds, suitable as MOCVD precursors has been and still is an important challenge. The requirements set for these compounds are that they are volatile and thermally stable up to evaporation temperatures , both in the liquid/solid and in the vapour phase, and relatively easy to synthesize. Potentially suitable types that have been extensively studied in recent years are fluorine-free and fluorine-substituted ß-diketonates, M(R'C(O)CHC(O)R")$_2$, which have a general structural formula as shown below.

$$M \left(\begin{array}{c} O-C \diagdown R' \\ CH \\ O-C \diagup R'' \end{array} \right)_2$$

In addition, much attention has been given to alkoxides M(OR)$_2$ [1,2], metal salt ligand complexes MX$_2$.nL [3] and substituted cyclopentadienyl derivatives (R$_n$Cp$_{5-n}$)$_2$M [4].
Investigations of fluorine-free ß-diketonates have concentrated on M(thd)$_2$ compounds (R'=R"= t-Bu), which are the most volatile within the series. Investigations on fluorinated ß-diketonates deal particularly with M(hfac)$_2$ compounds (R"=R"=CF$_3$) and to a lesser extend with the more volatile M(hfod)$_2$ compounds (R'=C$_3$F$_7$; R"=t-Bu). Recently, more highly fluorinated barium-ß-diketonates, R'=R"=C$_3$F$_7$ have also been investigated [5,6].
Monomeric Ba-, Sr- and Ca- ß-diketonates would be coordinatively unsaturated with a coordination number of 4, whereas coordination numbers of 6 up to 10 are usual. Therefore, in order to increase the coordination number, intermolecular association takes place resulting in the formation of clusters with reduced volatility. Volatility has been observed to increase with increased encapsulation of the metal ion e.g. upon replacing CH$_3$ by more bulky t-Bu groups.Upon introduction of fluorinated alkyl groups intermolecular interactions also decrease and consequently volatility increases. In the presence of free oxygen or nitrogen donor ligands complex formation results in dissociation of the clusters to form monomeric and dimeric species resulting in a notably increased volatility. Ba(thd)$_2$ (R'=R"=t-Bu), already known for 3 decades [7], used to be prepared in aqueous media. The purity of the product thus obtained is, however, questionable. Evaporation rate studies of Ba(thd)$_2$ obtained from different suppliers were found to be markedly different [8]. Consequently, inadequate purity and chemical instability of commercial source material has caused many problems in MOCVD experiments. Pure Ba(thd)$_2$, which appears to be a tetrameric cluster in the crystalline state [9,10], has been synthesized by the reaction of barium metal [10,11] or barium hydride, BaH$_2$ [11,12] with 2 equivalents of the free ß-diketone, Hthd. In contact with air substantial decomposition occurs due to hydrolysis and carbonate formation, so the product has to be stored in a closed vessel under dry nitrogen or air at low temperature [10-14]. Sievers et al [15] nicely demonstrated that Ba(thd)$_2$ as a result of hydrolysis is easily transferred into a Ba$_5$ cluster, Ba$_5$(thd)$_9$OH(H$_2$O)$_3$.

Application of pure Ba(thd)$_2$ as a barium precursor requires sublimation temperatures above 200°C, whereas the whole MOCVD system has to be kept at >250°C. Upon prolonged heating at these temperatures decomposition of Ba(thd)$_2$ occurs, resulting in the formation of non-volatile products and a significant decrease of the Ba-concentration in the deposited films. Evaporation of Ba(thd)$_2$ in the presence of oxygen or nitrogen donor ligands such as THF [16], CH$_3$O(CH$_2$CH$_2$O)$_n$CH$_3$ (n=1 or 2) [17], NH$_3$, NMe$_3$, NEt$_3$, pyridine, NH$_2$CH$_2$CH$_2$NH$_2$, Me$_2$NCH$_2$CH$_2$NMe$_2$ [18], has been found to have a positive effect on the thermal stability and the volatility of the barium precursors, resulting in a more constant and increased mass transport at lower evaporation temperatures.

Investigations into fluorine-substituted Ba-, Sr-, and Ca-ß-diketonates have dealt with various combinations of R' and R", either one or both being fluorine-substituted alkyl groups. These compounds sublime at temperatures around 200°C at 0.1 torr and have been applied as precursors for Ba-, Sr- and Ca-fluorides [19-22]. Like the fluorine-free ß-diketonates, fluorinated Ba-, Sr- and Ca-ß-diketonates appear to associate to oligomeric species, in particular Ba$_2$ and Ba$_4$ clusters have been detected in the vapour phase. Ca(hfac)$_2$.2H$_2$O (R'=R"=CF$_3$) is dimeric in the crystalline state [23], whereas the corresponding Ba(hfac)$_2$.1H$_2$O appears to be polymeric [23]. Thermally stable monomeric complexes of fluorine-substituted ß-diketonates with polyethers, primary found at TNO [3,17,24,25], are by far the most volatile Ba-, Sr- and Ca-precursors presently available. On the basis of cost-price versus performance the Ba(hfac)$_2$.L, Sr(hfac)$_2$.L and Ca(hfac)$_2$.L complexes, where L=triglyme, tetraglyme, hexaglyme or 18-crown-6, are the most interesting. The X-ray crystal structure determination of Ba(hfac)$_2$.tetraglyme (Fig.1) clearly shows a 9-coordinated barium atom encapsulated by the tetraglyme ligand and two ß-diketonate groups in opposite positions [24,25]. Meanwhile X-ray structures of the same [26] and analogous compounds, Ba(hfac)$_2$.(CH$_2$CH$_2$O)$_6$[27], and Ca(hfac)$_2$.CH$_3$O(CH$_2$CH$_2$O)$_4$CH$_3$ [28] have been reported by other groups.

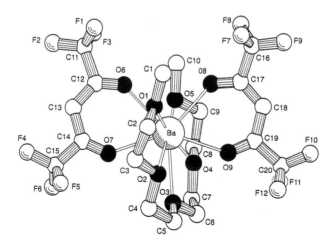

Figure 1. Structure of Ba(hfac)$_2$.tetraglyme [24,25]

The markedly enhanced volatility of Ba(hfac)$_2$.tetraglyme, Ba(hfac)$_2$.hexaglyme and Ba(hfac)$_2$.18-crown-6 is clearly demonstrated by the results of vapour pressure measurements. Vapour pressures are found to be in the same range as for the generally applied copper and yttrium precursors, Cu(thd)$_2$ and Y(thd)$_3$, (see Fig.2).

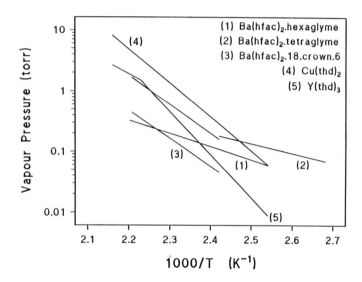

Figure 2. Vapour pressures of Barium, Yttrium and Copper MOCVD precursors

Fluorine-substituted Ba-, Sr- and Ca-ß-diketonate.polyether complexes in particular Ba(hfac)$_2$.tetraglyme, Sr(hfac)$_2$.tetraglyme, Ca(hfac)$_2$.tetraglyme and Ca(hfac)$_2$.triglyme have been successfully applied as MOCVD precursors for various types of yttrium [29], bismuth [30] and thallium [31] based superconducting thin films. Particularly the groups of Wessels and Marks et al. in numerous recent publications [29-31 a.o.] report these compounds to be superior Ba-, Sr- and Ca-MOCVD sources, for which the relatively high vapour pressures contribute to a greater control of the deposition process. A drawback of these precursors is that a hydrolysis procedure is required in order to transfer initially formed fluorides into oxides. As such they are excellent precursors for the deposition of Ba-, Sr- and Ca-fluorides.
Investigations into the development of fluorine-free precursors have been directed to the synthesis of a large series of complexes of Ba(thd)$_2$ with various types of oxygen and nitrogen donor ligands such as CH$_3$OH [4], Et$_2$O [32], 2,6-bis-t-butylphenol [33], tetraglyme [26,34,35], triglyme [28], diglyme [34], NH$_3$ [36], Me$_2$NCH$_2$CH$_2$NMe$_2$ [1], phenanthroline [37] and 2,2'-bipyridine [34]. These complexes, although stable at room temperature,

easily dissociate at elevated temperatures and reduced pressures and consequently they are not suitable as MOCVD precursors. This also holds for a series of Ba(thd)$_2$.L complexes with the tri-, tetra- and hexadentate cyclic aza compounds, 1,4,7-trimethyl-1,4,7-triazacyclononane (L$_1$), 1,4,8,11-tetrame-thyl-1,4,8,11-tetraazacyclotetradecane (L$_2$) and 1,4,7,10, 13,16-hexamethyl-1,4,7,10,13,16-hexaazacyclooctadecane (L$_3$) recently prepared at TNO [12].

$$L_1 = -[\overset{\overset{CH_3}{|}}{N}CH_2CH_2]-_3$$

$$L_2 = -[\overset{\overset{CH_3}{|}}{N}CH_2CH_2\overset{\overset{CH_3}{|}}{N}CH_2CH_2CH_2]-_2$$

$$L_3 = -[\overset{\overset{CH_3}{|}}{N}CH_2CH_2]-_6$$

New approaches into barium-ß-diketonates aiming at the development of fluorine-free, more volatile barium compounds comprise the use of alkoxyalkyl-substituted ß-diketones. Russian authors report that such compounds where R'= t-Bu, R"= C(CH$_3$)$_2$OCH$_3$ have enhanced stability compared to Ba(thd)$_2$ [38]. Rees et al. [39] report that a scorpiotail type complex R'=t-Bu, R"=CH$_2$CH$_2$OMe has been obtained. Vapour pressure data are lacking and there is no evidence that these compounds show increased performance as compared to Ba(thd)$_2$.

At TNO we looked into the properties of Ba(ß-diketonate)$_2$.18-crown-6 complexes with (A) chloro-(R'=CH$_3$, R"=CCl$_3$; m.p. 158-160°; decomp.) and (B) nitro-(R'=R"=p-NO$_2$C$_6$H$_4$) (thermally stable >200°C; non-volatile) substituted ß-diketonate groups [12]. However, these compounds were found to be thermally unstable and/or non-volatile.

(A) (B)

Recently Marks et al [40] reported on the synthesis of monomolecular bariumketoimines with a polyether group substituted at the nitrogen atom intramolecularly coordinating to the barium atom.

$$M \left\{ \begin{matrix} O - C \\ N - C \\ | \\ R^3 \end{matrix} \begin{matrix} R^1 \\ C-H \\ R^2 \end{matrix} \right\}_2$$

R^1=t-Bu, R^2=CH_3, R^3= -$(CH_2CH_2O)_2CH_3$ or
-$(CH_2CH_2O)_3C_2H_5$

According to the authors these complexes sublime with some decomposition at 80-120°C/10^{-5} torr. They are more volatile than Ba(thd)$_2$ but less volatile than the most volatile fluorinated ß-diketonate precursors. Whether these compounds show improved performance as barium precursors for MOCVD as compared to Ba(thd)$_2$ has not yet been reported.

Yttrium Precursors

At present Y(thd)$_3$ and its mono aqua adduct Y(thd)$_3$.H$_2$O are used, almost exclusively, as the volatile yttrium precursor in MOCVD deposition processes of yttrium containing thin films, such as the high Tc superconducting material YBa$_2$Cu$_3$O$_{7-x}$, yttrium stabilized zirconia and yttrium oxide.
At TNO investigations have been directed at the development of alternative yttrium precursors. Y(thd)$_3$ is coordinatively unsaturated, which implies that with neutral oxygen or nitrogen donor ligands complexes of the type Y(thd)$_3$.L are easily obtained, where L can be an amide, sulfoxide, amine-N-oxide, pyridine-N-oxide, phosphine oxide, mono-, di- and polyamine, pyridine, bipyridine or phenanthroline [41, 42]. Most complexes sublime without dissociation in the temperature range of 140-200°C at 0.01-0.05 torr. Y(thd)$_3$.L complexes with the 4-t-Bu and 4-Et substituted pyridine-N-oxides are of particular interest in that as a result of their low melting points they can be applied as liquid yttrium MOCVD precursor at temperatures >100°C. Y(thd)$_3$.4-Et-PyNO, mp 93-98°C, sublimes without dissociation at 140-150°C/0.01-0.05 torr. Simultaneous Thermal Analysis (STA) at atmospheric pressure reveals melting to occur at 106°C with subsequently a start of evaporation at 149°C. At the end of evaporation at 297°C a residue of 1.97 wt % is left. Similarly, Y(thd)$_3$.4-t-Bu-PyNO, mp 97-100°C, sublimes without dissociation at 140°C/0.01-0.05 torr, whereas STA at atmospheric pressure reveals melting to occur at 104°C with a start of evaporation at 146°C and a residue of 1.63 wt% at the end of evaporation at 306°C. Vapour pressures (Fig.3) appear to be somewhat higher than for Y(thd)$_3$. The use of these compounds as MOCVD precursors is currently investigated in the deposition of thin films of superconducting YBa$_2$Cu$_3$O$_{7-x}$ and of YSZ.

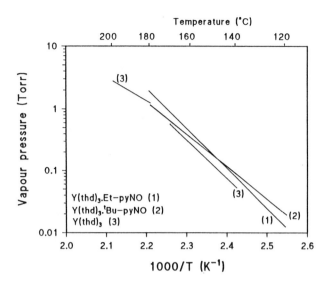

Figure 3. Vapour pressures of Yttrium MOCVD precursors.

CONCLUSIONS

Coordination chemistry plays a crucial role in the development of volatile early-main group and transition metal compounds that meet the requirements for MOCVD precursors. Thermally stable, volatile fluorine-containing Ba-, Sr- and Ca-compounds have become available in recent years. These compounds have been successfully applied for MOCVD of thin layers of HTcS, Ba-, Sr-, Ca-fluorides, $BaTiO_3$ and $Ba_{1-x}Sr_xTiO_3$. A second generation of fluorine free Ba-, Sr- and Ca-precursors for MOCVD with improved performance compared to $M(thd)_2$ (M=Ba, Sr, Ca) has not yet become available. Thermally stable $Y(thd)_3$.Ligand complexes, suitable as precursors for MOCVD, have been obtained. Complexes of $Y(thd)_3$ with 4-Et and 4-t-Bu substituted pyridine N-oxides are of particular interest as they act as liquid precursors at temperatures above 100°C.

ACKNOWLEDGEMENTS

For the investigations performed at TNO we acknowledge the support of the Commission of the European Communities under the Brite Euram Programme Contract No Breu/0438 and of our partners in this project, The Associated Octel Company Ltd, UK; Aixtron GmbH, D; NKT, DK and Strathclyde University, UK.

REFERENCES

1. L.G. Hubert-Pfalzgraf, Appl. Organometal. Chem. 6 627 (1992).
2. K.G. Caulton and L.G. Hubert-Pfalzgraf, Chem. Rev. 90 969 (1990)
3. K. Timmer and H.A. Meinema, Inorg. Chim. Acta 187 99 (1991)
4. T.P. Hanusa, Chem. Rev. 93 1023 (1993).
5. D.D. Gilliland, M.L. Hitchman, S.C. Thompson and D.J. Cole-Hamilton, J. Phys. III. 2 1381 (1992).
6. H. Sato and S. Sugawara, Inorg. Chem. 32 1941 (1993).
7. G.S. Hammond, D.C. Nonhebel and C.H.S. Wu, Inorg. Chem. 2 73 (1963).
8. E. Fitzer, H. Oetzmann, F. Schmaderer and G. Wahl, J. Phys. IV. C 2 713 (1991).
9. A. Gleizes, D. Medus and S. Sans-Lenain, in Better Ceramics trough Chemistry V. M.R.S. Symp. Proc. 271 919 (1992).
10. A.A. Drozdov and S.I. Trojanov, Polyhedron 11 2877 (1992).
11. A.D. Berry, R.T. Holm, M. Fatemi and D.K. Gaskil, J. Mater. Res. 5 1169 (1990).
12. K. Timmer, unpublished results
13. F.A. Kutznetsov, I.K. Igumenov and V.S. Danilov, Physica C 185-189 1957 (1991).
14. H. Busch, A. Fink, A. Müller and K. Samwer, Supercond. Sci. Technol. 6 42 (1993).
15. S.B. Turnipseed, R.M. Barkley and R.E. Sievers, Inorg. Chem. 30 1164 (1991).
16. S. Matsuno, F. Uchikawa and K. Yoshizaki, Jpn. J. Appl. Phys. 29 L947 (1990).
17. K. Timmer, C.I.M.A. Spee, A. Mackor and H.A. Meinema, Dutch Patent Appln 8901507; US Patent 5.248.787 (28 September 1993).
18. J.M. Buriak, L.K. Cheatham, R.G. Gordon, J.J. Graham and A.R. Barron, Eur. J. Solid State Inorg. Chem., 29 43 (1992).
19. R. Belcher, C.R. Cranley, J.R. Majer, W.I. Stephen and P.C. Uden, Anal. Chim. Acta 60 109 (1972).
20. A.W. Vere, K.J. Mackey, D.C. Rodway, P.C. Smith and D.M. Frigo, Angew. Chem. Adv. Mater. 101 1613 (1989).
21. R. Sing, S. Singha, P. Chou, N.J. Hsu, F. Radpour, H.S. Ullal and A.J. Nelson, J. Appl. Phys. 66 6179 (1989).
22. A.P. Purdy, A.D. Berry, R.T. Holm, M. Fatemi and D.K. Gaskill, Inorg. Chem. 28 2799 (1989).
23. D.C. Bradley, M. Hasan, M.B. Hursthouse, M. Motevalli, O.F.Z. Khan, R.G. Pritchard and J.O. Williams, J. Chem. Soc. Chem. Commun. 1992, 575.
24. K. Timmer, C.I.M.A. Spee, A. Mackor and H.A. Meinema,Inorg. Chim. Acta 190 109 (1991).
25. P. van der Sluis, A.L. Spek, K. Timmer and H.A. Meinema, Acta Crystallogr. Sect C 46 1741 (1990).
26. R.Gardiner, D.W.Brown, P.S.Kirlin and A.L.Rheingold, Chem. Mater. 3 1053 (1991).
27. J.A.T. Norman and G.P. Pez, J.Chem.Soc.Chem.Commun. 1991, 971.
28. S.R.Drake, S.A.S. Miller and D.J.Williams, Inorg.Chem. 32 3227 (1993).
29. C.I.M.A. Spee, E.A. van der Zouwen-Assink, K.Timmer, A.Mackor and H.A.Meinema, J.Phys IV C 2 295 (1991).
30. J.M.Zhang, B.W.Wessels, D.S. Richeson, T.J. Marks, D.C. DeGroot and C.R.Kannewurf. J.Appl.Phys. 69 2743 (1991).

31. D.L.Schulz, D.S.Richeson, G.Malandrino, D.Neumayer, T.J.Marks,D.C. DeGroot, L.J.Schindler, T.Hogan and C.R.Kannewurf, Thin Solid Films 216 45 (1992).
32. G.Rosetto, A.Polo, F.Benetollo, M.Porchia and P.Zanella, Polyhedron 11 979 (1992).
33. P.Miele, J.D.Foulon and N.Hovnanian, Polyhedron 12 209 (1993).
34. A.Drozdov, N.Kuzmina, S.Troyanov and L. Martynenko, Mater. Sci. Eng. B 18 139 (1993).
35. S.R.Drake, S.A.S. Miller, M.B.Hursthouse and K.M.A.Malik, Polyhedron 12 1621 (1993).
36. W.S.Rees, M.W. Carris and W.Hesse, Inorg.Chem. 30 4481 (1991)
37. R.Sato, K.Takahashi, M.Yoshino, H.Kato and S.Oshima, Jpn.J.Appl.Phys. 32 1590 (1993).
38. N.Snezhko, S.Moroz and P.Petchurova, Mater. Sci. Eng. B 18 230 (1993).
39. W.S.Rees, C.R. Caballero and W.Hesse, Angew.Chem. 104 786 (1992).
40. D.G.Schulz, B.J.Hinds, C.L.Stern and T.J.Marks, Inorg.Chem. 32 249 (1993).
41. K.Timmer and C.I.M.A. Spee, Dutch Patent Appln. 93.799 NL
42. K.Timmer, C.I.M.A.Spee, H.Linden and H.A.Meinema, to be published.

ADVANCES IN PRECURSOR DEVELOPMENT FOR CVD OF BARIUM-CONTAINING MATERIALS

BRIAN A. VAARTSTRA,* R. A. GARDINER,* D. C. GORDON,* R. L. OSTRANDER** and A. L. RHEINGOLD**
*Advanced Technology Materials Inc., 7 Commerce Dr., Danbury, CT 06810
**University of Delaware, Department of Chemistry, Newark, DE 19716

ABSTRACT

Barium titanate and barium-strontium titanate (BST) are high dielectric materials, likely to replace state-of-the-art capacitor materials for memory applications. Chemical Vapor Deposition (CVD) of these materials has been hampered, particularly by the lack of suitable precursors for barium. Although attempts to make volatile metal-organic barium compounds have met with some progress, a suitably stable, volatile barium source is still in demand. This paper will highlight recent developments at ATM, including syntheses and structures of polyamine and glycol ether adducts which have been designed to limit aggregation of barium diketonates, and stabilize the adducts with respect to ligand dissociation.

INTRODUCTION

Numerous reports have been published over the past few years as a result of attempts to synthesize volatile barium compounds, suitable for chemical vapor deposition of barium-containing ferroelectrics and high T_c superconducting oxides. Vapor transport of barium precursors is hampered by the tendency for barium to attain high coordination numbers, usually by forming multinuclear aggregates, which have poor volatility. Only those barium compounds with multidentate, sterically encumbered anions have proven useful for CVD appplications.

The air stability and moderate volatility of $Ba(thd)_2$ (thd = 2,2,6,6-tetramethyl-3,5-heptanedionate) have made it the most widely used barium source for MOCVD. Nevertheless, its vapor transport behavior is strongly dependent on synthesis conditions and purity of the compound, and decomposition occurs to varying degrees during sublimation. The determination that $Ba(thd)_2$ is tetrameric in the solid state[1] and that it is associated in the gas phase as well,[2] has suggested that further reduction in molecularity should increase volatility.

Addition of polyethers to Group IIA diketonates has proven to be effective in reducing aggregation through coordinative saturation. For example, $Ba(thd)_2(tetraglyme)$, is a crystalline colorless solid which is monomeric in the solid state.[3a] Although the nuclearity has been minimized in this compound, the tetraglyme dissociates upon attempted sublimation of the solid; a problem which has plagued many Lewis base adducts of barium diketonates.

Another strategy has been the incorporation of fluorine in the ligands surrounding barium, thus decreasing intermolecular van der Waals forces that also play a role in determining the vapor pressure of chemical species. Using both this strategy and the Lewis base adduct strategy, it has been shown that coordination of polyethers to fluoro-ß-diketonato barium compounds such as $Ba(CF_3COCHCOCF_3)_2$ yields ample volatility for MOCVD applications. However, thermally stable BaF_2 is obtained during deposition unless additional steps are taken to convert it to the desired oxide.[3] Non-fluorinated sources are therefore more desirable.

Nitrogen-donor ligands might be expected to make stronger bonds to barium diketonates than analogous oxygen-donor ligands, due to the increased basicity of amines compared to ethers. Ammonia, simple amines (NMe_3, NEt_3, pyridine) and even diamines ($H_2NC_2H_4NH_2$, $Me_2NC_2H_4NMe_2$) cause a decrease in the melting and sublimation temperatures of barium diketonates,[4] but the adducts do not have stable vapor pressures due to dissociation of the amine upon vaporization. With the exception of $[Ba(thd)_2(NH_3)_2]_2$,[5] adducts have not been characterized, and notably lacking is any data on non-fluorinated barium diketonates which have coordinated polyamines.[6]

An additional feature which has potential for increasing the bond strength of Lewis bases to barium diketonates is hydrogen bonding. Intermolecular hydrogen bonding, as observed for the methanol adduct, Ba(thd)$_2$(CH$_3$OH)$_3$,[7] increases aggregation and therefore would not be advantageous. However, intramolecular hydrogen bonding, between an acidic proton on a neutral donor ligand and oxygen atoms of a diketone, could strengthen the bonding of the Lewis base ligand to the barium diketonate, making it less prone to dissociation. In general, barium diketonates containing water or methanol have not been reported to exhibit intramolecular hydrogen bonding to diketonate oxygen atoms.[7,8] Likewise, the ammonia adduct, [Ba(thd)$_2$(NH$_3$)$_2$]$_2$,[5] does not appear to have this feature. One example of intramolecular hydrogen bonding in a barium diketonate has recently been reported by Hovnanian et al.,[9] wherein the X-ray structure of [Ba(thd)$_2$(HOAr)$_2$(THF)$_2$]$_2$ clearly revealed hydrogen bonding between the phenol and thd oxygens. We report here our initial study of the potential stability to be imparted to barium diketones by polyfunctional alcohols or amines through hydrogen bonding.

RESULTS

Polyamine adducts

The reactions of Ba(thd)$_2$ with 1,1,4,7,10,10-hexamethyltriethylenetetramine (hmtt) and 1,1,4,7,7-pentamethyldiethylenetriamine (pmdt) were carried out, producing the adducts Ba(thd)$_2$(hmtt) (**1**) and Ba(thd)$_2$(pmdt) (**2**). Reactions were performed under dinitrogen in anhydrous pentane or hexane as the solvent. NMR spectra were recorded at room temperature in benzene-d$_6$ on an IBM WP200SY spectrometer and peaks referenced to the protio impurity in the solvent at δ 7.15. In each case approximately 1.0 g of Ba(thd)$_2$ was partially dissolved in the solvent and to it was added one equivalent of the polyamine. Upon removal of solvent, colorless solids were obtained which were highly crystalline. In each case X-ray quality crystals were obtained by slow concentration of a pentane solution of the adduct. Elemental analyses (C, H, N) were satisfactory for both compounds.

Ba(thd)$_2$(hmtt) (1). Air stable crystals of the compound, Ba(thd)$_2$(hmtt) (**1**), crystallized in the space group P2$_1$/c with a = 10.833(6) Å, b = 20.442(12) Å, c = 19.404(9) Å, ß = 104.35(4) ° and Z = 4.[10] The molecular structure is shown in Figure 1. The metal center is eight-coordinate, with all four nitrogen atoms of the polyamine bound to barium. The thd ligands are oriented in a propeller-type geometry accompanied by a twist of the O-Ba-O planes: O(1)-Ba-O(4) = 86.2 °; O(2)-Ba-O(3) = 152.8°. This is a feature that is also observed in the polyether analog, Ba(thd)$_2$(triglyme).[11] Note that the hmtt ligand is substantially folded about the equator of the

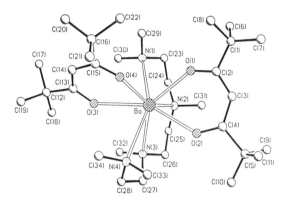

Figure 1: Molecular structure of Ba(thd)$_2$(hmtt)

coordination sphere, and is bound such that the methyl groups on the nonterminal amino groups extend in opposite directions. This would appear to minimize steric repulsion between methyl groups on the polyamine and on the thd ligands.

An interesting feature of this structure compared to related polyether adducts is the effect that additional methyl groups on the polyamine has on the Ba-O bond distances. Each of the thd ligands is involved in one shorter bond (2.61 Å and 2.58 Å) and one longer bond (2.68 Å and 2.72 Å, respectively) to the barium atom. As expected, the longer bonds are those which are closest to the coordinated polyamine. On the other hand, the Ba-N distances are rather consistent (within 3σ), between 2.986 and 3.060 Å.

The ^{1}H NMR spectrum of **1** revealed thd resonances at δ 5.84 (s, 2H) and δ 1.35 (s, 36H) and coordinated hmtt resonances at δ 2.61 (br), δ 2.17 (s, 12H), δ 2.08 (s, 6H), δ 1.70 (br). The thd methyl groups appear as a sharp singlet, representing a single methyl environment. This resonance is clearly shifted downfield from unadducted Ba(thd)$_2$, indicative of Lewis base coordination. Singlets are also observed for each of the terminal and nonterminal amino methylgroups with the expected integration in each case. The broad resonances for the ethylene protons of the polyamine suggest fluxionality in solution.

To compare the affinity of polyethers and polyamines toward Ba(thd)$_2$, NMR was used to monitor the effect of adding excess tetraglyme to a solution of Ba(thd)$_2$(hmtt). No ligand exchange was observed over 22 hrs, consistent with the expected greater basicity of the polyamine. On the other hand, addition of hmtt to a solution of Ba(thd)$_2$(tetraglyme) immediately produced free tetraglyme and Ba(thd)$_2$(hmtt). Excess hmtt exchanges with coordinated hmtt, suggesting that the fluxionality of Ba(thd)$_2$(hmtt) in solution may involve at least partial dissociation of hmtt.

The physical property of most interest from a CVD perspective is volatility. Sublimation of **1** was observed in the range 95-120 °C (50 mtorr). Unfortunately, the rate of sublimation was very poor at the low end of this range and hmtt dissociation became substantial as the temperature was increased. It is noteworthy that the tetraglyme adduct, Ba(thd)$_2$(tetraglyme), does not vaporize Ba(thd)$_2$ under the same conditions, despite the fact that it has an additional donor atom in the ligand compared to hmtt. The tetraglyme adduct merely undergoes ligand dissociation at 110 °C (0.1 torr).[3a] Since some transport of Ba(thd)$_2$ was observed in the sublimation experiment, one would expect that hmtt could be used as an effective additive to the carrier stream in a CVD process, much like the use of ammonia and simple amines by Barron et al.[4] We suggest that hmtt may have better capacity to transport Ba(thd)$_2$ than simple amines since it is bound more tightly, however, we have not quantified the rate of transport under similar conditions.

Ba(thd)$_2$(pmdt) (**2**). The study of an adduct containing less steric repulsion than hmtt, namely pmdt, was carried out under the suspicion that dissociation of hmtt from **1** might be due to the steric repulsions that were obvious in the solid state structure. Air stable crystals of the compound, Ba(thd)$_2$(pmdt) (**2**) crystallized in the space group P2$_1$/c with a = 10.577(3) Å, b = 23.547(7) Å, c = 15.963(5) Å, β = 105.21(2) ° and Z = 4. The X-ray structure is shown in Figure 2. The barium ion is seven-coordinate, having all three nitrogen atoms of the polyamine coordinated to the metal center. The diketonate ligands are pinched toward one another in order to accommodate the polyamine, which is bound in a meridional fashion between the thd ligands. Unlike compound **1**, the distances from Ba to the thd-oxygen atoms are similar (avg. 2.62 Å), as are the Ba-N bond distances (avg. 2.97 Å), with the distance to the "internal" nitrogen (N2) being only slightly longer than the others. Once again, the methyl groups on the polyamine are clearly oriented to minimize steric interactions with the tert-butyl groups of the thd ligands.

The ^{1}H NMR spectrum of **2** revealed thd resonances at δ 5.83 (s, 2H) and δ 1.32 (s, 36H), and coordinated pmdt resonances at δ 2.55 (br), δ 2.25 (s, 12H), δ 1.94 (s, 3H) and δ 1.89 (br). As with **1**, the ethylene resonances of **2** are broad and suggest fluxionality in solution.

Compound **2** sublimed over a slightly higher temperature range (120-150 °C / 50 mtorr) than compound **1**, again with dissociation of the polyamine. At the end of the attempted sublimation, both the sublimate and residue contained Ba(thd)$_2$, with very little pmdt present. The higher temperature required here suggests that the pmdt adduct is more thermally stable than the hmtt adduct. This is likely due to the steric repulsions discussed earlier for compound **1**, since pmdt would otherwise be expected to dissociate easier, owing to its lower boiling point and fewer donor atoms (ie. fewer bonds to barium).

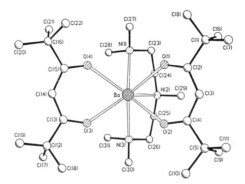

Figure 2. Molecular structure of Ba(thd)$_2$(pmdt)

Glycol ether adducts

Ba(thd)$_2$(mee) (3). The incorporation of additional bonding potential via hydrogen bonding was first attempted using the trifunctional alcohol, 2-(2-methoxyethoxy)ethanol (mee). Addition of either one or two equivalents of the alcohol to Ba(thd)$_2$ yielded the same product: a colorless crystalline solid which had the formula Ba(thd)$_2$(mee) (3), based on NMR and elemental analysis. Crystals suitable for an X-ray diffraction study were obtained by slow evaporation of a pentane solution of the compound. The molecular structure is shown in Figure 3.

The molecule is a centrosymmetric dimer, which is held together by two bridging thd groups. The bridging thd groups use only one oxygen atom to form the bridge, while the other is involved in a hydrogen bond to the alcohol proton of the mee ligand on the other barium atom. In this way, the dimer is therefore also held together by two hydrogen bonds. Each barium atom is eight-coordinate, with all three oxygen atoms of any given mee ligand bound to a single barium center.

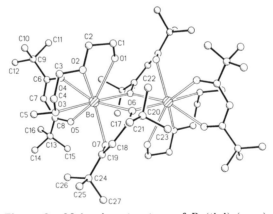

Figure 3. Molecular structure of Ba(thd)$_2$(mee)

Although the alcohol proton was not located crystallographically, the hydrogen bonding in this molecule is clear from the distance separating O(1) from O(7)'. Furthermore, the thd ligand involved in the hydrogen bond is obviously twisted in order to form the hydrogen bond. This is evident from the Ba-O(7)-C(19) angle of 119°, compared to more typical angles near 140° for terminal thd groups bound to a metal center. This twisting is much the same as that observed for [Ba(thd)$_2$(HOAr)$_2$(THF)$_2$]$_2$, where a similar hydrogen bond is formed between an alcohol proton and the non-bridged oxygen of a bridging thd group.[9]

The Ba-O bonds involving the mee ligand have noteably different lengths. The bond to the hydroxy oxygen atom, O(1), being the shortest, at 2.78 Å. The bonds to the ether oxygens, O(2) and O(3), are 2.93 Å and 2.85 Å, respectively. Although the latter two are in the range that is typical for Ba-O distances found in polyether adducts, the short Ba-O(1) distances is surely a result of the hydrogen bonding.

Compound 3 was also characterized in solution by NMR. The ^{1}H NMR spectrum in benzene-d$_6$ revealed thd resonances at δ 5.81 (s, 2H) and δ 1.30 (s, 36H) and coordinated mee resonances at δ 3.96 (br, 1H), δ 3.73 (m, 2H), δ 3.33 (m, 2H), δ 3.19 (m, 2H), δ 3.14 (s, 3H), δ 3.11 (m, 2H). The coordinated mee resonances were clearly shifted compared to a spectrum of free mee. The NMR spectrum is inconclusive about whether or not the dimeric structure is maintained in solution. Although it does not possess the number of methyl group resonances required by the solid state structure, a fluxional process may average these environments in solution.

The hydrogen bonding in 3 is also evident in the infrared spectrum of the compound. The IR spectrum of the solid (KBr pellet) exhibits a broad O-H stretch at 3171 cm^{-1}. This is extremely low, even for hydrogen bonded alcohols, but is comparable to other examples of intramolecular hydrogen bonded O-H stretching.[12]

Attempted sublimation of 3 was carried out at 70 mtorr. The compound melted in the range 125-130 °C and began to vaporize at 160 °C, recondensing as a solid on the cold finger of the sublimation apparatus. In the range 160-170 °C the compound vaporized with some decomposition, as mee-deficient product was collected and analyzed by NMR spectroscopy. Likewise the remaining residue was found to contain a mixture of Ba(thd)$_2$(mee) and Ba(thd)$_2$.

Although there was obvious decomposition of Ba(thd)$_2$(mee) during vaporization, it appeared to be significantly more difficult to remove mee from Ba(thd)$_2$ than any other neutral ligand we have encountered. It is therefore likely that we are observing a significant affect of hydrogen bonding upon the thermal stability of Ba(thd)$_2$ adducts.

Ba(thd)$_2$(^{3}egme) (4). The consequences of adding additional donors to the glycol ether ligand was examined by the reaction of Ba(thd)$_2$ with triethylene glycol monomethyl ether (^{3}egme). This tetrafunctional alcohol {CH$_3$(OCH$_2$CH$_2$)$_3$OH} was expected to exhibit similar hydrogen bonding to Ba(thd)$_2$ as mee, but perhaps discourage dimerization by an increase in coordination.

Addition of one equivalent of ^{3}egme to Ba(thd)$_2$ resulted in a colorless crystalline compound after removal of the reaction solvent. The product was confirmed to be a 1:1 adduct, Ba(thd)$_2$(^{3}egme) (4), by ^{1}H NMR spectroscopy. The resonances of the ^{3}egme were clearly shifted from that of free ligand and, as with the other adducts, only a single environment was observed for each type of proton in the thd groups.

Upon attempted sublimation of 4 (80 mtorr), the compound melted at 116-120 °C, which was accompanied by distillation of the glycol ether. This behavior was much like Ba(thd)$_2$(tetraglyme) which also loses the neutral ligand without transport of Ba(thd)$_2$.

Infrared spectroscopy suggested that there was hydrogen bond formation as evidenced by a broad O-H stretch at 3208 cm^{-1}, although this is indeed at slightly higher frequency than in the mee adduct. Unfortunately, we have not yet been able to obtain crystals that are suitable for X-ray studies, but it is obviously desirable to examine the solid state structure for an explanation of the poor thermal stability of this adduct compared to 3.

Polyamine adduct with potential hydrogen bonding

$Ba(thd)_2(deta)_2$ (5). In an attempt to benefit from both the added bond energy of polyamines and the adhesive forces of hydrogen bonding, an adduct was synthesized using diethylenetriamine (deta), $H_2NCH_2CH_2N(H)CH_2CH_2NH_2$. In this case it was found that a 2:1 adduct is formed on reaction of two equivalents of deta with $Ba(thd)_2$. Again a highly crystalline colorless product was obtained after removal of the reaction solvent. Integration of the ^1H NMR spectrum confirmed the formulation, $Ba(thd)_2(deta)_2$. Again, only a single thd environment was observed: δ 5.73 (s, 2H) and δ 1.31 (s, 36H), and the polyamine ligand appeared symmetrically bound on the NMR timescale: δ 2.53 (m, 8H), δ 2.40 (m, 8H), δ 1.19 (br, 10H).

Sublimation of this compound began at 140 °C (50 mtorr) with no melting. At 160 °C most of the sample eventually sublimed, although the sublimate was found to contain approximately 35% of its original deta. These observations are consistent with relatively high thermal stability of the adduct - although some ligand dissociation occurs. Although we have no firm evidence for intramolecular hydrogen bonding in this compound, an infrared spectrum revealed a broad N-H stretch at 3328 cm^{-1}, which appears consistent with hydrogen bonding. Nevertheless, hydrogen bonds involving amines are generally weaker and less distinguishing than those involving hydroxyl groups. It may be more important here that steric repulsions are minimized by the absence of alkyl groups on the donor atoms.

This compound seems to have gained us yet another incremental advantage in $Ba(thd)_2$ transport. An X-ray structure determination of this compound has revealed that it is mononuclear, and full details of its structure will be published at a later date.

ACKNOWLEDGEMENTS

We gratefully acknowledge ARPA (contract DASG60-92-C-0025) for financial support of this work.

REFERENCES

1. (a) A. Gleizes, S. Sans-Lenain, D. Medus, C. R. Acad. Sci. Paris, **313 II**, 761 (1991). (b) A. A. Drozdov, S. I. Trojanov, Polyhedron, **22**, 2877 (1992).
2. R. Belcher, C. P. Cranley, J. R. Majer, W. I. Stephen, P. C. Uden, Anal. Chim. Acta, **60**, 109 (1972).
3. (a) R. Gardiner, D. W. Brown, P. S. Kirlin, A. L. Rheingold, Chem. Mater., **3**, 1053 (1991). (b) K. Timmer, C. I. M. A. Spee, A. Mackor, H. A. Meinema, Eur. Pat. Appl. 0405634A2 (1991). (c) G. Malandrino, D. S. Richeson, T. J. Marks, D. C. DeGroot, J. L. Schindler, C. R. Kannewurf, Appl. Phys. Lett., **58**, 182 (1991). (d) J. M. Zhang, B. W. Wessels, D. S. Richeson, T. J. Marks, D. C. DeGroot, C. R. Kannewurf, J. Appl. Phys., **69**, 2743 (1991). (e) N. Hamaguchi, R. Gardiner, P. S. Kirlin, R. Dye, K. M. Hubbard, R. E. Muenchause, Appl. Phys. Lett., **57**, 2136 (1990).
4. J. M. Buriak, L. K. Cheatham, R. G. Gordon, J. J. Graham, A. R. Barron, Eur. J. Solid State Inorg. Chem., **29**, 43 (1992).
5. W. S. Rees, Jr., M. W. Carris, W. Hesse, Inorg. Chem., **30**, 4479 (1991).
6. An example of a polyamine coordinated to a fluorinated barium diketonate has been reported recently: S. R. Drake, M. B. Hursthouse, K. M. A. Malik, S. A. S. Miller, D. J. Otway, Inorg. Chem., **32**, 4464 (1993).
7. A. Gleizes, S. Sans-Lenain, D. Medus, Mat. Res. Soc. Symp. Proc., **271**, 919 (1992).
8. S. B. Turnipseed, R. M. Barkley, R. E. Sievers, Inorg. Chem., **30**, 1164 (1991).
9. P. Miele, J.-D. Foulton, N. Hovnanian, Polyhedron, **12**, 209 (1993). (HOAr = 3,5-di-tert-butylphenol)
10. Details of this and subsequent crystal structure studies will be published elsewhere.
11. S. R. Drake, S. A. S. Miller, D. J. Williams, Inorg. Chem., **32**, 3227 (1993).
12. B. A. Vaartstra, J. C. Huffman, P. S. Gradeff, L. G. Hubert-Pfalzgraf, J.-C. Daran, S. Parraud, K. Yunlu, K. G. Caulton, Inorg. Chem., **29**, 3126 (1990).

PRECURSOR DELIVERY FOR THE DEPOSITION OF SUPERCONDUCTING OXIDES: A COMPARISON BETWEEN SOLID SOURCES AND AEROSOL

O. THOMAS*, F. WEISS, J. HUDNER**, R. HAASE, C. DUBOURDIEU, E. MOSSANG, N. DIDIER, J.P. SENATEUR
Laboratoire des Matériaux et du Génie Physique, URA CNRS 1109, ENSPG, BP46, 38402 St Martin d'Hères, France

ABSTRACT

Epitaxial thin layers of $YBa_2Cu_3O_{7-x}$ (T_c = 90 K, J_c (0T, 77 K) = 2 10^6 A cm^{-2}) were synthesised by thermal decomposition (750 - 830 °C) of tetramethylheptanedionates of yttrium, barium and copper in the presence of oxygen. Three different precursor delivery systems have been used: In the first, the precursor's surface is flushed with the carrier gas (Ar), in the second, argon is directly injected through the powder leading to a much higher transport rate, and in the third, a liquid (diketonates in a solvent) is nebulized via a piezoelectric transducer. The transport and deposition rates obtained in these three different reactors will be compared.

INTRODUCTION

$YBa_2Cu_3O_{7-x}$ films deposited by MOCVD now have properties (T_c, J_c, R_s) which are comparable [1,2,8] with those obtained by other deposition techniques. The major problem with the chemical vapor deposition of superconducting cuprates from tetramethylheptanedionates (thd) is the instability of the barium complex [3]. Indeed the low vapor pressure of Ba(thd)$_2$ which requires the use of high sublimation temperatures (over 200 °C) together with the poor thermal stability of this compound leads to serious reproducibility problems. This imposes the use of reduced sublimation temperatures which in turn lead to low $YBa_2Cu_3O_{7-x}$ deposition rates. Ingenious solutions have been devised to overcome this limitation. The idea is always the same: a continuous sampling of the precursor, kept at room temperature, and complete volatilization at rather high temperature. The metalorganic complexes can either be in a solid form [4,5] or in solution [6].

This communication will address these problems, which are crucial to the development of a high deposition rate process. Three different deposition reactors with three different sources designs will be compared.

EXPERIMENTAL

Deposition Reactors

The three different setups used in this study will be labeled as H (for horizontal), V (for vertical) and P (for pyrosol). Reactor H is a horizontal reactor in which the reactive gases are flowing parallel to the substrate (in fact with an angle of 17 ° to account for gas phase depletion). It

---

* Present adress: MATOP Case 151, Faculté des Sciences et Techniques de St Jérôme, Av. de l'Escadrille Normandie Niémen, 13397 Marseille Cedex, France
** Permanent adress: Department of Solid State Electronics, Royal Institute of Technology, P.O. Box 1298, S-164 28 Kista, Stockolm, Sweden

has been fully described [7] earlier. The geometry of the sources is described in Figure 1. The precursor's surface is flushed with the carrier gas (Ar).

Reactor V has been especially designed for deposition on large areas with a sample holder of 120 mm in diameter. A more thorough description is given in [8]. The reactive mixture impinges on the horizontal substrate from the bottom up. Concerning the source geometry (see Figure 1), argon is directly injected through the powder which is inserted between two pieces of porous metal.

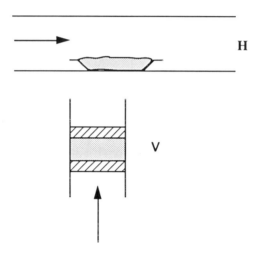

Figure 1 - The two different solid sources used in this study. The arrow indicates the direction of the carrier gas. The shaded area represents the organometallic precursor.

Reactor P has been described in [9]. It is a vertical reactor in which the gas flow is parallel to the substrate. The source principle is fundamentally different from the solid sources discussed above. The three thds are dissolved in a solvent (we used diglyme - diethylene glycol dimethyl ether - but others such as benzylalcohol are suitable). This source solution is nebulized by a piezoelectric transducer and the mist is transported by argon to a preheating zone. After evaporation of the solvent and of the precursors the gas phase is mixed with oxygen to form the reactive mixture as in a conventional MOCVD reactor.

Reactors H and V are multisources setups where the gas phase composition is monitored via the three temperatures of the sources. On the other hand reactor P is a single source device, in which the composition is fixed by the solution composition (typically a molar ratio of 1:2.6:2.3 is used).

The $YBa_2Cu_3O_{7-x}$ films have been deposited on MgO (100) in a temperature range of 750 - 850 °C and a total pressure range of 6.66 hPa - 13.33 hPa.

Characterization techniques

While the amount of evaporated precursor is easily deduced from the weight loss, the quantity of deposited material is less straightforward to measure. Films deposited by CVD tend to be off stoichiometric with CuO precipitates on their surfaces [2] (at least this is the region of the diagram where the best superconducting properties are encountered). Direct measurements of the deposited thickness, by a mechanical profilometer, is therefore unrepresentative of the condensed quantities of the different elements. In these copper-rich layers, however, most of the barium is involved in the $YBa_2Cu_3O_{7-x}$ matrix. The Ba component of the Rutherford Backscattering Spectrum is therefore related to the $YBa_2Cu_3O_{7-x}$ growth rate. Another way to estimate the rate is to measure the intensity of the Mg K_α emission peak by Energy Dispersive X-ray analysis (layers are deposited on MgO). This intensity I as a function of the thickness x, measured either by RBS or by alpha-step for stoichiometric films, has been fitted to an absorption law $\frac{I}{I_0} = \exp(-\frac{x}{a})$. This method allows the determination of the thickness between the precipitates.

RESULTS AND DISCUSSION

In Figure 2 the deposition rates (as measured by RBS) of individual oxides is plotted versus the evaporation rate of the corresponding tetramethylheptanedionates. All the layers have been deposited at the same temperature (825°C) and at the same pressure (6.66 hPa) on MgO in reactor H. It is clear that the maximum deposition rate for $YBa_2Cu_3O_{7-x}$ will be controlled by the maximum allowable $Ba(thd)_2$ temperature. For this particular case the maximum evaporation temperature of 218°C that was used for $Ba(thd)_2$ would lead to a maximum $YBa_2Cu_3O_{7-x}$ growth rate of 0.3 µm/h. The optimization of barium transport rate through a better design of the source is therefore of major importance.

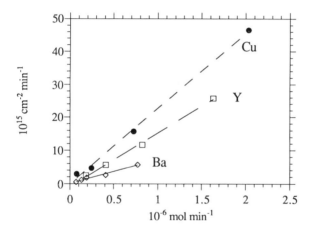

Figure 2 - Deposition rate of individual oxides at 825 °C on MgO versus evaporation rates of precursors.

When considering solid sources two important points need to be considered:
- Both source geometry and gas flow optimizations are needed for complete saturation of the carrier gas by organometallic vapor.
- Equilibrium vapor pressure of the solid at the source temperature sets an upper limit to the actual transport rate.

The optimization of source geometry is clearly shown in Figure 3 where the $Ba(thd)_2$ evaporation rates are compared for the two types of solid sources (H and V). The increase in the precursor's surface exposed to the carrier gas results in a roughly three times higher evaporation rate for the same source temperature. This results in a higher $YBa_2Cu_3O_{7-x}$ growth rate as shown in Figure 5.

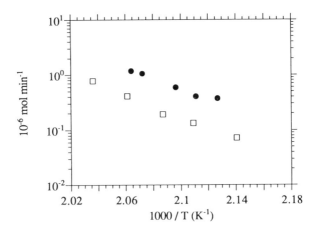

Figure 3 - Evaporation rates of Ba(thd)$_2$ as a function of the source temperature for two solid source geometries (H: open squares; V: solid circles).

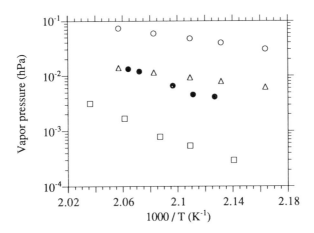

Figure 4 - Ba(thd)$_2$ vapor pressure as a function of temperature. Open squares and full circles: calculated from evaporation rates in sources H and V respectively. Open triangles: data from [11]. Open circles: data from [12].

According to Temple [10] an optimized source should saturate the carrier gas with a partial pressure of precursor equal to its vapor pressure at the source temperature. Assuming full saturation of the carrier gas one can extract vapor pressure values from evaporation rates [10]. This is done in Figure 4 where the hypothetical pressure values are calculated using the actual total pressures in the sources and the carrier gas flow rates (H: 50 sccm, 8.38 hPa; V: 40 sccm, 18.4 hPa). The large scattering in the vapor pressure values [11,12] of Ba(thd)$_2$, probably related to different preparation and/or storage conditions, makes saturation difficult to assess.

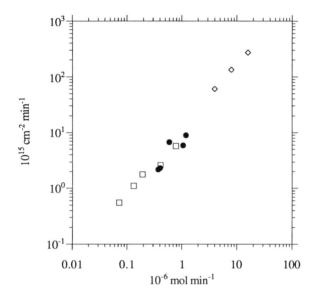

Figure 5 - Ba deposition rate versus transport rate for the 3 sources investigated (H: open squares, V: solid circles, P: open diamonds).

The nebulization of a liquid source, maintained at room temperature, lifts the restriction on the Ba source temperature and allows therefore very high deposition rates. This is clearly shown in Figure 5 where the Ba deposition rate is plotted as a function of the transport rate of the precursor for the three different sources compared here (note the log-log scale). Comparing sources H and V, the increased evaporation efficiency results in a higher deposition rate, this for a similar precursor's temperature (see Figure 3). In reactor P the evaporation rate is now no longer controlled by the source temperature but by the solution concentration and by the feed rate of the solution which depends on the carrier gas flow and on the ultrasonic excitation power (typical concentrations and feed rate are: 0.02 mol (Ba(thd)$_2$) l^{-1} and 0.4 ml / min). The cold liquid source makes YBa$_2$Cu$_3$O$_{7-x}$ growth rates as high as 15 µm/h possible. This process still needs optimization but this opens the way for thick film deposition. The properties of thin films (≤ 0.5 µm) deposited at high growth rates (6 µm/h) are good [13]: $T_c = 89$ K, J_c (0 T, 77 K) = 10^6 A cm^{-2} on perovskite substrates.

CONCLUSION

Thin $YBa_2Cu_3O_{7-x}$ films deposited by MOCVD on MgO from solid sources have properties ($T_c = 90$ K, J_c (0T, 77 K) = $2\ 10^6$ A cm^{-2}, R_s (77 K, 87 GHz) = 25 mΩ) comparable to the layers obtained by physical vapor deposition techniques. The growth rate is, however, rather low (≤ 0.5 µm/h). The use of a cold liquid source gives rise to high growth rates (up to 15 µm/h). A detailed study of the properties of the thick films obtained in these conditions remains to be done.

ACKNOWLEDGMENTS

This study has been funded by an EEC ESPRIT contract, "SUPERMICA" n° 6113.

REFERENCES

1. H. Yamane, H. Kurosawa, T. Hirai, K. Watanabe, H. Iwasaki, N. Kobayashi, Y. Muto, Supercond. Sci. Technol. **2**, 115 (1989).
2. J. Hudner, O. Thomas, E. Mossang, P. Chaudouet, F. Weiss, D. Boursier, J.P. Senateur, M. Östling, A. Gaskov, J. Appl. Phys. **74**, 4631 (1993).
3. S.B. Turnispeed, R.M. Barkley, R.E. Sievers, Inorg. Chem. **30**, 1164 (1991).
4. W.J. Lackey, W.B. Carter, J.A. Hanigofsky, D.N. Hill, E.K. Barefield, G. Neumeier, D.F. O'Brien, M.J. Shapiro, J.R. Thompson, A.J. Green, T.S. Moss, R.A. Jake, K.R. Efferson, Appl. Phys. Lett. **56**, 1175 (1990).
5. R. Hiskes, S. Dicarolis, J. Young, S.S. Laderman, R.D. Jacowitz, R.C. Taber, Appl. Phys. Lett. **59**, 606 (1991).
6. S. Matsuno, F. Uchikawa, S. Utsunomyia, S. Nakabayashi, Appl. Phys. Lett. **60**, 2427 (1992).
7. O. Thomas, A. Pisch, E. Mossang, F. Weiss, R. Madar, J.P. Senateur, J. Less Comm. Met. **164&165**, 444 (1990).
8. C. Dubourdieu, G. Delabouglise, O. Thomas, J.P. Senateur, D. Chateigner, P. Germi, M. Pernet, S. Hensen, S. Orbach, G. Müller, presented at EUCAS'93 - European Conference on Applied Superconductivity, October 4-8 1993, Göttingen.
9. F. Weiss, K. Fröhlich, R. Haase, M. Labeau, D. Selbmann, J.P. Senateur, J. de Physique IV C3 **3**, 321 (1993).
10. D. Temple, A. Reisman, J. Electrochem. Soc.**136**, 3525 (1989).
11. E. Waffenschmidt, J. Musolf, M. Heuken, K. Heime, J. of Supercond. **5**, 119 (1992).
12. P. Tobaly, G. Lanchec, J. chem. therm. **25**, 503 (1993).
13. F. Weiss, K. Fröhlich, R. Haase, M. Labeau, D. Selbmann, J.P. Senateur, O. Thomas, presented at EUCAS'93 - European Conference on Applied Superconductivity, October 4-8 1993, Göttingen.

STRUCTURE AND KINETICS STUDY OF MOCVD LEAD OXIDE (PbO) FROM LEAD BIS-TETRAMETHYLHEPTADIONATE (Pb(thd)$_2$)

WARREN C. HENDRICKS, SESHU B. DESU, AND CHING YI TSAI
Department of Materials Science and Engineering, Virginia Polytechnic Institute & State University, Blacksburg, VA 24061

ABSTRACT

Lead bis-tetramethylheptadionate (Pb(thd)$_2$) is an extremely useful precursor for the preparation of lead-based thin films such as PZT, lead titanate, etc. In this paper, lead oxide was deposited from Pb(thd)$_2$ in a hot-walled CVD reactor using oxygen as a reactive species and diluent gas. XRD and SEM were used to determine the structure of the material deposited by the CVD process. The CVD process consistently produced the monoxide of lead which was found to consist of a mixture of orthorhombic PbO with small tetragonal PbO platelets. TEM was used to determine the orientation of the individual platelets which was found to be consistently normal to the <201> family of zone axes. Deposition rates were determined and simulated using an FEM computer model to determine the rate constants for the overall deposition process.

INTRODUCTION

Metallorganic Chemical Vapor Deposition (MOCVD) is an important technique for the synthesis of thin films for the microelectronic industry. Many of these films which are members of the lead-based perovskites including lead titanate, PZT, PLZT, etc. are finding application as ferroelectric, pyroelectric, piezoelectric and optoelectronic devices.

Lead bis-tetramethylheptadionate (Pb(thd)$_2$) has been found to be one of the best candidates for a lead-based metallorganic precursor for the CVD process [1,2]. Some of the important properties considered in selecting a precursor include volatility, thermal and environmental stability, decomposition behavior, ease of handling, and safety considerations[1]; Pb(thd)$_2$ is adequate or superior in all of these categories. Thus, a knowledge of the kinetics and deposition characteristics of the most promising metallorganic lead source, namely Pb(thd)$_2$, is extremely useful for these applications and for this reason was chosen for our study.

In a typical hot-walled CVD tube furnace, it is often not practical to achieve a perfectly uniform temperature profile across the length; however, it has been shown that despite this nonuniform profile, it is possible to extract useful data regarding the kinetics of the overall deposition process using experimentally measured deposition rates across the length of the reactor in conjunction with a Finite Element Method (FEM) computer program to model the process.

EXPERIMENTAL PROCEDURE

Approximately 1.0 g of fresh Pb(thd)$_2$ precursor was placed inside a stainless steel bubbler at the start of each deposition. Using customized heaters, the bubbler and its contents were heated to the bubbler temperature, T_B to within ±1°C; in order to ensure sufficiently high vapor pressure the bubbler temperature was fixed at 165 °C. Oxygen was used both as the diluent and reactive species and was varied between 250 and 750 sccm; nitrogen was sent through the bubbler as the carrier gas and was fixed at 75 sccm. For each deposition experiment, eight polycrystalline alumina wafer pieces or sapphire disks were placed at equal intervals along a thirty centimeter length centered about the hot zone of the reactor. Deposition was continued for 30 minutes before closing off the bubbler and allowing the furnace to cool.

Deposition rates were determined by weight measurements. X-ray diffraction (XRD) and Transmission Electron Microscopy (TEM) were used to determine the crystal structure of the films; Scanning Electron Microscopy (SEM) was used to investigate the morphology.

RESULTS AND DISCUSSION

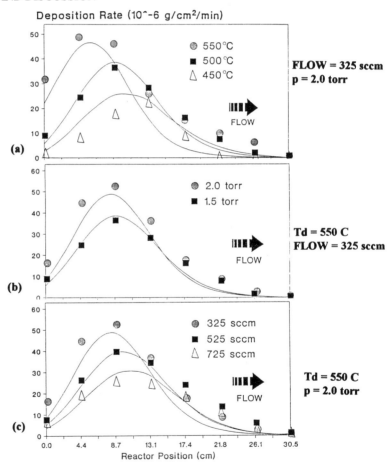

Figure 1: Simulated (curves) and experimental (symbols) data for deposition rate profiles demonstrate the effect of: **(a)** temperature, **(b)** pressure, and **(c)** total flow rate

Upon examining the deposition profiles that resulted from variations in the three experimental variables, namely temperature, pressure, and flow rate in Figs. 1(a), (b), and (c), respectively, it becomes obvious that no meaningful study can be conducted by examining the deposition rate at only a single point. The deposition rate for each point in the reactor depends not only on the temperature and pressure in the immediate vicinity but also on the flux of reactant to that point. The precursor reactant flux, in turn, depends on the overall amount of depletion occurring in earlier parts of the reactor; this is complicated considerably by the nonuniform temperature profile. Therefore, it is impossible to propose a single equation which will describe the deposition rate as a function of position, temperature, pressure, etc. Consequently, it becomes necessary to introduce a model using Finite Element Methods (FEM) which can be solved by an appropriate computer program.

FEM Modeling of the MOCVD Process

There are many possible side reactions that occur during the CVD process; however, the model used here assumes that the final surface reaction occurring on the substrate is the rate limiting step in the overall process. Thus, the reaction rate constant, k, is identical to the surface reaction rate constant, k_s, which can be expressed as follows:

$$k = k_s = k_0 \exp\left(\frac{-E_a}{RT}\right) \quad (1)$$

where k_0 is the preexponential constant and E_a is the activation energy for the process; first order reaction kinetics are being assumed.

For each given point in the reactor the total mass entering and leaving that point by diffusion, convective transport and surface reaction must be accounted for; hence the following continuity equation is employed:

$$\left.\frac{dP}{dt}\right|_x = -D\frac{d^2P}{dx^2} + U\frac{dP}{dx} + \left(\frac{k_s}{2\pi r}\right)P = 0 \quad (2)$$

In eq. 2 above, P respresents the partial pressure of $Pb(thd)_2$, x represents the distance along the reactor (cm), D is the diffusion coefficient of the precursor (cm/s^2), T is the temperature (K), U is the convective transport coefficient (cm/s) and $2\pi r$ is the circumference of the reactor (cm). The deposition rate at any given point in the reactor is the product of the local rate constant and reactant concentration. Assuming that oxygen is in excess, only the concentration of the $Pb(thd)_2$ precursor need be considered in determining the expression for the deposition rate, D.R.:

$$D.R. = k_s\, C_{Pb(thd)2} = k_0 \exp\left(\frac{-E_a}{RT}\right)\left(\frac{P_{Pb(thd)2}}{RT}\right) \quad (3)$$

It is possible to combine the results of Eqs. 2 & 3 simultaneously by an iterative technique to determine the deposition rate, D.R. as a function of position, x, once we have the appropriate kinetic parameters, k_0 and E_a. Alternatively, the kinetic parameters can be obtained using several experimentally determined deposition profiles; this was the approach taken here. The equations can not be solved explicitly, as mentioned earlier, because of the nonuniform temperature profile. Consequently, a sixteen element finite element grid was used with each substrate position representing a node in the grid. For each experiment, we can find the temperature and deposition rate at each node. These data can be entered into the finite element program we have developed which will determine the deposition profile for the specified kinetic parameters. By choosing suitable kinetic parameters which are fixed for all experiments once chosen, deposition profiles can be obtained which will fit all of the experimental data quite well. The preexponential rate constant, k_0, was found to be 33 g/cm^2/min (0.15 mol/cm^2/min) while the activation energy was found to be 82 kJ/mol. By substituting these values into Eq. 3 it is possible to obtain an expression for the deposition rate dependent on only the local partial pressure and temperature.

Crystal Structure and Morphology

In every case, regardless of oxygen partial pressure, only the monoxide form of lead, PbO, was observed; litharge (tetragonal α-PbO) was found to be deposited at the lower temperatures while a mixture of litharge and massicot (orthorhombic β-PbO) was deposited at the higher temperatures. Metastable massicot was found in detectable amounts on substrates deposited at temperatures as low as 420 °C despite the fact that the equilibrium temperature for the α and β forms to coexist is 550 °C [3]. The XRD patterns for a typical set of films from a single batch deposited at T_d = 500 °C is given in Fig. 2.

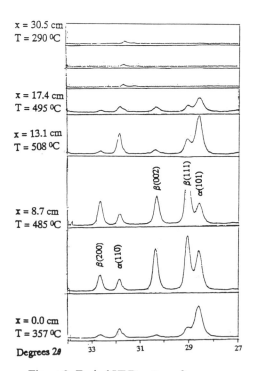

Figure 3: SEM micrograph of typical platelet structure observed

Figure 2: Typical XRD patterns for samples deposited at different locations, x, in the reactor for which Td = 500 C; local temperatures, T, are also given.

An SEM photo of the typical lead oxide morphology is shown in Fig. 3. The morphology of the deposited PbO was found to be dependent on the temperature and deposition rate. A smooth film was observed only for very low deposition rates; as the deposition rate became higher a structure consisting of randomly oriented plates was found to occur. These plates became significantly larger as the temperature and deposition rate was increased. The presence of these lath-shaped crystals is indicative of a large anisotropy in growth rate kinetics with respect to crystallographic orientation.

In general, the formation of well defined crystal planes during the growth of a crystal results from substantially slower growth rates in the direction of these planes. Thermodynamically, we would predict the (001) planes to have the lowest growth rate since they have the lowest surface energy due to the weak van der Waals bonding in this plane[4]. This would consequently determine the crystal habit; the simulated crystal structure of these nearly close-packed planes is illustrated in Fig. 4. For crystals grown hydrothermally, this is indeed reported to be the case [3, 5].

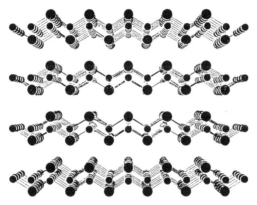

Figure 4: Simulated crystal structure of α-PbO looking in the <010> direction

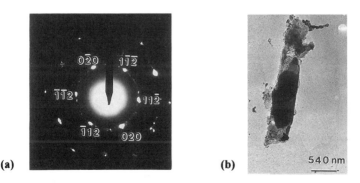

Figure 5: (a) EDP of α-PbO platelet showing <201> zone axis; (b) corresponding bright-field image

To determine the orientation of the plates grown in this experiment, the coating from one of the sapphire substrates which was found to consist of large PbO plates was scraped free and prepared for TEM investigation. Electron diffraction patterns were obtained for several platelets; in all cases, the phase of the platelets was tetragonal and the orientation was found to be normal to the <201> family of zones. Electron diffraction patterns and corresponding bright field photographs are given for a typical platelets in Fig. 5. While <201> is probably not the thermodynamically favored orientation for tetragonal PbO platelets, it seems that in the particular CVD process studied here this growth morphology was favored from a kinetic standpoint; interestingly, <201> oriented island growth was reported elsewhere for the growth of lead oxide on the (111) surface of metallic lead by the incorporation of oxygen[6].

SUMMARY AND CONCLUSIONS

Pb(thd)$_2$ precursor was used to deposit PbO over a range of experimental conditions and was found to conform well to the Finite Element Model that had been proposed to simulate the deposition rate profile. The films deposited on sapphire and alumina were found from XRD to consist of a mixture of tetragonal and orthorhombic lead oxide, the proportion of which depended on both the deposition temperature and the deposition rate. Using SEM, it was observed that the tetragonal form of PbO tends to grow in the form of platelets randomly oriented with respect to the substrate which were about 1 micron in diameter and about 100 Å or less in thickness. The orientation of the each individual platelet was confirmed by TEM electron diffraction patterns to be the normal to the <201> zone.

ACKNOWLEDGEMENTS

Special thanks to Chien Peng for assistance in the design and operation of the CVD equipment and May Nyman for the use of static decomposition data on Pb(thd)$_2$. I also wish to thank C. Chiu for help in taking the SEM photographs and in interpreting the TEM electron diffraction patterns. Gang Chen and J. K. Chen have provided invaluable assistance with their help in the TEM work and I would also like to acknowledge Prof. G. V. Gibbs for the use of his computer program for generating the crystal structure images.

REFERENCES

1. Peng, C. H., PhD Thesis, Virginia Polytechnic Institute & State University, 1992.

2. R. C. Mehrotra, R. Bohra, D. P. Gaur, <u>Metal Beta-Diketonates and Allied Derivatives</u>, Academic Press, Inc., London, pp. 58-69, 1978.

3. C. J. M. Rooymans and W. F. Th. Langenhoff, Journal of Crystal Growth, **3** (4) (1968).

4. B. G. Hyde, Sten Andersson, <u>Inorganic Crystal Structures</u>, John Wiley & Sons, Inc., New York, 1989.

5. A. A. Chernov, <u>Modern Crystallography III, Crystal Growth</u>, Springer Verlag, New York, p. 383, 1984.

6. J. W. Matthews, C. J. Kircher, and R. E. Drake, Thin Solid Films, **47**, 95 (1977).

LIQUID DELIVERY OF LOW VAPOR PRESSURE MOCVD PRECURSORS

R. A. GARDINER, P. C. VAN BUSKIRK and P. S. KIRLIN
Advanced Technology Materials Inc., 7 Commerce Dr., Danbury, CT 06810

ABSTRACT

A significant limitation in MOCVD processing of advanced materials is the ability to deliver low volatility precursors, in a repeatable fashion, to the deposition reactor. Apparatus for delivering low volatility precursors in gaseous form, wherein a precursor source liquid is flash vaporized at elevated temperature has been developed. Stable, reproducible delivery of liquid precursors for TiN and Ta_2O_5 has been demonstrated. For multicomponent ceramic material systems simultaneous delivery of multiple cation species via a single source liquid has been achieved. The effect of flash vaporization on precursors for PZT was investigated. Pb, Zr and Ti precursors were dissolved in an organic medium with a defined cation ratio. The source solution was vaporized, the precursors transported as vapor then collected and analyzed by ^1H and ^{13}C nmr. No decomposition of the precursors was observed post vaporization and source solution stoichiometry was maintained in the collected material. The use of this flash vaporization technique has already been particularly successful for MOCVD of $BaTiO_3$, $MgAl_2O_4$, $YBa_2Cu_3O_{7-x}$, YSZ, $LaSrCoO_3$ and Cu metal.

INTRODUCTION

Many of the potential application of materials for 'next generation' electronic materials require their use in thin film, coating, or layer form. Next generation materials include the recently discovered high temperature super conducting (HTSC) materials including $YBa_2Cu_3O_{7-x}$[1] and BiSrCaCuO.[2] Barium titanate $(BaTiO_3)$[3] and barium strontium titanate $(Ba_xSr_{1-x}TiO_3)$[4] have been identified as ferroelectric and photonic materials with unique and potentially very useful properties. Barium strontium niobate $(Ba_xSr_{1-x}Nb_2O_6)$[5] is a photonic material whose index of refraction changes as a function of electric field and also as a function of the intensity of light upon it. Lead zirconate titanate $(PbZr_{1-x}Ti_xO_3)$ (PZT) is a ferroelectric material whose properties are extremely important.[6,7,8] Ferroelectric ceramics such as PZT are highly acclaimed materials due to their applications in electro-optic, piezoelectric and pyroelectric devices. In bulk form they have been widely used as piezoelectric elements and in capacitors because of their high dielectric constant. Non-volatile random access computer memories (RAM) are likely the most significant application for ferroelectric materials in today's market. Such materials are ideal in this role since they retain polarization (i.e. charge storage) even after power is removed, and they can be switched rapidly (in hundreds of nanoseconds). Ferroelectric memories are particularly attractive for military and space applications because they are inherently radiation hard. Another highly promising application of ferroelectric thin films is in pyroelectric infrared sensors.

Tantalum oxide (Ta_2O_5)[9] is seeing expanded use in the microelectronics industry as a promising dielectric for storage capacitors in scaled down memory cells and as a gate insulator of metal-oxide-semiconductor devices. Titanium nitride (TiN)[10] is of use a low resistance contact and diffusion barrier in interconnect metallization schemes. For the described applications the materials have to be deposited in thin film or layer form. The use of 'next generation' electronic materials in microelectronics applications has been limited by materials processing and compatibility problems. As a result, it has proven difficult to deposit high quality thin films at temperatures compatible with standard processing technology. Improvements in thin film technology for these materials are therefore in great demand.

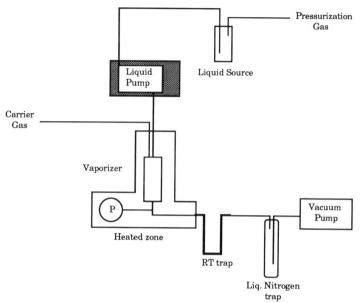

Figure 1. Liquid delivery system schematic

Metaloorganic chemical vapor deposition (MOCVD) is a particularly attractive method for forming thin films or layers because it is readily scaled up to production runs and because the electronic industry has a wide experience and an established equipment base in the use of MOCVD technology which can be applied to new MOCVD processes. The problem of controlled delivery of liquid MOCVD reagents into deposition reactors has previously been addressed.[11] The delivery of reagents into the deposition chamber in vapor form is accomplished by providing the reagent in a liquid form, neat or solution, and flowing the reagent liquid from a liquid reservoir onto a flash vaporization matrix structure which is heated to a temperature sufficient to flash vaporize the reagent source liquid. A carrier gas may optionally be flowed by the flash vaporization matrix structure to form a carrier gas mixture containing the flash vaporized reagent source liquid.

This work describes apparatus and a method for delivering low volatility reagents to an MOCVD tool. Precursors for the Ta_2O_5, TiN and PZT material system were delivered by a technique developed at Advanced Technology Materials using room temperature liquid delivery of the precursors to a flash vaporization zone using a system represented in Figure 1

EXPERIMENTAL

PZT Precursor Synthesis and Purification

Synthesis and handling of all organometallics was performed under nitrogen atmosphere using standard Schlenk techniques or in a Vacuum Atmospheres inert atmosphere glove box. Cryogenic nitrogen was purified using Ridox® and molecular sieves. Lead bis(2,2,6,6-tetramethyl-3,5-heptandionate) [$Pb(thd)_2$] and zirconium tetrakis(2,2,6,6-tetramethyl-3,5-heptandionate) [$Zr(thd)_4$] (Strem Chemicals) were purified by reduced pressure sublimation. Titanium bis(isopropoxide) bis(2,2,6,6-tetramethyl-3,5-heptandionate) [$Ti(O^iPr)_2(thd)_2$] was synthesized at ATM and also purified by reduced pressure sublimation. Titanium isopropoxide (Strem Chemicals) was distilled prior use. Tantalum ethoxide and tetrakis(diethylamido)titanium (Prochem) were filtered (0.5μ) and used as received. Pentane was refluxed over purple sodium benzophenone ketyl and distilled prior to use. 2,2,6,6-tetramethyl-3,5-heptanedione (Strem Chemicals), isopropanol (Aldrich), tetrahydrofuran (Aldrich) and tetraethylene glycol dimethyl ether (Aldrich), were dried over 4A molecular sieves. Nuclear magnetic resonance spectra were obtained on an IBM WP200SY spectrometer. ^{1}H and ^{13}C NMR spectra were measured at 200 MHz with an IBM WP200SY spectrometer. Chemical shifts are reported in ppm on a d scale with residual resonances from the NMR solvents serving as internal standards [^{1}H (C_6D_6 δ = 7.15 ppm), ^{13}C (C_6D_6 δ 128.0 ppm)].

Preparation of Ti(O^{i}Pr)$_2$(thd)$_2$. Titanium isopropoxide (54.0 g, 0.19 moles) was added, via syringe to a stirring solution of H(thd) (69.0 g, 0.38 moles) in 250 ml of freshly distilled pentane. A slight exotherm was noted on addition of the titanium isopropoxide. The pentane was then removed by a N_2 purge to leave a white solid. The crude material was then purified by reduced pressure sublimation at 150°C and 10^{-2} Torr to yield a white solid, 75.0 g, 75%. ^{1}H NMR (C_6D_6) δ 1.06 (s, 18H), 1.26 (s, 18H), 1.34 (d, 12H), 4.95 (m, 2H), 5.85 (s, 2H). ^{13}C NMR δ 25.7 (CH(CH$_3$)$_2$), 28.6 (C(CH$_3$)$_3$), 40.6 (C(CH$_3$)$_3$), 77.6 (CH(CH$_3$)$_2$), 92.3 (CH), 200.5 (C=O).

Purification of Pb(thd)$_2$ and Zr(thd)$_4$. 25.0g of Pb(thd)$_2$ was purified by reduced pressure sublimation at 140°C and 10^{-2} Torr to yield a white, crystalline solid, 24.1g, 96%. ^{1}H NMR (C_6D_6) δ 1.23 (s, 36H), 5.66 (s, 2H). ^{13}C NMR δ 28.8 (C(CH$_3$)$_3$), 42.0 (C(CH$_3$)$_3$), 93.1 (CH), 198.5 (C=O). 25.0g of Zr(thd)$_4$ was purified by reduced pressure sublimation at 140°C and 10^{-2} Torr to yield a white, crystalline solid, 24.9g, 99%. ^{1}H NMR (C_6D_6) δ 1.24 (s, 36H), 5.89 (s, 2H). ^{13}C NMR δ 28.7 (C(CH$_3$)$_3$), 40.5 (C(CH$_3$)$_3$), 91.7 (CH), 196.9 (C=O).

Investigation of Pb(thd)$_2$, Zr(thd)$_4$ and Ti(O^{i}Pr)$_2$(thd)$_2$ in solution.

For single source liquid delivery of PZT precursors to be a commercially viable MOCVD process the precursors chosen must not undergo ligand exchange when mixed (either as liquid or vapor) which may produce nonvolatile species. ^{1}H and ^{13}C NMR spectroscopy were used to analyze mixtures of Pb(thd)$_2$, Zr(thd)$_4$ and Ti(O^{i}Pr)$_2$(thd)$_2$ in solution. The spectra of individual compounds as compared with that of a mixture to identify if any chemical reactions (e.g. ligand exchange) had occurred. The NMR time scale generally refers to lifetimes of the order of 1 to 10^{-6} seconds. If the exchange is greatly in excess of the NMR time scale the mixture spectrum will be a superposition of the individual component spectra. If the exchange is fast with respect to the NMR time scale the mixture spectra will be considered as due to a single species whose parameters (i.e. chemical shifts) are the relevant averages of those for the individual components. ^{1}H and ^{13}C NMR proton shifts for a mixture of Pb(thd)$_2$, Zr(thd)$_4$ and Ti(O^{i}Pr)$_2$(thd)$_2$ are shown in Table I.

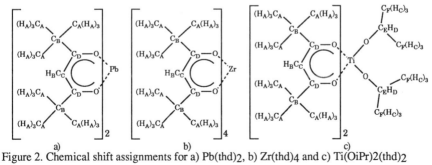

Figure 2. Chemical shift assignments for a) Pb(thd)$_2$, b) Zr(thd)$_4$ and c) Ti(OiPr)2(thd)$_2$

Table I. ^{1}H and ^{13}C chemical shifts of a mixture of Pb(thd)$_2$, Zr(thd)$_4$ and Ti(O^{i}Pr)$_2$(thd)$_2$

^{1}H NMR		^{13}C NMR	
Chemical Shift in solution (d ppm)	Identity	Chemical Shift in solution (d ppm)	Identity
1.07	H$_A$ - Ti(O^{i}Pr)$_2$(thd)$_2$	198.6	C$_D$ - Pb(thd)$_2$
1.23	H$_A$ - Pb(thd)$_2$	196.9	C$_D$ - Zr(thd)$_4$
1.24	H$_A$ - Zr(thd)$_4$	93.1	C$_C$ - Pb(thd)$_2$
1.27	H$_A$ - Ti(O^{i}Pr)$_2$(thd)$_2$	92.4	C$_C$ - Ti(O^{i}Pr)$_2$(thd)$_2$
1.36	H$_C$ - Ti(O^{i}Pr)$_2$(thd)$_2$	91.8	C$_C$ - Zr(thd)$_4$
1.38	H$_C$ - Ti(O^{i}Pr)$_2$(thd)$_2$	77.7	C$_E$ - Ti(O^{i}Pr)$_2$(thd)$_2$
multiplet, 4.97	H$_D$ - Ti(O^{i}Pr)$_2$(thd)$_2$	42.0	C$_B$ - Pb(thd)$_2$
5.66	H$_B$ - Pb(thd)$_2$	40.6	C$_B$ - Ti(O^{i}Pr)$_2$(thd)$_2$
5.87	H$_B$ - Ti(O^{i}Pr)$_2$(thd)$_2$	40.5	C$_B$ - Zr(thd)$_4$
5.90	H$_B$ - Zr(thd)$_4$	28.8	C$_A$ - Pb(thd)$_2$
		28.7	C$_A$ - Zr(thd)$_4$
		28.3	C$_A$ - Ti(O^{i}Pr)$_2$(thd)$_2$
		25.7	C$_F$ - Ti(O^{i}Pr)$_2$(thd)$_2$

The assignment of chemical shifts is given in Figure 2 and should be compared to the shifts of the pure compounds individually in solution described in the subsection 'PZT Precursor Synthesis and Purification'. The chemical shifts of the PZT precursors in solution are identical with those for the individual compounds in solution. No ligand exchange occurs under these experimental conditions.

PZT precursor Liquid Delivery

Handling of the PZT precursors was performed under nitrogen atmosphere using standard Schlenk techniques or in a Vacuum Atmospheres inert atmosphere glove box. The solids were dissolved in a mixture of thf, isopropanol and tetraethylene glycol dimethyl ether in a volumetric ratio of 8 : 2 : 1. The solutions were then filtered (0.45µ) and introduced into the liquid delivery system reservoir via syringe. Liquid delivery experiments were carried out using a solution of total cation concentration of 0.40 moles/liter (ratio Pb:Zr:Ti = 1 : 5 : 5) or 0.57 moles/liter (ratio Pb:Zr:Ti = 2 : 1 : 1). Liquid flow rates of 0.50 and 0.90 ml/min were used.

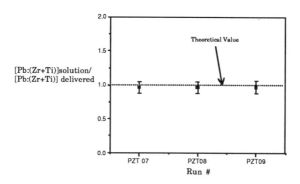

Figure 3. Graph of accuracy and repeatability of PZT precursor delivery

The vaporization temperature was 200°C and carrier gas flow rates of 200 to 400 sccm of nitrogen. The system pressure varied between 1.6 and 6.9 torr. The vaporized precursors were collected in the room temperature trap shown in Figure 1. The cation ratio [Pb/(Zr+Ti)] of the trapped solid was then measured by ICP analysis (Galbraith). This ratio was compared to that in the starting solution. The results are plotted in Figure 3.

The graph shows that across a number of runs the ratio of [Pb/(Zr+Ti)] solution divided by [Pb/(Zr+Ti)] collected was identical, and well within the experimental error of the theoretical value.

Tantalum Ethoxide Liquid Delivery

Tantalum ethoxide was filtered into the liquid delivery system reservoir via syringe. The pure liquid was delivered to the vaporization zone at a rate of 0.10 ml/min. The vaporization temperature was 165°C and a carrier gas flow rate of 50 sccm was used. The system base pressure was 1.1 torr. As the tantalum ethoxide was introduced to the system the pressure increased stabilizing as the delivery rate equilibrated. The variation in system pressure gave an indication of the stability of the tantalum ethoxide vaporization and vapor flow. The stability of tantalum ethoxide flow versus time is shown in Figure 4. Tantalum ethoxide was delivered consistently with a variation of ± 1.5% of set point.

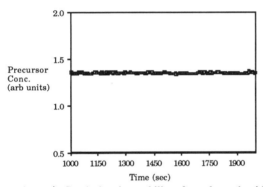

Figure 4. Graph showing stability of tantalum ethoxide delivery

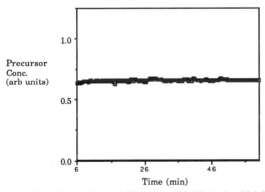

Figure 5. Graph showing stability of tetrakis(diethylamido)titanium delivery

Tetrakis(diethylamido)titanium Liquid Delivery

Tetrakis(diethylamido)titanium was filtered into the liquid delivery system reservoir via syringe. The pure liquid was delivered to the vaporization zone at a rate of 0.05 ml/min. The vaporization temperature was 165°C and no carrier gas was used. The system base pressure was 0.1 torr. The variation in system pressure gave an indication of the stability of the tetrakisdiethylamido titanium vaporization and vapor flow. The stability of tetrakis(diethylamido)titanium flow versus time is shown in Figure 5. Tetrakis(diethylamido)titanium was delivered consistently with a variation of ± 1.1% of set point.

CONCLUSIONS

The stable delivery of tantalum ethoxide and tetrakis(diethylamido)titanium have been demonstrated using a liquid delivery technology developed at Advanced Technology Materials. The technology can also be applied to the delivery of solid MOCVD precursors dissolved in solution. The repeatable delivery of lead, zirconium and titanium compounds has been accomplished using a single solid source solution technique. The apparatus and method described in this work have also been used to deliver precursors in the MOCVD of a number of materials systems including $BaTiO_3$,[12] $MgAl_2O_4$,[13] $YBa_2Cu_3O_{7-x}$,[14] YSZ,[12] $LaSrCoO_3$[15] and Cu metal.[16]

REFERENCES

[1] J. Zhang, R. A. Gardiner and P. S. Kirlin, Mater. Res. Soc. Proc. **275**, San Francisco, CA, 1992, pp419-424.

[2] T. Sugimoto, N. Kubota, Y. Shiohara and S. Tanaka, Appl. Phys. Lett **60** (11), 1387 (1992).

[3] H. Funakobo, Y. Inagaki, K. Sinosaki and N. Miautani, j. Chem. Vap. Dep. **1**, 73 (1992).

[4] K. Koyama, T. Sakuma, S. Yamamichi, H. Watanabe, H. Aoki, S. Ohya, Y. Miyasaka, T. Kikkawa, IEDM Technical Digest, 823-826 (1991).

[5] Y. Xu, C. J. Chen, R. Xu and J. D. Mackenzie, Phys. Rev. B **44** (1), 35 (1991).

[6] K. Iijima, R. Takayama, Y. Tomita, and I. Ueda, J. Appl. Phys., **60** (8), 2914 (1986).

[7] B. A. Tuttle, J. A. Voight, D. C. Goodnow, D. L. amppa, T. J. Headley, M. O. Eatough, G. Zender, R. D. Nasby, and S. M. Rodgers, J. Am. Ceram. Soc. **76** (6), 1537 (1993).

[8] M. Okada, K. Tominaga, T. Araki, S. Katayama, and Y. Sakashita, J. J. Appl. Phys. **29** (4), 718 (1990).

[9] H. Treichel, A. Mitwalsky, N. P. Sandler, D. Tribula, W. Kern, and A. P. Lane, Adv. Mat. Opt. Electr. **1**, 299 (1992).

[10] N. Yoshikawa and A. Kikuchi, J. Cryst. Growth **130**, 578 (1993)

[11] Peter S. Kirlin, Robin L. Binder and Robin A. Gardiner, U.S. Patent No. 5,204,314 (20 April 1993).

[12] P. C. VanBuskirk, R. A. Gardiner, P. S. Kirlin and S. Nutt, J. Mater. Res. **7** (3), 542 (1992).

[13] G. T. Stauf, P. C. VanBuskirk, P. S. Kirlin, W. P. Kosar and S. Nutt, presented at the 1993 MRS Spring Meeting, San Francisco, CA, 1993 (unpublished).

[14] J. Zhang, R. A. Gardiner, P. S. Kirlin, R. Boerstler, and J. Steinbeck, J Appl. Phys. Lett., **61**, 2884 (1992).

[15] J. Zhang, Guang-Ji Cui, D. Gordon, P. C. VanBuskirk, and J. Steinbeck,, presented at the 1993 MRS Spring Meeting, San Francisco, CA, 1993 (unpublished).

[16] G. A. Petersen, T. R. Omstead, P. M. Smith and M. F. Gonzalez, presented at the 1993 Electrochemical Society Meeting, Honolulu, Hawaii, May 1993.

SYNTHESIS AND DECOMPOSITION OF A NOVEL CARBOXYLATE PRECURSOR TO INDIUM OXIDE

ALOYSIUS F. HEPP,* MARIA T. ANDRAS,*,‡ STAN A. DURAJ,** ERIC B. CLARK,* DAVID G. HEHEMANN,***,‡‡ DANIEL A. SCHEIMAN,† AND PHILLIP E. FANWICK††
*NASA Lewis Research Center, Photovoltaic Branch, M.S. 302-1, Cleveland, OH 44135
**Department of Chemistry, Cleveland State University, Cleveland, OH, 44115
***School of Technology, Kent State University, Kent, OH 44242
†Sverdrup Technology, Inc., 2001 Aerospace Parkway, Brook Park, OH 44142
††Department of Chemistry, Purdue University, West Lafayette, IN 47907

ABSTRACT

Reaction of metallic indium with benzoyl peroxide in 4-methylpyridine (4-Mepy) at 25 °C produces an eight-coordinate mononuclear indium(III) benzoate, $In(\eta^2-O_2CC_6H_5)_3(4-Mepy)_2 \cdot 4H_2O$ (I), in yields of up to 60%. The indium(III) benzoate was fully characterized by elemental analysis, spectroscopy, and X-ray crystallography; (I) exists in the crystalline state as discrete eight-coordinate molecules; the coordination sphere around the central indium atom is best described as pseudo-square pyramidal. Thermogravimetric analysis of (I) and X-ray diffraction powder studies on the resulting pyrolysate demonstrate that this new benzoate is an inorganic precursor to indium oxide. Decomposition of (I) occurs first by loss of 4-methylpyridine ligands (100°-200°C), then loss of benzoates with formation of In_2O_3 at 450°C. We discuss both use of carboxylates as precursors and our approach to their preparation.

INTRODUCTION

Our interest in oxygen-containing gallium and indium complexes derives from efforts to produce precursors for deposition of thin-film materials for solar cell fabrication [1]. Oxides are deposited or grown on solar cells to provide electrical insulation, to decrease surface recombination, or to produce anti-reflective coatings [2]. A major issue surrounding technological applications of InP is the deposition of insulators on the surface that are chemically stable with good electrical and interfacial properties. With this goal in mind, we are searching for easy to prepare and handle chemical vapor deposition (CVD) precursors to In_2O_3. Recently, we reported the preparation and characterization of the first main group oxo-centered trimeric carboxylate, $[Ga_3(\mu_3-O)(\mu-O_2CC_6H_5)_6(4-Mepy)_3][GaCl_4]$ [3]. In an attempt to prepare a monomeric indium benzoate, we reacted indium metal with benzoyl peroxide in 4-methylpyridine at 25 °C and obtained $In(\eta^2-O_2CC_6H_5)_3(4-Mepy)_2$ (4-Mepy = 4-methylpyridine) (I) in good yields.

The existence of indium and gallium carboxylates is well-documented [4,5]. Numerous homoleptic polynuclear indium(III) carboxylates, $[In(O_2CR)_3]_x$ (R = H, CH_3, C_2H_5, n-C_3H_7, $(CH_3)_2CH$, and $(CH_3)_3C$) [6], as well as polynuclear organoindium(III) carboxylates, $[R_2In(O_2CR')]_x$ (R= CH_3, C_2H_5, R'=CH_3, C_2H_5 or R = n-C_4H_9, R' = C_2H_5) are known [6-10]; however, there is a void in the literature on analogous indium(III) benzoates. To the best of our knowledge, $Cl_2In(\eta^2-O_2CC_6H_5)py_2$ (py = pyridine), a six-coordinate mononuclear species, is the only structurally characterized indium(III) benzoate complex to date [11]. We describe the synthesis, structure and mass spectral analysis of the eight-coordinate mononuclear indium(III) benzoato complex (I). Thermogravimetric analysis (TGA) of (I) and X-ray diffraction powder (XRD) studies on the resulting pyrolysate demonstrate that this new benzoate is a precursor to indium (III) oxide, In_2O_3.

‡ - National Research Council/NASA Lewis Research Center Resident Research Associate.
‡‡ - Senior Research Fellow/NASA Lewis Research Center Resident Research Associate.

EXPERIMENTAL

All operations of moisture- and air-sensitive materials were performed under an inert atmosphere using standard Schlenk techniques and a double-manifold vacuum line. Solids were manipulated in a Vacuum Atmospheres Co. drybox equipped with a HE-493 dri-train. Solvents were freshly distilled from sodium benzophenone ketyl prior to use. Solutions were transferred via stainless steel cannulae and/or syringes. Indium powder (Aldrich) was used without additional purification. Benzoyl peroxide was deareated under vacuum at room temperature. Elemental analyses were performed by Galbraith Microanalytical Laboratories, Inc. (Knoxville, TN). Thermogravimetric analyses were performed under an atmosphere of nitrogen using a Perkin Elmer TGS-II. Powder X-ray diffraction (XRD) data was collected using monochromated Cu K_α radiation on a Scintag PAD V and a Phillips APD diffractometer. Electron impact mass spectra were recorded on a Finnigan TSQ–45 mass spectrometer. X-ray diffraction data were collected at 20 ± 1 °C on a 0.38 x 0.38 x 0.31 mm crystal using an Enraf-Nonius CAD-4 diffractometer.

$In(\eta^2-O_2CC_6H_5)_3(4\text{-Mepy})_2$ was prepared by reaction of 150-mesh indium powder (0.50 g, 4.35 mmol) and benzoyl peroxide (1.58 g, 6.52 mmol) in 35 mL of 4-methylpyridine at ambient temperature for several days. The mixture was filtered, and the resulting off-white solid washed with three 25 mL aliquots of hexane and dried under vacuum for 2 h. Hexane, 150 mL, was added to the bright yellow filtrate to further precipitate the white solid. The supernatant was decanted, and the white solid was washed with two 25 mL aliquots of fresh hexane and dried under vacuum for 2 h. The solids were combined, recrystallized from 4-methylpyridine/hexane (v/v 40/70) and dried under vacuum for 18 h, the yield is 53-60%. The analytical data was consistant with the single crystal X-ray structure and is detailed in a prior publication [12].

RESULTS AND DISCUSSION

Oxidation of indium metal by benzoyl peroxide in 4-methylpyridine produces the first mononuclear indium(III) benzoate in yields of up to 60% (Eq. 1):

$$2\ In^0\ +\ 3\ (C_6H_5CO)_2O_2 \xrightarrow[25\ °C]{4\text{-Mepy}} 2\ In(\eta^2-O_2CC_6H_5)_3(4\text{-Mepy})_2 \quad (1)$$

$In(\eta^2-O_2CC_6H_5)_3(4\text{-Mepy})_2$ is very stable; it can be stored under an inert atmosphere at room temperature for extended periods of time. TGA studies show that it is thermally stable up to 100 °C, at this temperature loss of 4-methylpyridine occurs. In contrast, pyridine adducts of indium(III) acetate and formate are unstable, losing pyridine slowly at room temperature [13].

Colorless single crystals, suitable for X-ray diffraction studies, were grown by slow interdiffusion of hexane into a 4-methylpyridine solution of (I), the compound crystallized as the tetrahydrate (I)·4H$_2$O, no doubt from trace amounts of water in the coordinating solvent. Crystallographic data are summarized in Table 1, a detailed description of the structural analysis is given in [12]. Single crystal X-ray diffraction analysis reveals that (I)·4H$_2$O is composed of an ordered array of discrete mononuclear eight-coordinate molecules positioned on a crystallographic two-fold rotation axis. The solid-state molecular structure of (I) is shown in Figure 1. The immediate coordination sphere around the central indium(III) atom is best described as a pseudo-square pyramid with each bidentate benzoate assuming a single position. The In atom is bound to six oxygen atoms from three equivalent (vide infra) bidentate benzoate groups. The In-O bond distances range from 2.225(6) to 2.413(5) Å. Within the symmetrically independent benzoato ligand, the In-O bond lengths are not equivalent. The In-O(22) bond length, 2.413(5) Å, is slightly longer (by 0.19 Å) than the In-O(21) bond length, 2.225(6) Å. Unsymmetrical bonding of chelating carboxylate groups to an indium(III) center is not unusual [14].

TABLE 1.
Crystallographic Data for In(η^2-O$_2$CC$_6$H$_5$)$_3$(4-Mepy)$_2$·4H$_2$O.

chemical formula InO$_{10}$N$_2$C$_{33}$H$_{37}$	formula weight 736.49
a = 11.7195(8) Å	space group C2/c (No. 15)
b = 11.995(1) Å	T = 20 °C
c = 25.407(2) Å	λ = 0.71073 Å
β = 94.177(6)°	ρ_{calc} = 1.373 g cm^{-3}
V = 3562.0(8) Å^3	μ(Mo Kα) = 7.04 cm^{-1}
Z = 4	R(F$_o$)^a = 0.059
	R$_w$(F$_o$)^b = 0.079

^{a} R(F$_o$) = Σ||F$_o$| - |F$_c$||/Σ|F$_o$|; ^{b} R$_w$(F$_o$) = [Σw|F$_o$| - |F$_c$|]^2/Σw|F$_o$|^2]$^{1/2}$; w = 1/σ^2(|F$_o$|).

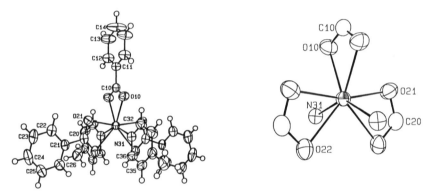

Figure 1. ORTEP drawing of the In(η^2-O$_2$CC$_6$H$_5$)$_3$(4-Mepy)$_2$ molecule showing 50% thermal ellipsoids and the atomic labeling scheme. Compound (1) InO$_6$N$_2$ core on right.

A common reactivity/bonding characteristic of indium(III) complexes is the expansion of the indium(III) atom coordination sphere through polymerization or adduct formation [6-15]. In the case of indium(III) carboxylates (of which (I) is an example), the coordination number of the indium(III) atom generally increases to 6 or 8 via polymerization of the [In(OOCR)$_3$] units - oxygen atoms from adjacent carboxylate molecules bridge the units creating infinite [In(OOCR)$_3$]$_n$ chains. In (I), the presence of the two methylpyridine ligands prevents such polymerization by coordinatively saturating the indium(III) atom, resulting in the formation of the first mononuclear eight-coordinate indium(III) benzoato species. The tendency to associate can be seen by the complete lack of parent ions in mass spectra with 25 and 70 eV ionization. The most intense peaks in the spectra, with m/e values: 105 and 77, correspond to loss of O and CO$_2$ from the benzoate group to form C$_6$H$_5$$^+$ and C$_6$H$_5$(CO)$^+$ ions. While much less intense, the strongest metal-containing ion peaks, in decending intensity order are m/e (for ^{115}In): 357, 313, 115, 269, 435, 479, these peaks can be assigned to, respectively: In(O$_2$CC$_6$H$_5$)$_2$$^+$, C$_6H_5$In(O$_2CC_6H_5$)$^+$, In$^+$, (C$_6H_5$)$_2In^+$, (C$_6H_5$)In(O$_2CC_6H_5$)$_2$$^+$, and In(O$_2CC_6H_5$)$_3$$^+$. While the volatility of (I) is low at decomposition, the compound readily produces In$_2$O$_3$.

Thermal decomposition of $In(\eta^2-O_2CC_6H_5)_3(4\text{-Mepy})_2$ was followed by TGA in both air and nitrogen, and the composition of the final pyrolysate determined by XRD. No attempt was made to identify the intermediate pyrolysates produced during this analysis. The first two steps in the thermogram (see Figure 2) correspond to the sequential loss of the two 4-methylpyridine ligands. The same results were obtained in air and nitrogen. This lack of effect on the formation of oxide is expected due to the InO_6N_2 coordination environment around indium, it is also consistant with the ready decomposition of benzoate seen in the mass spectra.

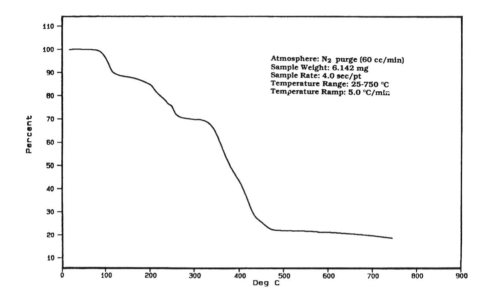

Figure 2. Thermogravimetric analysis of $In(\eta^2-O_2CC_6H_5)_3(4\text{-Mepy})_2$ under an atmosphere of nitrogen. The theoretical values for weight changes are: 86%, $[In(O_2CC_6H_5)_3(4\text{-Mepy})_2 - (4\text{-Mepy})]$; 72%, $[In(O_2CC_6H_5)_3(4\text{-Mepy})_2 - 2(4\text{-Mepy})]$; 21%, $In_2O_3/In(O_2CC_6H_5)_3(4\text{-Mepy})_2$.

The final weight loss corresponds to complete decomposition of $In(\eta^2-O_2CC_6H_5)_3(4\text{-Mepy})_2$ to In_2O_3, as demonstrated by weight loss and XRD pattern, Table 2 [16,17]. The most interesting aspect of the TGA data is the relatively low temperature (475 °C) for the stabilization of weight loss. By contrast, the polymeric $In(\eta^2-O_2CCH_3)_n$ [13], did not reach a stable mass until between 1000 and 1100 °C, above the sublimation temperature of In_2O_3 of 850 °C [18]. No attempt was made to characterize the intermediate materials. The morphology of material produced during a typical thermal analysis run is shown in a scanning electron micrograph, figure 3. As can be seen, melting has occurred in the material, as the temperature of the sample reached 750 °C; this is still 300 °C below the temperature needed to fully convert the polymeric acetate [13].

TABLE 2.

X-ray Diffraction (XRD) Powder Pattern for Pyrolysate, In_2O_3, Between 1.20 and 5.00 Å.

Angle, 2θ	d, Å	I/I_{max}, %
21.522	4.126	1.18
30.750	2.905	100.00
35.510	2.53	35.19
37.725	2.383	3.48
41.895	2.155	11.72
45.710	1.983	6.64
49.302	1.847	3.11
51.115	1.786	65.64
52.745	1.734	2.99
56.075	1.638	7.54
59.200	1.56	8.30
60.675	1.525	58.17
52.222	1.491	13.16
63.700	1.46	12.67
65.198	1.43	4.85
68.030	1.377	8.11
69.475	1.351	5.16
73.745	1.284	5.63
75.102	1.264	8.50
76.272	1.247	7.92

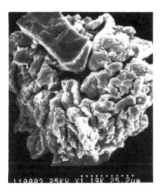 

Figure 3. Scanning electron micrographs of typical samples of In_2O_3 produced during TGA experiments and charcterized by XRD, see text.

CONCLUSIONS

We have demonstrated a simple and direct one-step route to $In(\eta^2-O_2CC_6H_5)_3(4-Mepy)_2$, the first mononuclear eight-coordinate indium(III) benzoate. Our approach to the synthesis of indium(III) carboxylates differs significantly from previously reported methods [4-11]. The presence of 4-methylpyridine at the initial stages of reaction virtually eliminates all probability of $In(O_2CR)_3$ polymerization by coordinatively saturating the indium(III) center as it is formed. In addition, we have established that (I) is a stable inorganic precursor to indium oxide.

ACKNOWLEDGEMENT

A.F.H. (Director's Discretionary Fund), M.T.A. (Postdoctoral Fellowship, National Research Council/NASA Lewis Research Center), S.A.D. (NASA Cooperative Agreement NCC3-162), D.G.H. (NASA Cooperative Agreement NCC3-318), and P.E.F (NASA Cooperative Agreement NCC3-246) acknowledge support from NASA Lewis Research Center. We thank Ms. Ruth Cipcic (XRD, NASA LeRC) and Mr. Frederick K. Oplinger (NMR, CSU) for assistance.

REFERENCES

1. I. Weinberg, Solar Cells **29**, 225 (1990).

2. J. van de Ven, J.J.M. Binsma, and N.M.A. de WIld, J. Appl. Phys. **67**, 7568 (1990).

3. M.T. Andras, S.A. Duraj, A.F. Hepp, P.E. Fanwick, and M.M. Bodnar, J. Am. Chem. Soc. **114,** 786 (1992).

4. R.C. Mehrotra and R. Bohra, *Metal Carboxylates*; (Academic Press, London, 1983) pp. 121-129, 178-181.

5. D.G. Tuck, in *Comprehensive Organometallic Chemistry*; *Vol. 1*, edited by G. Wilkinson, F.G.A. Stone, E.W. Abel, (Pergamon Press, New York, 1982) pg. 719.

6. V.W. Lindel and F.Z. Huber, Anorg. Allg. Chem. **408**, 167 (1974).

7. H.D. Hausen, J. Organometal. Chem. **39**, C37 (1972).

8. F.W.B. Einstein, M.M. Gilbert, and D.G. Tuck, J. Chem. Soc., Dalton Trans., 248 (1973).

9. J.J. Habeeb and D.G. Tuck, Can. J. Chem. **52**, 3950 (1974).

10. R. Nomura, S. Fujii, K. Kanaya, and H. Matsuda, Polyhedron **9**, 361 (1990).

11. M.A. Khan, C. Peppe, and D.G. Tuck, Acta Cryst. **C39**, 1339 (1983).

12. M.T. Andras, A.F. Hepp, S.A. Duraj, E.B. Clark, D.A. Scheiman, D.G. Hehemann, and P.E. Fanwick, Inorg. Chem. **32**, 4150 (1993).

13. J.J.Habeeb and D.G. Tuck, J. Chem. Soc., Dalton Trans., 243 (1973).

14. For example, $In(O_2CMe)_3L$ where L is a bidentate N, N'-donor molecule, i.e., 2,2' bipyridine or 1,10-phenanthroline; see V.H. Preut and F.Z. Huber, Anorg. Allg. Chem. **450**, 120 (1979).

15. R. Kumar, H.E. Mabrouk, and D.G. Tuck, J. Chem. Soc., Dalton Trans., 1045 (1988).

16. The diffraction pattern was matched with In_2O_3 in the JCPDS International Center for Diffraction Data 1989. Powder Diffraction File No. 6-0416.

17. Chemical analysis showed total C, H, and N content below detection limit.

18. *Handbook of Chemistry and Physics*, edited by R.C. Weast, 55th ed. (Chemical Rubber Company Press, Cleveland, OH, 1974) pg. B-96.

LOCAL EQUILIBRIUM PHASE DIAGRAMS FOR SiC DEPOSITION IN A HOT WALL LPCVD REACTOR

CHIEN C. CHIU[†], SESHU B. DESU[††], ZHI J. CHEN[†], AND CHING YI TSAI[*]
[‡]Department of Materials Science and Engineering and [*]Department of Engineering Science and Mechanics, Virginia Polytechnic Institute and State University, Blacksburg, VA 24061–0237
†: Author to whom correspondence should be addressed.

ABSTRACT:

The traditional CVD phase diagrams are only valid for cold wall reactors because of the neglecting of the depletion effects in hot wall reactors. Due to the depletion effects along the reactor, the traditional CVD phase diagrams can not accurately predict the phases deposited on the substrate in a hot wall CVD system. In this paper, a new approach to calculate the local equilibrium CVD phase diagrams in a hot wall reactor is presented by combining the depletion effects with the equilibrium thermodynamic computer codes (SOLGASMIX–PV). In this study, the deposition of SiC using methyltrichlorosilane (MTS)–hydrogen (H_2) was chosen to verify this new approach. The differences between the new CVD phase diagrams and the traditional phase diagrams were discussed. The calculated CVD phase diagrams were also compared with the experimental data. The single phase of SiC predicted by this approach is much better than the traditional phase diagrams. The experimental regions for depositing single phase SiC are larger than the calculated local phase diagrams because of the higher linear velocity of the gas flux under low pressure and the polarity of the Si carrying intermediate species.

INTRODUCTION:

Chemical vapor deposition (CVD) working in a hot wall reactor and a low pressure is currently used in semiconductor industry to manufacture microelectronic devices because of the high throughput, improved thickness homogeneity, and the reduced contamination from susceptors by using very low deposition pressure [1]. Although the CVD phase diagrams have been studied using lots of models, all of the calculated CVD phase diagrams are based on the cold wall assumptions [2–4]. Since, for a hot wall reactor, the temperature profile is continuously increasing from the inlet of the reactor to the substrates, it is expected that the precursors will undergo some chemical reactions (*i.e.* depletion effects) along the reactor. These effects exhaust the precursor concentrations and generate some intermediate gaseous species. Some other kinetic factors were also believed to play a role in the depletion effects. Therefore, with the constant changes in the reactive gaseous species along the length of hot wall CVD reactor because of the depletion effects, it is apparent that the local gaseous species concentrations at the substrate are significantly different from the input concentrations. This phenomenon will be more obvious under higher deposition temperature [5]. Thus the current phase diagrams are not suitable to study the CVD processes in hot wall reaction conditions [2–4]. Although various models have been reported to study the hot wall CVD processes [6–9], no study for the local CVD phase diagrams at the substrate under various processing conditions was reported, which is of fundamental importance in understanding and optimizing the CVD processes.

For this study, we present a new approach to the local CVD phase diagrams in a hot wall reactor, which couples the depletion effects to the equilibrium thermodynamic calculation. The schematic diagram for this approach is depicted in Fig. 1; the depletion effects were considered before performing the equilibrium thermodynamic calculation to obtain local CVD phase diagrams. The detailed analytical method is described in the next section. The deposition of silicon carbide (SiC) from MTS–H_2 in a hot wall reactor was chosen to study the local equilibrium CVD phase diagrams. Comparisons between the CVD phase diagrams for depositing SiC developed based on both the hot wall and cold wall assumptions are emphasized. Experimental results were also used to verify the calculated results.

METHOD OF CALCULATION:

For a hot wall CVD, the reactive species are consumed or generated along the reactor. In order to include this hot wall factor into the calculations of the CVD phase diagrams for SiC deposition, a kinetic model for this deposition process was used to calculate the spatial dependence of the gaseous species concentrations [10,11].

As shown in Fig. 1, in the kinetic model, the steps involved in the deposition of SiC from MTS–H_2 precursors are assumed to be: (i) the decomposition of MTS molecules into two major intermediates, one containing silicon (IP_{si}) and the other containing carbon (IP_c), as well as gaseous byproducts (BP), (ii) adsorption of intermediate species onto the surface sites of the growing films, and (iii) reaction of the adsorbed intermediates to form silicon carbide. Detailed description of the model was shown elsewhere [10].

For the present study, the kinetic analysis of silicon carbide deposition, in a hot wall reactor, is first introduced to identify the partial pressures of the intermediate species (IP_{si} and IP_c) and gaseous byproducts (BP) over the position where the substrate is located. Then, equilibrium thermodynamic calculations were performed at this position to calculated the CVD phase diagram for this hot wall reactor.

The thermodynamic analysis was conducted with a modified SOLGASMIX–PV computer program to calculate the equilibrium composition of the system [12]. The procedures are based on free–energy minimization under given conditions of temperature, total pressure, and the input gas concentrations. Detailed descriptions of the calculation were reported by Eriksson [13] and White [14].

For a hot wall CVD reactor, the MTS precursor decomposes into some gaseous intermediate species when it transports from the entrance of the furnace to the substrates [15–17]. It was also reported that the concentrations or the partial pressures of the gaseous species in a hot wall reactor can be predicted from the equilibrium calculations, assuming no condensed phases present [1,16]. In these calculations, the reactants are the same as those at the entrance of the furnace (*i.e.* no intermediate species and gaseous byproducts). However, these equilibrium calculations can only predict the partial pressures for silicon bearing (IP_{si}), carbon bearing (IP_c) species, and the gaseous byproducts (BP). The experimentally determined concentrations of MTS along the reactor were much larger than the calculated values, especially at high temperatures (T > 1200 K) [16]. Thus the analysis of the CVD system by merely using the equilibrium thermodynamic analysis is insufficient to understand the partial pressures for each gaseous species in a hot wall reactor without considering the kinetic factors.

For the present study, the partial pressures of MTS, the intermediate species (IP_{si} and IP_c), and the gaseous byproducts (BP) in the gas phase above the substrate were obtained from the kinetic analysis stated above. Because of the surface controlled reactions under the experimental conditions [11], it is also assumed that, above the substrate, the partial pressures of the gas species (MTS, IP_{si}, IP_c, and BP) in the boundary layer are the same as those in the gas phase, and the reactions in the boundary layer above the substrate come to a rapid equilibrium. Before the thermodynamic analysis with the presence of condensed phases, the orders and the ratios for the intermediate species (IP_{si} and IP_c), and the gaseous byproducts (BP) above the substrate in the gas phase were obtained by assuming no condensed phases formed under equilibrium conditions and with the initial H_2/MTS ratio at the entrance of the furnace. Since the kinetic analysis can only give the total pressures of IP_{si}'s, IP_c's, and BP's, the individual partial pressure of each IP_{si}, IP_c and BP was obtained from the kinetic analysis and from the ratios of equilibrium thermodynamic calculation, assuming no condensed phases. Then the equilibrium calculations were performed with the presence of the condensed phases in obtaining the CVD phase diagram in a hot wall reactor. In this

The approach to calculate local phase diagram in a hot wall reactor

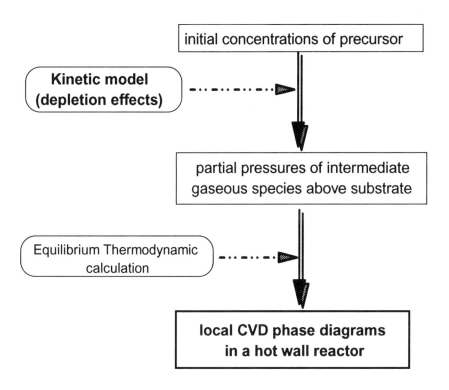

Figure 1: The schematic diagram of calculation.

study, 59 gas species and 6 condensed phases were considered. The free energies, entropies, and enthalpies of formation of the species are available in SOLGASMIX–PV [12], JANAF thermochemical table [18], and Ref. [19]. Since the cubic form, β–SiC, was more stable than the hexagonal polytypes for the temperatures considered, only one polymorph of SiC (*i.e.* cubic form β–SiC) was considered in the calculation. In most cases the results are presented as CVD phase diagrams as a function of temperature and H_2/MTS ratios in the input gas for a specific system pressure. The calculated CVD phase diagrams were compared with the data from our own experimental results and from the literatures for the hot wall type of reactor beneath 10 Torr and from 1000 to 1500 K. Detailed experimental procedure was described elsewhere [11,20,21].

RESULTS AND DISCUSSION:

The experimental results of the present study together with the calculated CVD phase diagrams without the depletion effects from a hot wall reactor are shown in Fig. 2. As shown in Fig. 2, the calculated CVD phase diagrams indicated that the deposit compositions at thermodynamic equilibrium were graphite+β–SiC under small H_2/MTS ratios for the considered deposition pressures (1.8, 5, and 8 Torr). The region for graphite+β–SiC expands toward higher values of H_2/MTS ratios above 1100 K with decreasing deposition pressure. Thus the equilibrium phase diagrams without the depletion effects in a hot wall reactor can only predict the deposit composition with higher deposition pressures, higher H_2/MTS ratio ,and lower temperature than the present study.

In contrast to Fig. 2, the CVD phase diagrams for MTS–H_2 system in a hot wall reactor and the comparison with the experimental results are given in Fig. 3. In comparison with the results from the cold wall CVD reactors (Fig. 2), the boundaries between graphite+β–SiC and β–SiC, in a hot wall reactor, are not affected by the depletion factors at temperatures lower than 1100 K. This is because the depletion effect was not prominent at lower temperatures. However, it is clear that the boundaries between graphite+β–SiC and β–SiC move toward smaller H_2/MTS ratios with increasing deposition temperature (T > 1100K) and pressure. The regions for Si+β–SiC and/or Si also appear at higher deposition pressures and higher temperatures (*e.g.* Figs. 3(b) and 3(c)). With increasing temperature, the compositions of the deposit seem to be more dependent on deposition pressure than those for cold wall reactor (Fig. 2). As shown in Fig. 3, the regions containing the deposition or codeposition of Si move toward lower temperatures and low H_2/MTS ratios with increasing deposition pressure. This is because of the depletion effect from the hot wall reactor, which is more significant at higher deposition temperature and pressure. Therefore, β–SiC was expected to be codeposited with Si at higher deposition pressure in a hot wall CVD reactor. The codeposition of Si with SiC is also suggested by the experimental results obtained from the literatures using MTS–H_2 gas system as the SiC precursor in a hot wall reactor [22–25].

Since the calculated boundaries between graphite+β–SiC and β–SiC move toward smaller values of H_2/MTS as shown in Fig. 3, the experimental data at 1.8, 5, and 8 Torr agree much better than those without the consideration of depletion factors, especially under high H_2/MTS ratios (Fig. 2). However, for lower deposition pressures of 1.8 and 5 Torr (Figs. 3(a) and 3(b)), the experimental boundaries between the regions of graphite+β–SiC and β–SiC appear to be located at lower values of H_2/MTS ratios and temperatures. The boundary between β–SiC and Si+β–SiC also seems to be moved toward the region of calculated region for Si+β–SiC at 8 Torr (Fig. 3(c)). These indicate that the deposition of single–phase β–SiC is not only influenced by the thermodynamic equilibrium factor but also by the kinetic ones. According to Fischman et al. [2], the expansion of single phase β–SiC toward the region of graphite+β–SiC at lower H_2/MTS ratios can be explained by higher rate of deposition of silicon than carbon because the primary intermediate species carrying Si are polar molecules, whereas the primary molecules bearing carbon are nonpolar. On the other hand, the shift of the boundaries between β–SiC and Si+β–SiC toward the region for Si+β–SiC was believed to be due to

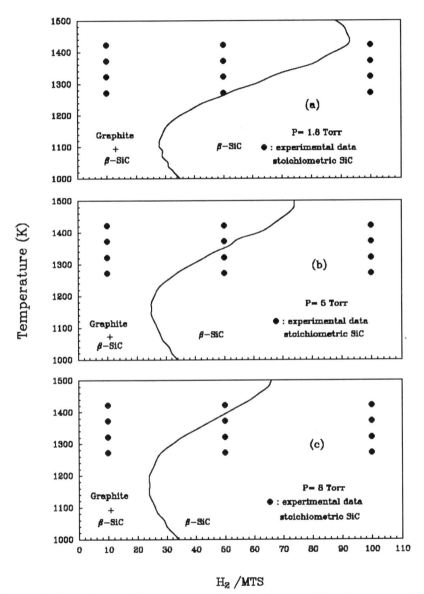

Figure 2: Comparison of the experimental results from a cold wall reactor with the calculated CVD phase diagrams without considering the depletion effect for (a) 1.8 Torr, (b) 5 Torr, and (c) 8 Torr.

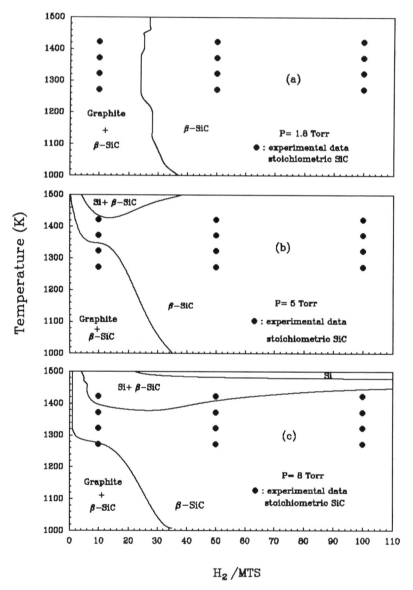

Figure 3: Comparison of the experimental results from a hot wall reactor with the calculated CVD phase diagrams with the consideration of the depletion effect for (a) 1.8 Torr, (b) 5 Torr, and (c) 8 Torr.

be due to the suppression of the deposition of free Si under high linear velocity of gas flux at low deposition pressures and high temperature [29]. Therefore, the experimental regions for single phase of β–SiC should be larger than those from the calculated CVD phase diagrams in a hot wall reactor.

SUMMARY:

A new model considering both the depletion effects and equilibrium thermodynamics was proposed to calculated the CVD phase diagrams in a hot wall reactor. For the deposition of SiC from MTS–H_2 gas system, in comparison with the CVD phase diagrams without considering the depletion effects, the calculated equilibrium CVD phases diagrams in a hot wall CVD reactor showed better agreement between the calculated and experimental results and depicted that the composition of the deposit was strongly influenced by the deposition pressure. It is also shown that Si starts to deposit or codeposit with SiC with increasing deposition temperature and pressure because of the depletion effects. The discrepancies between the calculated and experimental results were probably attributed to a high linear velocity of the gas flux at low pressure and polarity of the Si carrying molecules.

REFERENCES:

1. F. Langlais, F. Hottier, and R. Cadoret, J. Cryst. Growth, **56**, 659 (1982).
2. G.S. Fischman and W.T. Petuskey, J. Am. Ceram. Soc., **68**, 185 (1985).
3. A.I. Kingon, L.J. Lutz, P. Liaw, and R.F. Davis, J. Am. Ceram. Soc., **66**, 558 (1983).
4. K. Minato and K. Fukuda, J. Nuclear Mater., **149**, 233 (1987).
5. S.A. Gokoglu, in *Chemical Vapor Deposition of Refractory Metal and Ceramics II*, edited by T.M. Besmann, B.M. Gallois, and J.W. Warren (Mater. Res. Soc. Symp. Proc. 250, Mater. Res. Soc., Pittsburgh, PA, 1992), p. 18.
6. F. Langlais, C. Prebenge, B. Tarride, and R. Naslain, J. de Physique, Colloque c5, supplément au n° 5, Tome 50, C5–93 ~ C5–103 (1989).
7. D. Neuschütz and F. Salehomoum, in *Chemical Vapor Deposition of Refractory Metal and Ceramics II*, edited by T.M. Besmann, B.M. Gallois, and J.W. Warren (Mater. Res. Soc. Symp. Proc. 250, Mater. Res. Soc., Pittsburgh, PA, (1992), p. 41.
8. F. Langlais and C. Prebende, in *Proc. 11th Int. Conf. on CVD*, edited by K.E. Spear and G.W. Cullen (The Electrochem. Soc., Pennington, NJ, 1990), p. 686.
9. S.A. Gokoglo and M.A. Kuczmarski, in Proc. of the 12th Int. Symp. on Chemical Vapor Deposition 1993, edited by K.F. Jeusen and G.W. Cullen (The Electrochem. Soc., Inc., Pennington, NJ, Proc. vol. 93–2), p. 392
10. C.Y. Tsai, S.B. Desu, and C.C. Chiu, J. Mater. Res., **9**, (1994) (in press).
11. C.C. Chiu, S.B. Desu, and C.Y. Tsai, J. Mater. Res, **8**, 2617 (1993).
12. B.W. Sheldon, in Solgasmix–PV for the PC, ORNL, Oct., 1989.
13. G. Eriksson, Acta Chem. Scand., **25**, 2651 (1971).
14. W.B. White, W.M. Johnson, and G.B. Dantzig, J. Chem. Phys., **28**, 751 (1958).
15. T.M. Besmann, B.W. Sheldon, T.S. Moss III, and M.D. Kaster, J. Am. Ceram. Soc., **75**, 2899 (1992).
16. T.M. Besmann and M.L. Johnson, in *Proc. 3rd Int. Symp. on Ceramic Materials and Components for Engine*, Las Vegas, NV, 443–456 (1988).
17. J.N. Burgess and T. Lewis, Chem. and Industry, **19**, 76 (1974).
18. JANAF Thermochemical Tables, 3rd Edition, J. Phys. Chem. Reference Data, Vol. 14, (1985).
19. Z.J. Chen and S.B. Desu, unpublished paper.
20. C.C. Chiu and S.B. Desu, J. Mater. Res., **8**, 535 (1993).
21. C.C. Chiu, S.B. Desu, G. Chen, C.Y. Tsai, and W.T. Reynolds, Jr., submitted to J. Mater. Res.
22. T.M. Besmann, B.W. Sheldon, and M.P. Kaster, Surface and Coating Technol.,

43/44, 167 (1990).
23. W. Schintlmeister, W. Wallgram, and K. Gigl, High Temp.–High Pressures, **18**, 211 (1986).
24. D.H. Kuo, D.J. Cheng, W.J. Shyy, and M.H. Hon, J. Electrochem. Soc., **137**[11], 3688 (1990).
25. S. Motojima and M. Hasegawa, J. Vac. Sci. Technol., **A8**, 3763 (1990).

DEVELOPMENT AND IMPLEMENTATION OF LARGE AREA, ECONOMICAL ROTATING DISK REACTOR TECHNOLOGY FOR METALORGANIC CHEMICAL VAPOR DEPOSITION

G.S. TOMPA, P.A. ZAWADZKI, M. MCKEE, E. WOLAK, K. MOY, R.A. STALL, A. GURARY, AND N.E. SCHUMAKER
EMCORE Corportion, 35 Elizabeth Avenue, SomersetT, NJ 08873.

ABSTRACT:

The vertical, high speed, rotating disk reactor (RDR) has, in recent years, found broad application in the Metalorganic Chemical Vapor Deposition of a variety of material systems. These applications include epitaxial films of III-V and II-VI compound semiconductors, oxides (such as YBCO superconductors/ferroelectrics and SiO_2, amongst others), Group IV materials (such as diamond and SiC), and metals (such as copper and tungsten). As production of these material systems increases, so too does the need for economical, high yield equipment capable of producing these materials with high levels of uniformity and repeatability. We have used computational fluid dynamic modeling to investigate the complex flow and thermal dynamics required for scaling existing RDRs (as large as a 7.25" diameter disk handling up to 3x3" wafers) to larger dimensions (11" and 12" diameter disks for multiple 4" and 15.5" diameter disk for 3x6" wafers). The scaling parameters predicted by the modeling codes are reviewed and correlate well with experimental results. Materials results on GaAs films using TBAs, TMGa, and TMAl for the 11" diameter system routinely demonstrate within wafer thickness uniformities of <1.1% for 3x4" wafers, as well as for 6" or 8" diameters, wafer to wafer uniformities <1% and run to run repeatabilies within 1%. These results are verified by SEM analysis, as well as with GaAs/AlGaAs Bragg reflectors. The excellent results on the 11" and 15.5" diameter platters combined with modeling indicated that 4x4" wafers on a 12" diameter platter would produce ideal films which, indeed, is the case. The 11" diameter results have been surpassed, demonstrating <0.9% for >9" diameters (4x4" wafers) on a 12" diameter susceptor. With high reactant efficiencies (>36%), short cycle times between growths using the loadlock, and minimal maintenance requirements, the costs per wafer in a cost of ownership model are found to be dramatically less than in competitive technologies.

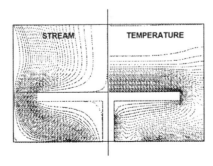

Fig. 1: Models Showing Contours of Stream and Temperature in an RDR

INTRODUCTION:

Several reactor types have been investigated for the epitaxial deposition of compound semiconductors. Molecular Beam Epitaxy (MBE) and Metalorganic Chemical Vapor Deposition (MOCVD) have evolved to be the predominant deposition technologies. MOCVD has become the preferred method for production due to its versatility, scalability and economical operation.

Several forms of MOCVD reactors have evolved, with horizontal tube technology (including planetary reactors) and vertical (including high speed rotating disk reactors (RDR)) being the preeminent technologies. We will report on the developments of the vertical, RDR, which is the reactor technology we have historically developed[1]. The advantage of vertical RDR's over horizontal tubes include: uniform isotherms with sharp temperature gradients, which maximizes thermophoresis effects[2] (minimizing particulate incorporation) promoting uniform composition; forced convection which eliminates buoyancy or recirculation effects and generates higher reactant efficiencies; and the ability to give all the reactants an equivalent process history, which allows advanced chemical processes to be utilized and minimizes premature reactions.

THEORY:

Several authors have experimentally and computationally investigated the rotating disk reactor systems [1-5]. Table I (after ref. 3) shows the important parameters and equations associated with RDRs and the scaling of them. The important features are that the scaling equations are dimensionless, and that there are no fundamental limits to scaling of RDRs. Experimentally, we have verified this for rotating disks as large as 15.5" in diameter. Figure 1 shows the results of computational fluid dynamical modeling of a RDR. The result shows laminar, non-recirculating contours of stream and uniform, sharply graded temperature isotherms. These models are for

DISCOVERY SERIES

ENTERPRISE SERIES

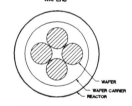

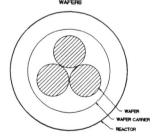

FIG. 2 HISTORICAL PROGRESSION OF EMCORE Turbo Disk (RDRs)

the 12" diameter wafer carrier operating at 15 Torr and rotating at 700 rpm. Figure 2 shows the historical progression of EMCORE *Turbo*Disc RDRs.

	Theory	Practice
$Re \propto \dfrac{r_d^2 \omega}{v_\infty}$	>500	~500
$\dfrac{Gr}{Re^{3/2}} = \dfrac{g(1-\bar{\rho}_\omega)}{\bar{\rho}_\omega \omega \sqrt{\omega\, v_\infty}} < 1$	~1	
$\mu_T = .75 \sqrt{\omega\, v_\infty}$		~40

Re = disk Reynolds number
Gr = Mixed Convection Parameter
Re³/²
r_d = disk radius
ω = disk spin rate (rad/sec)
v_∞ = kinematic viscosity
  (evaluated at reactor inlet)
g = gravitational acceleration
$\bar{\rho}_\omega$ = dimensionless density
  (evaluated at disk surface) =ρ_ω/ρ_∞

Table I: Scaling Parameters

EXPERIMENT:

The operating principles of RDRs have been described by several authors and will be reviewed here briefly. In an RDR the carrier gas and reactants are injected uniformly from the top. The high speed rotating disk imparts a viscous drag to the gas, pumping it to the surface and then out toward the exhaust. The viscous drag negates bouyancy effects, and, in combination with the total flow, is used to eliminate recirculation cells. The result is uniform and sharply graded isotherms and smooth, non-mixing streamlines. The specific mix of reactants is independently controlled through a reactant distribution manifold and generally the hydrides are separated from the alkyls.

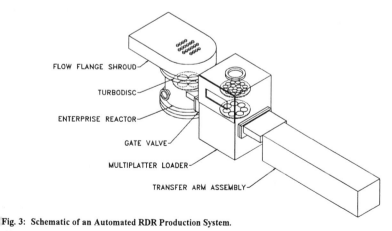

Fig. 3: Schematic of an Automated RDR Production System.

Our specific configuration for this series of experiments was as follows. An oversized stainless steel reactor, similar to the reactor shown schematically in fig. 3, was used to allow us to investigate growth on platters up to 15.5" in diameter. In combination with the oversized reactor we used internal shrouds to mimic ideal reactor diameters as predicted by the modeling. Thermal and flow studies also indicted that the use of a shroud would have no significant impact on the system fluid or thermal dynamics. The rotating platter is radiatively heated by a fixed resistive heater mounted beneath it. Optical ports are located in the horizontal plane and at high angles of incidence. The temperature is monitored by thermocouples and pyrometers and was calibrated by observing the AlSi eutectic on a witness wafer. The stainless steel construction is compatible with loadlocking for single or multiple reactor dockings to the platter transfer sytem. A standard exhaust system is utilized to effectively pump the system and remove solid and gaseous toxic effluents.

In all cases, the runs were performed at ~700°C, 15-20 Torr and rotation rates of ~700rpm. The total flow was varied from ~30 to 50 slm. We used AsH_3, TBA, TMGa, and TMAl as the reactants for these studies. Standard high speed, run / vent / idle, pressure balanced gas switching was used for the reactant transport. For the uniformity measurements, a half 2" wafer was used in combination with the 4" wafers to generate the full uniformity maps.

RESULTS:

With the 11" diameter platter we were able to achieve <1.1% GaAs/AlAs thickness uniformity across the wafers as shown in fig. 4a. Based on these results, we then increased the platter diameter to 12" and the inner shroud to 16" and achieved GaAs/AlAs thickness uniformities of <0.9% as shown in figure 4b. With both platter sizes we went on to produce GaAs/AlAs Bragg reflectors. The reflectivity uniformity repeated the SEM thickness uniformity. Fig. 5 shows a uniformity plot for one of 4x4" wafers from the 12" diameter platter. We repeated these experiments, but for peak reflections at blue, yellow, green and red wavelengths, with equivalent results.

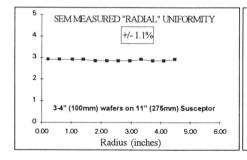 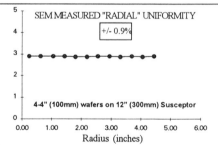

Fig. 4a) 11" diameter platter uniformity, b) 12" diameter platter Uniformity.

For both platter diameters, the efficiency of the deposition rate controlling species was calculated and found to be >36% over the flow range examined. The efficiency is calculated as {moles of reactant deposited}/{moles of reactant consumed}. The system throughput efficiency is enhanced through the use of a loadlock. The loadlock allows us to prepare up to 10 platters for deposition and maintain them under vacuum prior to deposition. Thus we can safely remove hot (>300°C) platters and load new ones into the reactor without fear of contamination. It is also our experience that literally hundreds of microns of epitaxy can be deposited before any cleaning of the reactor is required. These features result in a low cost per wafer and a high throughput as shown in figure 6. The curves shown in the figure depend primarily on the deposition time, since we have found that the wafers can be cycled to the loadlock at 350-400°C, and hence heat-up, cool-down and transfer times are not generally significant (<10-15 min.).

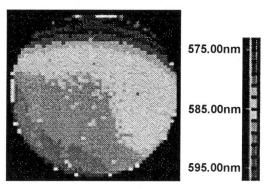

Fig. 5: Reflectivity Spectra of 1 of 4-4" Wafers Showing <0.9% Variation.

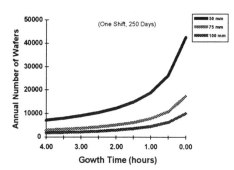

Materials Demonstrated in *TurboDisc* Systems

III-V	GaAs, AlGaAs, InP, InGaAsP, InGaP, InGaAlP, InSb, GaN
II-VI	ZnSe, CdSe, CdTe, HgCdTe, ZnTe, CdZnTe, CdSeTe
IV-IV	SiC, Diamond, SiGe
OXIDES	YBCO, BaTiO$_3$, MgO, ZrO,SiO$_2$, ZnO, ZnSiO, SrTiO$_3$, NdGaO$_3$, LaAlO$_3$, CeO$_2$, MgAl$_2$O$_4$, NdGaO$_3$, LaAlO$_3$
METALS	Al, Cu, In, W

Table II:

Fig. 6. 12' diameter throughput based on deposition time.

CONCLUSION:

We have utilized computational and experimental results to produce a series of rotating disk reactors which address research, pilot production and production needs. The systems have been scaled from 3" to 15.5" diameter susceptors and are used for deposition on single and multiple substrates. The versatility of RDR systems is exemplified by the wide range of materials it has been successfully applied to, as shown in Table II. We have recently focused on the 12" diameter system, which holds 4x4" or 17x2" wafers, for production needs. The process parameters have experimentally scaled in accordance with theory, consistently transferring process parameters from reactor to reactor as we increase the dimensions. At all sizes, efficiencies, repeatabilities, and uniformities are more than suitable for economic device production. With the automation of these systems we truly have compound semiconductor production tools at a level normally ascribed to "Si world" production tools.

ACKNOWLEDGMENTS:

The authors thank W.G. Brieland, G.H. Evans and P. Esherick of Sandia National Laboratories for many useful technical discussions, and Peter Broskie of EMCORE for his assistance with the preparation of this document.

REFERENCES:

[1] G.S. Tompa, M.A. McKee, C. Beckham, P.A. Zawadzki, J.M. Colabella, P.D. Reinert, K. Capuder, R.A. Stall, and P.E. Norris, J. Crystal Growth 93, 220 (1988)
[2] R.W. Davis, E.F. Moore and M.R. Zachariah, J. of Crystal Growth 132 (1993)
[3] W.G. Breiland and G.H. Evans, J. Electrochem. Soc. 138 (1991) 1806.
[4] J.F. Brady and L. Durlofsky, J. Fluid Mech. 125 (1887) 363; and D.G. Kotecki and S.G. Barbee, J. Vac. Sci. Technol. A 10(4) (1992) 43.
(5) D.I. Fotiades, A.M, Kramer, D.R. McKenna and K.F. Jensen, J. of Crystal Growth 85 (1987) 154; and C.R. Biber, C.A. Wang, and S. Motakef. J. of Crystal Growth 123 (1992) 545.

PART VI

High T_c Superconductors

SOME CONSIDERATIONS OF MOCVD FOR THE PREPARATION OF HIGH T_c THIN FILMS

MICHAEL L. HITCHMAN*, SARKIS H. SHAMLIAN*, DOUGLAS D. GILLILAND*, DAVID J. COLE-HAMILTON+, SIMON C. THOMPSON+, STEPHEN L. COOK# AND BARBARA C. RICHARDS#
*Department of Pure and Applied Chemistry, University of Strathclyde, Glasgow GB-G1 1XL
+Chemistry Department, University of St. Andrews, St. Andrews GB-KY16 9ST
#The Associated Octel Co. Ltd., PO Box 17, Ellesmere Port, South Wirral, GB-L65 4HF

ABSTRACT

In this paper we first give a brief overview of the MOCVD of high temperature superconductors and consider the general requirements of a precursor. We then present results for deposition from Y containing precursors and compare apparent enthalpies of sublimation and deposition activation energies with values in the literature. The problem of Ba containing precursors are briefly reviewed and results given for the sublimation of and deposition from a new stable and volatile precursor. Finally, some comments are made about possible deposition mechanisms.

INTRODUCTION

After the discovery of high temperature superconductors (HTS) by Mueller and Bednorz in 1986 [1] and the subsequent preparation a year later [2] of $YBa_2Cu_3O_{7-x}$ (YBCO) with a value of T_c above the b.p. of liquid nitrogen, the first paper describing the preparation of thin films of YBCO by CVD appeared in 1988 [3]. Subsequently, the discoveries of an HTS containing Bi, Sr, Ca and Cu (BSCCO) and one containing Tl, Ba, Ca and Cu (TBCCO) in 1988 were rapidly followed by methods describing the CVD of these materials [4,5]. Since that time there have been more than 200 papers on the CVD of YBCO, over 40 papers on BSCCO CVD, and more than 10 papers on TBCCO CVD. The recent announcement [6] of a Hg, Ba, Ca, Cu containing HTS with the highest T_c yet reported of 133K will no doubt be followed by a paper on the CVD of this latest HTS.

At the same time as the research of the new HTS materials has developed there has been intense commercial interest in them. In 1988 it was predicted [7] that the first commercial applications of HTS would be in electronic devices, and this was fulfilled in 1990 with a description of the use of a superconductor transistor in microwave amplifiers, oscillators, phase shifters and signal mixers [8], and with the first superconductor-CMOS semiconductor circuits being described in 1993 [9]. By the year 2005 it is estimated that the world market for electronic devices based on HTS will be in excess of $300 M [10].

In order to build electronic HTS devices it will be necessary to produce thin films of HTS by techniques which are compatible with fabrication lines, which allow close control of layer properties, which produce conformal thin films over large areas and have a high level of intra- and inter-sample uniformity, and which give a high sample throughput. CVD is just such a technique, and in addition it has the features of allowing a range of oxidants to be used (e.g. O_2, N_2O, O_3), of permitting low temperature deposition (e.g. PECVD, photoCVD), and of maintaining a high partial pressure of oxidant, which is often desirable [11]. However, the C in CVD emphasises that the technique is very dependent on precursor chemistry, deposition chemistry, and reaction parameters. This rapidly becomes apparent if one reviews the results in the literature on the CVD of HTS [12,13]. A wide range of layer properties have been reported. For example, values of critical temperature (T_c) for YBCO from ca. 20K to ca. 92K have been obtained, with values of critical current density (J_c at 77K and zero applied field) from 2×10^4 A cm^{-2} to 6×10^6 A cm^{-2}. Factors affecting both T_c and J_c have been found to include layer composition, layer rugosity, crystalline quality, epitaxial orientation, and thickness. These factors are, in their turn, very dependent on deposition conditions for which there is also a large gamut. For example, a large number of precursors has been used showing a wide variety of volatility, chemical stability, and reproducibility, but with limited information being available of the physical and chemical characteristics of many of the precursors. Their deposition temperatures reported have ranged from ca. 400°C to ca. 1000°C and deposition pressures have been from below ca. 40 mTorr up to ca. 760 Torr. The different types of oxidants and process activation have already been mentioned above while the number of annealing procedures described is approximately equal to the number of publications mentioning this topic. Finally, many different substrates have been used including oxides, various buffer layers on silicon, and metals.

Clearly, the CVD of YBCO and other HTS, like many CVD processes, is a multiparameter problem and much of what has been reported has been of an empirical nature. There is a need for understanding both precursor and deposition chemistry as well as the effect of deposition parameters on layer properties. In this paper we consider some aspects of the chemistry associated with organometallic precursors of Y and Ba and with the CVD of thin films from such precursors. This allows us to obtain some particular and general insights into the use of MOCVD for the preparation of HTS. The experimental system and conditions used have been described elsewhere [14,15] and are omitted from the text here because of the limitation on space.

PRECURSORS

One can specify a number of general ideal requirements for any CVD precursor. These requirements include a high vapour pressure (e.g. > 1 Torr at 100°C), efficient mass transport at low temperature (e.g. < 200°C), a low m.p., thermal stability, chemical stability under ambient conditions (e.g. non-oxidisable, non-hydrolysable, non-photolysable), user "friendliness" (e.g. easy and reproducible synthesis, ease of handling, high purity, non-toxic, "green"), and clean and controllable decomposition under the required deposition conditions. For a HTS precursor all of these requirements also apply, but in addition there is the need to introduce oxygen into the deposit. This can be done with an external oxygen source, such as O_2, N_2O or O_3 as has already been mentioned, or with an internal oxygen source, such as is found with an alkoxide, or with a combination of both these approaches. For the MOCVD of HTS the combination approach is most often used with the external oxygen source not only being used as a source of oxygen for controlling layer stoichiometry but also as a means of reducing carbon incorporation. However, alkoxides have not, in general, been found to be very suitable precursors for CVD of HTS because of their tendency to oligomerise and to become too involatile [12]. Alternative precursors have therefore been sought and the most widely investigated have been β-diketonates, with the general molecular formula shown in Fig. 1 and with a range of substituent groups as illustrated in Table 1. The coordination via both oxygen atoms blocks some sites which might otherwise be used for bridging, while the ring substitutents reduce the likelihood of intermolecular hydrogen bonding and also act as bulky groups to sterically hinder oligomerisation. In addition the use of an electronegative substituent element such as fluorine will reduce the negative charge density on the oxygen atoms, which again will lower the tendency to form the intermolecular linkages associated with poor volatility. All of these factors assist too in minimising the air and moisture sensitivity of β-diketonate complexes.

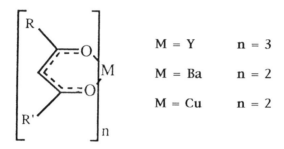

Fig. 1 General molecular formula for metal β-diketonates

Table 1 Some substitutional groups for metal β-diketonates

R	R'	Abbreviation
Me	Me	ACAC
^{t}Bu	^{t}Bu	TMHD or DPM
^{t}Bu	CF_3	TFHD or TPM
^{t}Bu	C_2F_5	PFHD or PPM
^{t}Bu	C_3F_7	HFOD
CF_3	CF_3	HFAC
CF_3	C_3F_7	DFHD
C_2F_5	C_2F_5	DFD
C_3F_7	C_3F_7	TDFND

'Y' DEPOSITION

For the deposition of Y_2O_3 the most widely used precursor is $Y(TMHD)_3$; the methyl substituted precursor, $Y(ACAC)_3$, is involatile because van der Waals forces between molecules produce high lattice energies [16]. The molecular structure has been described by Rees et al [17], and the STA (Fig. 2) shows two weak endotherms (i) probably corresponding to loss of free H-TMHD ligand (present as an impurity) at ca 70°C and residual moisture at ca. 90°C; in very carefully purified precursor these two endotherms are not present. Precursor volatilisation (ii) occurs up to and through the m.p. of ca. 170°C (iii), and there is volatilisation without any signficant signs of decomposition above ca. 200°C; the small residue of 1.6% is close to the detection limits of the TGA. The lack of any thermal events in the DTA which could be associated with precursor decomposition strongly indicates the suitability of $Y(TMHD)_3$ as a MOCVD precursor, as indeed is found, and shows the usefulness of STA as a technique for precursor screening. However, it is not just a question of precursor stability and volatility, but also of knowing the temperature dependence of that volatility in order to control carry-over and deposition rates. From experiments where $Y(TMHD)_3$ was sublimed under reaction conditions and collected in a cold trap just before the reactor is was possible to determine an effective enthalpy of vaporisation ($\Delta H_{eff,v}$) of 159 kJ mol^{-1}. This value is compared with other values reported in the literature in Table 2. The wide range of values arises since they are not all necessarily true thermodynamic values, having been determined, as in our case, under non-equilibrium conditions. In such instances $\Delta H_{eff,v}$ obtained can clearly be a function of the method used, but in addition it can be dependent on precursor pot geometry [18,19] and purity of the sample, as is discussed further below.

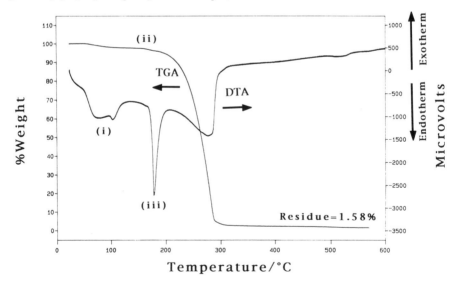

Fig. 2   Simultaneous thermal analysis (STA) of $Y(TMHD)_3$

Therefore literature values of "enthalpies of vaporisation" should be treated with great care since the majority of them are far from true thermodynamic values and they should not be quoted as such. Nevertheless, the relatively high values of $\Delta H_{eff,v}$ do show the need for careful temperature control of the precursor; a variation of $1°C$ would give rise to ca. 10% variation in carry-over rate.

Fig. 3 shows the dependence of growth rate on deposition temperature (T_d) of Y_2O_3 from $Y(TMHD)_3$ in the form of an Arrhenius plot. The presence of kinetic and transport controlled regions is evident. However, as has been pointed out elsewhere [26], there is not a sharp transition between these two regions and a true kinetic activation energy can only be obtained by allowing for the contribution of transport controlled growth at low temperatures. This is done using the simple relationship

Table 2   Comparison of effective enthalpies of vaporisation for $Y(TMHD)_3$

Reference	$\Delta H_{eff,v}$/kJ mol^{-1}	Method of Determination
[20]	130	Weight loss
[21]	98	Weight loss
[22]	61	Deposition rate
[23]	65	Deposition rate
[24]	157	Knudsen diffusion
[25]	120	Static vapour pressure
This work	159	Sublimate collection

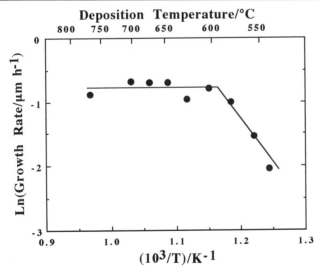

Fig. 3   Dependence of growth rate on deposition temperature of Y_2O_3 from $Y(TMHD)_3$

$(1/j_T) = (1/j_k) + (1/j_L)$ where the fluxes (j) are respectively the overall growth rate, the kinetically controlled growth rate, and the transport limited growth rate. Taking an average value of j_L of 0.47 μm h^{-1} from Fig. 3 the kinetic growth rates for T_d < 600°C can be calculated. An Arrhenius plot of these j_k values gives a kinetic activation energy of ca. 300 kJ mol^{-1}. This is significantly higher than the uncorrected value of 114 kJ mol^{-1} obtained from Fig. 3 and also that reported in the literature [20] of 120 kJ mol^{-1}, which was also obtained directly from an uncorrected Arrhenius plot. The effect of mass transport which, in turn, depends on reactor geometry and gas flow characteristics is very apparent as is the need to correct for it in order to obtain true kinetic parameters.

From the point of view of material quality, XRD analysis of films grown either in the transport controlled region (e.g. T_d = 700°C) or well into the region where kinetics play a significant role (e.g. T_d = 535°C) showed that crystalline Y_2O_3 was produced in both cases [27]. The main effect of depositing at higher temperatures was to promote c-axis oriented crystal growth and to increase grain size with a resultant decrease in film smoothness. For HTS films based on YBCO it has been found [28] that for high current densities epitaxial films are required, and so, based on the results for Y_2O_3 deposition, a high deposition temperature is indicated. Also rougher layers have been shown [29] to lead to flux pinning with a resulting slower decrease in critical current with the applied field, again suggesting higher temperature deposition is desirable. However, if layers become too rough then photolithography for device fabrication would become more difficult and so crystalline growth cannot be allowed to occur too extensively. Other parameters such as partial pressure and carrier gas flow could be adjusted to influence layer morphology [27].

The results presented above have provided some basic information about the use of Y(TMHD)$_3$ as a precursor for CVD of Y_2O_3 and, by implication, for YBCO. As already indicated, it is widely used for this purpose. What has not been said, and is also practically never mentioned in publications on MOCVD with Y(TMHD)$_3$, is the variability in growth rates under seemingly identical conditions. Fig. 4 shows the spread of results obtained with different samples of Y(TMHD)$_3$ obtained over a period of several months. Studies [27] of the underlying causes of the lack of reproducibility have strongly suggested that the principal factor is variations in sublimation rate. For samples of Y(TMHD)$_3$ from different sources this could result from differences in the degree of purity; this problem is discussed further in the context of barium precursors below. For identical samples of Y(TMHD)$_3$ from the same batch it was found that reproducibility of deposition was acceptable (± 10%) on a short time basis (8-10 hours), but it deteriorated (± 30%) when the precursor pot was allowed to cool down and then reheated (e.g. from day-to-day), and especially when the pot was refilled with precursor sample. This deterioration is unlikely to be due to chemical changes in the sample because of its thermal stability (cf. Fig. 2) but is most probably attributable to changes in surface area on thermal cycling of the precursor exposed to the carrier gas and also resulting differences in thermal contact between the sample and the container. These results emphasise the need for the possible development of a precursor container capable of achieving stable sublimation rates and/or the use of a technique for in situ monitoring and control of sublimation rates [30]. An alternative strategy is to use a precursor which is a liquid under the sublimation conditions. Such a precursor for the MOCVD of Y_2O_3 has been synthesised [31] with the ligand ^{t}Bu-pyNO complexed with Y(TMHD)$_3$.

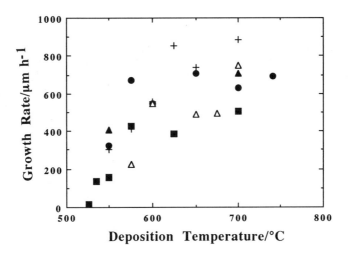

Fig. 4 Growth rate of Y_2O_3 from $Y(TMHD)_3$ as a function of temperature under otherwise "identical" conditions

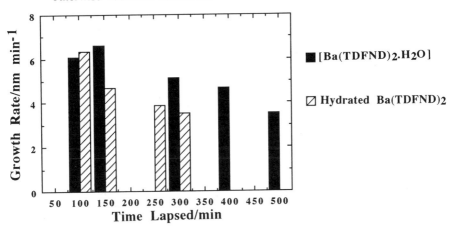

Fig. 5 Growth rate of BaF_2 from $[Ba(TDFND)_2 \cdot H_2O]$ and hydrated $Ba(TDFND)_2$

'Ba' DEPOSITION

Although there are problems associated with 'Y' deposition these pale into insignificance compared with those encountered in obtaining a suitable Ba precursor and reproducible results for 'Ba' deposition. The most commonly used barium precursors are hydrates of β-diketonates (Fig. 1) with R = R' = CF_3 (HFAC) or R = ^tBu and R' = C_3F_7 (HFOD). All of these compounds are oligomeric and of low volatility and they decompose close to their sublimation temperatures; this is clearly shown in their STAs [14]. This means that the reproducibility of the barium content of the gas phase is poor, and that fresh samples are required for each growth run. Several approaches have been proposed to overcome the problem [15], but the one that we have employed has been to increase the fluorine content of the ring substituents to the extent of all four groups being C_3F_7; i.e. $Ba(TDFND)_2$ [14,15]. In spite of a formula weight increase of nearly three times over that of $Ba(TMHD)_2$ the STA of the monohydrate of this compound showed it [14] to be more volatile than $Ba(TMHD)_2$ and to sublime completely, after loss of water, without decomposition at one atmosphere pressure. This was the first barium complex to show stable and complete volatilisation at ambient pressure. Thus the intermolecular forces of any oligomeric structures are weaker than for other barium complexes and suggest that it should be a good MOCVD precursor. However, when used at 160°C for the CVD of BaF_2 it was found that [15] whilst there was a very high initial deposition rate there was a sequential and marked drop in the rate when the same precursor material was used for a series of depositions. Apart from a slight discoloration of the precursor sample during use there was no evidence of significant decomposition of the precursor, in agreement with the STA result, and in fact ^1H nmr and ^{19}F nmr spectra run on fresh and used samples were essentially identical [14,15]. Also microanalysis results for C and H for hydrated and anhydrous samples both before and after use closely correlated with expected values showing only the loss of water during use. Thus there is no evidence of any significant compositional changes on heating the precursor to just below its m.p. in order to volatilise it for CVD. There was, though, a slight increase in the m.p. of the precursor from 187°C before use to 196°C after use. This we have attributed [15] to structural changes in the solid and/or melt which lead to stronger intermolecular forces and a resulting slow decrease in volatility with increased use. This concept is given some credence by a comparison of growth rates of BaF_2 from a sample of monohydrate $[Ba(TDFND)_2H_2O]$ and hydrated $Ba(TDFND)_2$ - Fig. 5. There is a striking similarity in the fall in growth rate with usage time for both samples. Furthermore, XRDs for a hydrated sample of $Ba(TDFND)_2$ before and after use as a precursor showed evidence of crystal growth occurring on heating. Certainly changes in crystalline structure would be expected to lead to differences in evaporation rate.

Additional support for the effect of loss of water of crystallisation was given by a series of 19 growth runs done under identical conditions over a period of two months with an anhydrous precursor sample. This gave very constant and stable deposition rates: 2.95 ± 0.09 nm min^{-1}. However, even here the situation was not completely straightforward. That result was obtained using the same sample of precursor throughout. A second sample of anhydrous $Ba(TDFND)_2$ also gave self-consistent results, but different from the first sample: 2.11 ± 0.04 nm min^{-1} for six runs. Yet another anhydrous sample even showed a fall off in growth rate (3.1 to 2.5 nm min^{-1}) with time, although the percentage decrease (19%) was less than that for a hydrated sample; e.g. 40% for $[Ba(TDFND)_2.H_2O]$, cf Fig. 5. Since all the samples had been prepared by an identical synthetic route, the differences in the results for the rate of deposition suggested varying levels of an impurity in the original samples, possibly acting as a Lewis base and giving rise to varying degrees of intermolecular bonding and volatility. This would manifest itself in a similar effect to that observed for the presence of water of crystallisation already described, with a decrease in volatility and hence deposition rate with increasing use of an impure sample as the impurity is lost by evaporation; indeed, the loss of water, itself acting as a Lewis base, from a hydrated sample may not just result in crystallinity changes but also in enhanced intermolecular bonding and oligomerisation as well. A test of this idea of Lewis base adducts being formed from impurities is shown in Fig. 6. This shows growth rates of BaF_2 from samples of anhydrous $Ba(TDFND)_2$. Carefully purified precursor (a) shows no variation of deposition with time while a sample which was slightly "wet" due to retention of unreacted ligand HTDFND (b) shows a high initial deposition rate and then a gradual fall off with usage; i.e. as the more volatile HTDFND is lost by evaporation. Taking the pure sample (a) and deliberately contaminating it with HTDFND (sample (c)) shows behaviour remarkably similar to that of the impure sample. This result underlines the comment made earlier about lack of reproducibility of growth rates resulting from variations in precursor sublimation rates and highlights the importance of the need to take great care in purifying precursor samples.

However, even if very pure samples are used there is still the possibility with solid precursors of crystalline changes occurring on repeated heating and cooling leading to different degrees of volatility and carry over rates; the XRDs of anhydrous $Ba(TDFND)_2$ before and after use show evidence of annealing and crystallite growth similar to that found for the initially hydrated samples receiving the same treatment. Therefore, as with the case of Y precursors, a Ba precursor which is a liquid at the temperature of sublimation would be preferable. We have recently reported [19] on the tetraglyme complex of $Ba(TDFND)_2$ which has a m.p. of $70^{\circ}C$ and is the most volatile and stable barium precursor under atmospheric conditions yet reported. Of course, for YBCO preparation use of either $Ba(TDFND)_2$ or the tetraglyme complex requires in situ or post deposition hydrolysis of the fluoride which is deposited rather than the oxide. This is a drawback, but no non-fluorine containing barium precursors, to our knowledge, show the same degree of stability, volatility and reproducibility that we have found for the TDFND complexes. The BaF_2 which is deposited from both precursors has been shown [19,27] to be of a high purity and highly orientated, provided that growth is carried out at a high enough temperature ($>450^{\circ}C$) corresponding to mass transport controlled growth. Used in conjunction with $Y(TMHD)_3$ and $Cu(TMHD)_2$, HTS layers of YBCO have been grown [19,27]. Therefore the precursors are extremely interesting and promising candidates for not just MOCVD of BaF_2 but for HTS YBCO as well.

DEPOSITION MECHANISMS

In this section we restrict our comments primarily to deposition from Ba precursors. First, we note that whereas deposition from $Ba(TDFND)_2$ produces BaF_2 films for which the refractive index (n) has a value equal to the bulk value of 1.474 and is independent of deposition temperature, for deposition from the tetraglyme precursor only films deposited under mass transport controlled conditions have n = 1.474; at lower temperatures the value of n decreases quite markedly [19]. This we have attributed to the fact that under kinetic controlled deposition not all of the precursor reaching the surface will undergo decomposition and unreacted, or partially reacted, species could be incorporated into the layer, hence leading to a low n value. This concept is supported by the fact that as the partial pressure of the precursor was

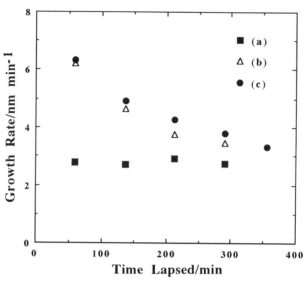

Fig. 6 Growth rate of BaF_2 from samples of $Ba(TDFND)_2$ of varying purity (a) Dry (b) Slightly wet due to retention of HTDFND (c) As in (a) but with 10% HTDFND added

lowered, at a temperature corresponding to mixed transport and kinetic control at a higher precursor partial pressure, the refractive index rose until it became equal to 1.474 again. This is understandable in terms of the shape of a curve for growth rate as a function of reciprocal temperature, where as reactant partial pressure decreases the transport controlled region extends to lower and lower temperatures [26]. The fact that films grown by deposition from $Ba(TDFND)_2$ do not show this effect is probably associated with the higher activation energy for the decomposition of the tetraglyme complex as compared with that for the decomposition of $Ba(TDFND)_2$ alone [19].

The second point to comment on is the fact that deposition from either of the Ba precursors gives BaF_2 even though oxygen is present in the reactor. This is also true for the case of MOCVD from $Ba(HFAC)_2$. On the other hand, deposition from $Cu(HFAC)_2$ in the presence of oxygen gives CuO [32]. It has been shown [33] that there is considerable intermolecular bonding in $Ba(HFAC)_2$ in the solid state, in particular involving the F atoms of the CF_3 groups of one molecule forming a bridging bond to the Ba atom of an adjacent molecule. If this type of bonding is retained in the vapour phase, and there is evidence for this from mass spectrometry [14], then when thermal molecular decomposition starts to occur there is already a link for the formation of the Ba-F bond, which apparently is dominant even though the Ba is also attached to O atoms in the ring structure. One might expect similar behaviour for the TDFND complexes. The reason why the intermolecular bonding occurs so readily in the case of Ba precursors is because of the ability of Ba to increase its coordination sphere and form oligomeric clusters. In contrast, copper complexes in general, and β-diketonate complexes in particular, are usually 4 coordinate and square planar [34]

Any tendency to higher coordination is provided by long range intermolecular interactions resulting in less stable complexes. Therefore there is less likelihood of a Cu-F bond being formed on decomposition of the precursor than there is in the case of the Ba complexes. Even when $Ba(TDFND)_2$ is complexed with tetraglyme, when one would expect on the basis of structural studies of the analogous [$Ba(HFAC_2$.tetraglyme] [35] the Ba to possibly be already at least 9 coordinate and to have a globular appearance with the Ba atom well shielded from neighbouring molecules, BaF_2 is still formed. This is probably because although the tetraglyme assists volatility it comes off before the TDFND complex decomposes; mass spectral evidence supports this idea. Therefore the Ba atom is no longer protected from intermolecular coordination with F and BaF_2 is again formed. These and other aspects of deposition mechanisms are being studied at the moment with techniques for in situ analysis.

ACKNOWLEDGEMENTS

We acknowledge the support of the Commission of the European Communities under the BRITE/EURAM Programme, Contract No. BREU/0438. We also wish to thank The Associated Octel Co. Ltd for two studentships (SCT and DDG) and for other financial support, and Mr. R. P. McGinty and his colleagues for performing the thermal analyses.

REFERENCES

1. J. G. Bednorz and K. A. Mueller, Z. Phys. B. 64, 189 (1986)
2. M. K. Wu, J. R. Ashburn, C. J. Torng, P. H. Hor, R. L. Meng, L. Gao, Z. J. Huang, Y. Q. Wang and C. W. Chu, Phys. Rev. Letts. 58, 908 (1987)
3. A. D. Berry, K. D. Gaskill, R. T. Holm, E. J. Cukauskas, R. Kaplan and R. L. Henry, Appl. Phys. Lett. 52, 1743 (1988)
4. A. D. Berry, R. T. Holm, E. J. Cukauskas, M. Fatemi, D. K. Gaskill, R. Kaplan and W. B. Fox, J. Crystal Growth 92, 344 (1988)
5. D. S. Richeson, L. M. Tonge, J. Zhao, J. Zhang, H. O. Marcy, T. J. Marks, B. W. Wessels and C. R. Kannewurf, Appl. Phys. Lett. 54, 2154 (1989)
6. R & D in Brief, R. & D Magazine June 1993, p.5
7. W. Gosling in Superconductivity: The Opportunities, Implications and Applications, edited by A. Conway (IBC Technical Services Ltd, London, 1988), p.25
8. Science and the Citizen, Scientific American 264, 15 (1991)
9. Industry News, Semiconductor International, April 1993, p.18
10. News Item, Chem. Eng. P. Sept. 1989, p.44
11. J. Zhao, C. S. Chern, Y. Q. Li, D. W. Noh, P. Norris, P. Zawadzki, B. Kear and B. Gallois, J. Cryst. Growth 107, 699 (1991)
12. M. L. Hitchman, D. D. Gilliland, D. J. Cole-Hamilton and S. C. Thompson, in *New Materials and their Applications (Inst. Phys. Conf. Ser. III)*, edited by D. Holland (Institute of Physics, Bristol, 1991), p. 305.
13. M. Leskela, H. Molsa and L. Niinisto, Supercond. Sci. Technol. 6, 627 (1993)
14. S. C. Thompson, D. J. Cole-Hamilton, D. D. Gilliland, M. L. Hitchman and J. C. Barnes, Adv. Mat. Opt. Electron, 1, 81 (1992)

15. D. D. Gilliland, M. L. Hitchman, S. C. Thompson and D. J. Cole-Hamilton, J. Phys. III France 2, 1381 (1992).
16. R. E. Sievers and J. E. Sadlowski, Science 201, 217 (1978)
17. W. S. Rees, H. A. Luten, M. W. Carris, E. J. Doskocil and V. L. Goedken, MRS Symp. Proc. 1992, 141
18. E. Fitzer, H. Oetzmann, F. Schmaderer and G. Wahl in *Proc. Eighth Eur. Conf. CVD*, edited by M. L. Hitchman and N. J. Archer (Les Editions de Physique, Paris, 1991), p.C2-713
19. S. H. Shamlian, M. L. Hitchman, S. L. Cook and B. C. Richards, J. Mat. Chem., in press
20. F. Schmaderer, R. Huber, H. Oetzmann and G. Wahl, in *Proc. Eighth Eur. Conf. CVD*, edited by M. L. Hitchman and N. J. Archer (Les Editions de Physique, Paris, 1991), p.C2-539
21. S. H. Kim, C. H. Cho, K. S. No and J. S. Chun, J. Mat. Res. 6, 704 (1991)
22. A. Erbil, K. Zhang and E. P. Boyd, SPIE, 1187, 104 (1989)
23. C.I.M.A. Spee, E. A. Van der Zouwen-Assink, K. Timmer, A. Mackor and H. A. Meinema, in *Proc. Eighth Eur. Conf. CVD*, edited by M. L. Hitchman and N. J. Archer (Les Editions de Physique, Paris, 1991), p.C2-295
24. R. Amano, A. Sata and S. Suzuki, Bull. Chem. Soc. Jap. 54, 1368 (1981)
25. C.I.M.A. Spee and L. Saunders (private communication)
26. M. L. Hitchman, Progr. Crystal Growth 4, 249 (1981)
27. D. D. Gilliland, PhD thesis, University of Strathclyde, 1993
28. D. Caplin, Nature 335, 204 (1988)
29. Y. Q. Li, J. Zhao, C. S. Chen, P. Lu, B. Gallois, P. Norris, B. Kear and F. Cosandey, Physica C 195, 161 (1992)
30. E. J. Thursh, C. G. Cureton, J. M. Trigg, J. P. Stagg and B. R. Butler, Chemtronics 2, 62 (1987)
31. K. Timmer and C.I.M.A. Spee, Dutch Patent Application No. 9302030, 1993
32. K. Shinohara, F. Munakata and M. Yamanaka, Jap. J. Appl. Phys. 27, L1683 (1988)
33. D. C. Bradley, M. Hasan, M. B. Hursthouse, M. Motevalli, O. F. Z. Khan, R. G. Pritchard and J. O. Williams, J. Chem. Soc., Chem. Commun. 1992, 575
34. B. J. Hathaway, in *Comprehensive Coordination Chemisty, Vol. 5*, edited by G. Wilkinson, R. D. Gillard and J. A. McCleverty (Pergamon, Oxford, 1987), p.533
35. K. Timmer, C.I.M.A. Spee, A. Mockor and H. A. Meinema, Inorg. Chim. Acta. 190, 109 (1992)

PLASMA-ENHANCED MOCVD OF SUPERCONDUCTING OXIDES

Kenji EBIHARA*, Tomoyuki FUJISHIMA*, Masanobu SHIGA**, and Quanxi JIA***, *Department of Electrical Engineering and Computer Science, Kumamoto University, Kurokami, Kumamoto 860, Japan, **Dojindo Laboratories, Kumamoto Techno Research Park, Kumamoto 861-23, Japan, *** Department of Electrical and Computer Engineering, State University of New York at Buffalo, Bonner Hall, Amherst, New York 14260

ABSTRACT

The plasma-enhanced MOCVD is developed to prepare high-Tc oxide superconducting thin films. Plasmas generated by microwave and rf discharges decompose effectively the source materials (ß-diketonate chelates) into their elements and oxides. YBaCuO thin films were deposited on the MgO substrate of 500-650 °C at total pressure of 0.6-5 Torr with O_2 contents less than 30% . The as-grown films produced by two kinds of plasma enhancement were porous and consisted of crystalline grains, but showed the superconducting transition after heating procedure at around 800 °C. It is shown that the inherent crystalline orientations of the as-grown films determine the crystal structure of the post-annealed films. The film of the metal atomic ratios of 0.91 for Ba/Y and 2.64 for Cu/Y showed the superconducting properties with $T_{c(zero)}$ of 89 K and the critical current density (at 77 K) of 5×10^4 A/cm^2. Spectroscopic analysis showed that the plasmas are composed of many excited species such as Y, Y$^+$, Ba, Ba$^+$, Cu, YO, BaO, CuO. Formation of the metal-oxides through the gas phase reaction is essential for the high-quality YBaCuO superconducting thin film preparation in the PE-MOCVD.

INTRODUCTION

Plasma-enhanced metalorganic chemical vapor deposition (PE-MOCVD) has been widely applied to preparation of silicon based materials , metal oxides and GaAs. Since in this technique the source gases are decomposed by collisional excitation with activated species and high energy electrons, plasma properties such as ionization degree, plasma temperature and plasma density affect significantly growth process and physical and crystallographic properties of the deposited films. The discovery of the high-Tc superconductivity of the YBaCuO and BiSrCaCuO systems has stimulated the development of the advanced process technology suitable to the high-Tc superconducting thin film preparation. The PE-MOCVD has been attempted to fabricate the oxide superconducting thin films using various metalorganic precursors[1,2]. The well-controlled PE-MOCVD systems can provide high quality superconducting transition similar to the oxide superconducting thin films by other techniques like the plasma sputtering and the laser ablation depositions[3-7]. Much effort has been devoted to improving the film quality generated by the PE-MOCVD. Although the PE-MOCVD has possessed inevitable difficulty for high-Tc superconductor preparation, many attempts for his PE-MOCVD process have been made to prepare the high-quality superconducting films satisfying the requirement for the microelectronic devices. These studies include the new development of metalorganic precursors, processing plasma control, monitoring the plasmas, and relaxation of the processing conditions such as substrate temperature.

We have developed plasma-enhanced MOCVD for preparation of high-Tc oxide superconducting thin films[8-11]. In our processes several types of discharge plasmas have been applied to decomposition of a mixture of precursor and dilute gases. The YBaCuO films were deposited at various deposition conditions to clarify optimum parameters. Optical emission spectroscopy and mass spectroscopic analysis were used to identify the decomposed species during the film depositions.

PLASMA-ENHANCED MOCVD SYSTEMS FOR HIGH-Tc SUPERCONDUCTING FILMS

In our PE-MOCVD processes, the discharge plasmas were produced by an rf planar discharge and a microwave discharge. These plasmas provide the different decomposition of the source materials depending on the energy distribution of the decomposed species such as electrons and activated ions. The rf plasma enhanced MOCVD system was demonstrated previously[12]. An rf plasma was generated between a pair of electrodes connected to an rf generator (13.56 MHz). The substrates of MgO(100) were placed on the electrodes which were heated to 500-800 °C. Total pressure during the film deposition was in the range from 0.6 to 5 Torr and rf power of 100 W corresponding to a power density of about 0.5 W/cm^2 was supplied to the electrodes. Typical deposition conditions for the rf PE-MOCVD are summarized in Table 1.

Table 1. Deposition conditions of rf PE-MOCVD and microwave PE-MOCVD.

Process	rf PE- MOCVD		MicrowavePE-MOCVD
Frequency	13.56 MHz		2.45 GHz
Power(W)	100		100-400
Ambient gas	Ar + O$_2$		
Total pressure	0.6-4 Torr		0.6 - 5 Torr
O$_2$ pratial pressure	0 - 0.5 Torr (30 %)		0 - 0.5 Torr (30 %)
Substrate	MgO(100)		
Substrate temperature	500 - 650°C		500 - 650°C
Deposition time	45- 60 min.		20 - 45 min
Metalorganic materials	Y(dpm)$_3$	Ba(dpm)$_2$	Cu(dpm)$_2$
Sublimation temperature	120 - 190 °C	210 -270°C	120 - 200°C
Post-annealing	1 atm air, 800 - 850°C		as-grown

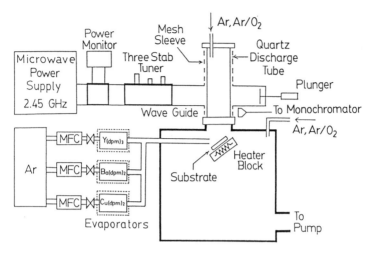

Fig. 1. A schematic diagram of the microwave PE-MOCVD system.

Figure 1 shows a schematic diagram of a downstream microwave plasma-enhanced MOCVD system[9, 10]. The microwave power from the 2.45 GHz oscillator was supplied to a quartz tube (dia. 50 mm) through a rectangular wave guide (WR430). Chelates of Y, Ba and Cu were used as precursors. Metalorganic compounds of $Y(dpm)_3$, $Ba(dpm)_2$, and $Cu(dpm)_2$ (dpm = dipivaloymethane, $C_{11}H_{19}O_2$) were sublimated under appropriate temperatures. The compounds were loaded into individual evaporators of cylindrical tubes passed by carrier gas. The evaporators were maintained at 120-190, 210-270, 110-200 °C for the Y, Ba, and Cu compounds, respectively. The carrier gases of helium, argon and a mixture of Ar/O_2 were fed at flow rates of 15-40 sccm for Y, 30-40 sccm for Ba and 15-40 sccm for Cu source. The gas mixture leaving the evaporators was transported in the heated tubes in order to avoid the precursor condensation and deposition in the gas transportation system. The MgO(100) substrates were located on a susceptor heated up 500-650°C. Table 1 also shows the deposition conditions for the microwave plasma-enhanced MOCVD system. The optical emission from the YBaCuO plasmas was measured using a 0.25-m monochromator equipped with a photomultiplier. The digitized signals were transferred to a microcomputer to investigate the plasma properties. The deposited films were characterized using x-ray diffraction, scanning electron microscopy (SEM), and resistivity versus temperature measurements. The composition of each metal element was measured by electron probe microanalysis (EPMA) which was calibrated by the $YBa_2Cu_3O_{7-x}$ superconductor.

RESULTS AND DISCUSSION

Optical emission spectroscopy for the processing plasmas

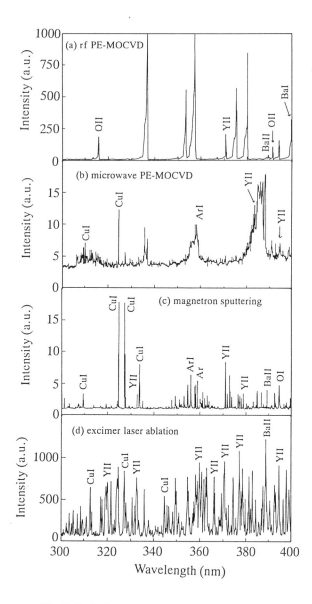

Fig. 2. Emission spectra of the process plasmas.

Optical emission spectroscopy has provided useful information about decomposition of precursors and film growth. We attempted to identify the species decomposed during the PE-MOCVD process. Figure 2 shows typical spectra of YBaCuO plasmas generated by rf and microwave discharges. In order to make clear the features of decomposition in the PE-MOCVD process, the spectra of a magnetron sputtering and an excimer laser ablation are also illustrated. The deposition conditions for the rf discharge plasma was the total pressure of 0.8 Torr (partial pressure of O_2 : 150 mTorr) and rf power of 100 W. The microwave plasma was generated at the conditions of total pressure of 0.8 Torr (partial pressure of O_2 : 60 mTorr) and microwave power of 400 W. These spectra of Fig. 2(a) and (b) indicate that Y, Y^+, Ba, Ba^+, Cu elements are generated by decomposing the precursors with rf and microwave plasmas. We also notice appearance of band spectra due to molecular species such as metal oxides (YO, BaO, CuO) and CH. On the other hand, in the physical vapor deposition like the magnetron sputtering and the excimer laser ablation, the observed spectral lines of the metal elements have strong intensity and narrow line width as shown in Fig. 2(c) and (d). We estimated the plasma temperatures by means of the relative spectral intensity method. Several spectral lines originated in Y, Ba, Cu elements were used to determine the excitation temperatures. The results showed that the excitation temperature of the PE-MOCVD is around 4000 K ,while the plasmas in the magnetron sputtering and the laser ablation have high temperatures of about 10000 K. In order to avoid the degradation of the film quality caused by activated ion bombardment in the PE-MOCVD, plasma temperature is required to maintain at appropriate temperatures by controlling the deposition conditions. It was found that the increase of oxygen partial pressures leads to decreasing the spectral intensity due to metal oxides formation in gas phase.

Surface morphology of the PE-MOCVD films

The films were prepared under various deposition conditions in the rf and the microwave PE-MOCVD. The as-deposited films with dull gray were electrically conducting. Figure 3 shows SEM micrographs of the YBaCuO film surface. The film by rf PE-MOCVD (Fig. 3(a))

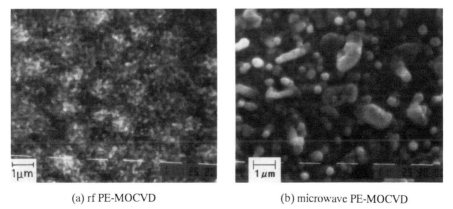

(a) rf PE-MOCVD (b) microwave PE-MOCVD

Fig. 3. SEM micrographs of the as-grown films by the rf and the microwave PE-MOCVD.

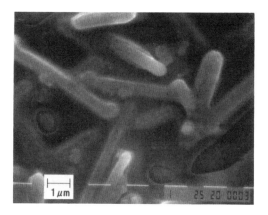

Fig. 4. SEM micrograph of the annealed film.

was prepared at the substrate temperature of 600 °C, the total pressure of 0.9 Torr (O_2 :0.15 Torr) and rf power of 100 W. The morphology of the film is granular-like with small grain sizes of 0.2 - 0.5 μm. X-ray diffractory provides weak signals due to polycrystalline structure. Figure 3(b) is the microstructure of YBaCuO films prepared by the microwave PE-MOCVD at substrate temperature of 620°C, the total pressure of 1 Torr (O_2 :0.2 Torr), and power of 100 W. The grain size is slightly larger than that of the rf PE-MOCVD films. The YBaCuO films at the reduced substrate temperatures around 600 °C did not show highly textured structures composed of needle-like grains. Post-annealing procedure of these as-grown films improved the morphology as shown in Fig. 4. This micrograph is the SEM of the post-annealed film prepared by the microwave PE-MOCVD. The as-grown film was annealed at 800 - 880°C for 30 min.

Crystallographic orientation

X-ray diffraction spectra from YBaCuO films prepared at various deposition conditions are shown in Fig. 5 and Fig. 6. Figures 5(a) and (b) are x-ray diffraction patterns for the as-grown films prepared by rf PE-MOCVD and microwave PE-MOCVD, respectively. Surface morphology and the deposition conditions for these samples are given in Fig. 3(a) and (b). The results show that the x-ray pattern for the microwave PE-MOCVD film has pronounced intensity of (00n) lines, indicating c-axis orientation perpendicular to the substrate surface. Polycrystal structure of the rf PE-MOCVD film with (112), (005) and (200) is due to crystal growth caused by the ion bombardment on the film surface. Figure 6 shows that the post-annealing procedure of the as-grown films enhances crystallization of the specified orientation which appeared in the as-grown film. Table 2 shows the c-axis length of the YBaCuO films which was estimated from the x-ray diffraction measurements.

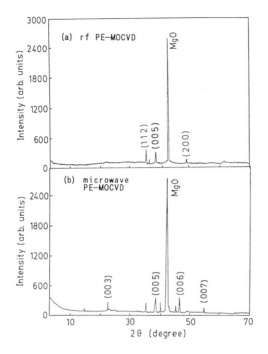

Fig. 5. X-ray diffraction spectra of the as-grown films.

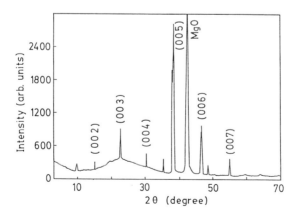

Fig. 6. X-ray diffraction spectrum of the annealed film.

Table 2 c-axis length of the YBaCuO thin films by PE-MOCVD.

Plane of Crystal	c-axis length (Å)			
	as-grown		annealed	
	rf	microwave	rf	microwave
(005)	11.615	11.725	11.730	11.675
(006)	---	11.724	11.724	11.676
(007)	---	11.711	---	11.676
The c-axis length Orthorhombic cell ($YBa_2Cu_3O_{0.85}$) : 11.669Å, Tetragonal cell ($YBa_2Cu_3O_6$) : 11.738Å.				

Post-annealing greatly improves the surface morphology and eliminates the porosity, providing good superconducting properties of high critical temperature and critical current density. It is clear from Table 2 that after the heat treatment the c-axis length of the microwave PE-MOCVD films is nearly equal to the c-lattice parameter (c_0 = 11.67 Å) of bulk ceramics. It has been reported that there is a close relation between critical temperature and c-lattice parameter for the YBaCuO superconducting thin films. It has been shown that the zero resistivity temperature of about 90K is attained in the range of c_0=11.70-11.73 Å.

Stoichiometry of the deposited films

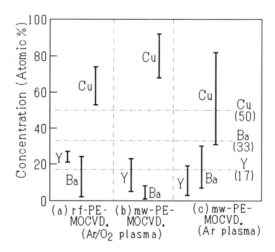

Fig. 7. Dependence of deposition processes on composition ratios of as-grown films.

The composition of the YBaCuO films deposited by various processing conditions was measured by EPMA. Figure 7 shows the composition of the deposited films. The stoichiometric $YBa_2Cu_3O_{7-x}$ film has elemental concentrations of Y=17%, Ba=33% (Ba/Y=2) and Cu=50% (Cu/Y = 3). We notice that the deposited films are deficient in Ba component and rich in Cu element. The concentration ratio of each element is affected by the mixture ratio of the carrier gases, the substrate temperature and the supplied power. In these film depositions the element concentration of the precursors was varied in the range of Y:Ba:Cu = 1: 2.6-6:1.5-3. The scattering of the compositional ratios in Fig. 7 is mostly due to the variety of mixture vapor composed of sublimated precursors. When only argon gas is used as carrier gas, high decomposition of source materials is attained without forming metaloxides. Although the deposited films under this condition provided stoichiometric compositions, the superconductivity was not attained. Since the films deposited without oxygen carrier gas are deficient in oxygen, the post-annealing in oxygen ambient is required to make phase change to the orthorhombic structure.

Superconducting properties

The temperature dependence of the resistivity was measured by the DC four probe method with Ag electrodes sputtered on the films. The current density for the resistivity measurement was 0.1 mA (about 10 A/cm^2). Most of the as-grown films showed high resistivity and non-superconducting properties like semiconducting behavior. Since these films were deposited at relative low substrate temperature about 600°C, crystalline growth and oxygen incorporation were insufficient for providing superconducting transition over liquid nitrogen temperature (77 K). Figure 8 shows the typical results of the temperature versus resistivity curves under various

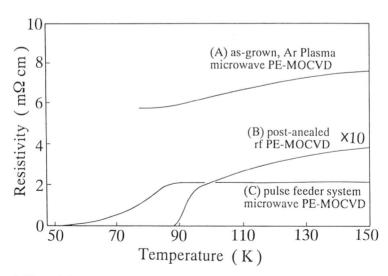

Fig. 8. The resistivity-temperature characteristics of the YBaCuO thin films by PE-MOCVD.

deposition conditions. The curve (A) for the as-grown film prepared at 650 °C shows semiconducting behavior. The superconducting transition was achieved by annealing the as-grown films at 880 °C in air. The post-annealed film as demonstrated in the curve (B) has zero resistivity critical temperature of 89 K and the critical current density of 5×10^4 A/cm^2 at 77 K and zero magnetic field.

It was found in this experiment about the PE-MOCVD process that the reduction of the substrate temperature leads to degradation of crystallinity, off-stoichiometry of compositional ratios, insufficient oxygen incorporation and poor reproducibility of the deposited film properties. In order to eliminate these problems in the PE-MOCVD process, we have developed new supply system of source materials. Sequential metalorganic source supply for the layer-by-layer film deposition was reported previously[13]. In the present PE-MOCVD system, the pulsed feeder system supplied in pulse each vapor of source material to the decomposition / reaction region. Figure 9(C) shows the temperature versus resistivity profile for the as-grown film which was prepared at substrate temperature of 670 °C, total pressure of 5 Torr (partial pressure of oxygen : 1.7 Torr) and microwave input power of 200 W. The film possesses the superconducting transition with the onset temperature Tc(onset) of 87 K and the zero resistivity temperature Tc(zero) of 53 K. We can notice the wide transition temperature ΔT of 30 K, indicating the insufficient crystal growth due to deficient oxygen, crystal defect, and grain boundary.

<u>Comparison with other YBaCuO thin films prepared by laser ablation technique and plasma sputtering technique</u>

The high Tc superconducting thin films have been prepared using the physical vapor deposition(PVD) such as plasma sputtering and laser ablation techniques. It is the essential difference between the PE-MOCVD and the PVD that the sources for the film deposition are the sublimated vapor for the PE-MOCVD and the sintered solid targets for the PVD process. A pulsed laser deposition (PLD) has shown great promise in producing high Tc oxide superconducting thin films with good epitaxy, and high critical current density and transition temperatures. In this comparative experiment, a pulsed KrF excimer laser (Lambda Physik LPX305icc, wavelength = 248nm, pulse duration=25 ns) was used to ablate a $YBa_2Cu_3O_{7-x}$ superconducting target in a deposition chamber[14]. The YBaCuO film was deposited on the MgO substrate heated at 710 °C in the oxygen ambient.

A magnetron sputtering technique has been widely used for the preparation of high Tc oxide superconducting thin films. In conventional low pressure magnetron sputtering, the composition of the deposited films is known to be quite different from that of the target due to resputtering of the film by negative ions. A facing targets magnetron sputtering (FTMS) process is one of approaches to reduce the resputtering and ion bombardment on the films. Our experimental system was illustrated somewhere[15].

Table 3 shows the superconducting properties of the films prepared using PE-MOCVD, the pulsed laser deposition and the facing targets magnetron sputtering. The microwave downstream plasma technique using metalorganic precursors has high possibility for an

Table 3. Superconducting properties of the films deposited by PE-MOCVD, PLD and FTMS.

Process	Plasma production	Pressure (Torr)	Substrate temperature(°C) (MgO)	Tc(onset) Tc(zero)	Jc (A/cm^2)	Other conditions
PE-MOCVD	rf discharge	0.4-2.5	600 (880)	93-90 87-85	10^4	Annealing:880°C Power:100W
	microwave discharge	5(in pulse)	670	87 53	10^3	as-grown, power:200W
Pulsed laser ablation (PLD)	plasma plume (laser induced discharge)	0.15(O_2)	710	92 91	5×10^5	as-grown, KrF laser energy density:2J/cm^2
Facing target magnetron sputtering	magneto plasma (DC and rf discharge)	0.06 (O_2:40%)	650	87 81	10^4	as-grown, rf power:50W

alternative process for high Tc superconducting thin films. Precious control of the sublimation process for the precursors and the purification of the source materials are required to improve the superconducting properties. The pulsed laser deposition provides high quality superconducting thin films with Tc(zero) of 91 K and Jc of 5×10^5 A/cm^2. But this PLD process also has problems such as deposition of small droplets on the film surface. Furthermore, new additional technology controlling the laser-induced plasma is necessary to obtain the large-area and uniform thin films when the deposited films will be applied to microelectronic devices.

CONCLUSIONS

We have developed the plasma enhanced MOCVD processes using rf capacitive and microwave downstream discharges. Spectroscopic measurements showed that the metalorganic precursors as the source materials in the PE-MOCVD were decomposed into their elements such as Y, Ba, Cu and their oxides. The as-grown films deposited at low substrate temperatures around 600°C did not demonstrate the superconducting transition over 77K. The post-annealing procedure was needed to make phase transition to the orthorhombic structure. Although precious process control and annealing procedure at around 800°C are necessary to fabricate high quality superconducting thin films, the PE-MOCVD has still remarkable advantages of large area processing and high deposition rate. Active approaches to overcome some obstacles have been continued[16].

ACKNOWLEDGMENTS

This research was supported in part by a Grant-in-Aid for Scientific Research in Priority Area from the Ministry of Education, Science and Culture of Japan. The authors wish to thank Dr.T.Ikegami, Dr.Y.Yamagata, Dr.S.Kanazawa, Y.Yoshimoto, K.Yufu and K.Suga for their help during this experiment. The authors also would like to acknowledge Dr.M.Ohno, Dr.J.Suzuki, Dr.K.Sakai, Hitachi VLSI Engineering Co.Ltd., and K.Harada, Denki Kogyo Co.Ltd., for their continued encouragement and support.

REFERENCES

1. G.R.Bai, W.Tao, R.Wag et. al, Appl. Phys. Lett., **55**, 194 (1989).
2. K.Kobayashi, S.Ichikawa and G.Okada, Jpn. J. Appl. Phys., **28**, L2165 (1989).
3. J.Zhao, D.W.Noh, C.S.Chern, Y.Q.Li, P.Norris, B.Gallois and B.Kear, Appl. Phys. Lett., **56**, 2342 (1990).
4. C.S.Chern, J.Zhao, Y.Q.Li, P.Norris, B.Kear and B.Gallois, Appl. Phys. Lett., **57**, 721 (1990).
5. C.S.Chern, J.Zhao, Y.Q.Li, P.Norris, B.Kear, B.Gallois and Z.Kalman, Appl. Phys. Lett., **58**, 185 (1991).
6. Y.Q.Li, J.Zhao, C.S.Chern, E.E.Lemonie, B.Gallois, P.Norris and Z.Kalman, Appl. Phys. Lett., **58**, 2300 (1991).
7. J.Zhao, C.S.Chern, Y.Q.Li, P.Norris, B.Kear, X.D.Wu and R.E.Muenchausen, Appl. Phys. Lett., **58**, 2839(1991).
8. K.Ebihara, S.Kanazawa, T.Ikegami and M.Shiga, J. Appl. Phys., **68**, 1151 (1990).
9. K.Ebihara, T.Ikegami, S.Kanazawa and M.Shiga, Mat. Res. Soc. Symp. Proc. Plasma Processing and Synthesis of Materials III, Vol. 190,155 (1991).
10. K.Ebihara, T.Fujishima, T.Ikegami and M.Shiga, IEEE Transactions on Applied Superconductivity, **3**, 976 (1993).
11. T.Fujishima, Y.Yoshimoto and K.Ebihara, Trans. IEEE of Japan(in Japanese), **113-A**, 540 (1993).
12. K.Ebihara, T.Ikegami, T.Fujishima and M.Shiga, Proceedings of 10th International Symposium on Plasma Chemistry, **3**, 2.4-8p.1 (1991).
13. K.Fijii, H.Zama, and S.Oda, Jpn. J. Appl. Phys., **31**, L787 (1992).
14. K.Ebihara, Y.Yamagata and T.Ikegami, Proceedings of 11th International Symposium on Plasma Chemistry Vol. 4, 1615 (1993).
15. M.Fukumoto, Y.Akasaka, T.Shigehisa, T.Ikegami and K.Ebihara, Proceedings of 6th Asia Conference on Electrical Discharge, (Oita, Japan, 1993) in press.
16. for example, International workshop on MOCVD of High Temperature Superconductors and Related Topics, (Organized by T.J.Marks, Evanston, Illinois, USA, 1992).

METAL-ORGANIC CHEMICAL VAPOR DEPOSITION OF EPITAXIAL $Tl_2Ba_2Ca_2Cu_3O_{10-x}$ THIN FILMS

BRUCE J. HINDS*, JON L. SCHINDLER**, BIN HAN**, DEBORAH A. NEUMAYER*, DONALD C. DEGROOT**, TOBIN J. MARKS*, AND CARL R. KANNEWURF**

Science and Technology Center for Superconductivity and the Materials Research Center, Northwestern University, Evanston, IL 60208-3118
*Department of Chemistry
**Department of Electrical Engineering and Computer Science

ABSTRACT

Superconducting thin films of $Tl_2Ba_2Ca_2Cu_3O_{10-x}$ (TL-2223) have been grown on single crystal (110) $LaAlO_3$ using a two-step process. $Ba_2Ca_2Cu_3O_x$ precursor films are deposited via metal-organic chemical vapor deposition (MOCVD) in a horizontal hot walled reactor. The second generation precursors $Ba(hfa)_2 \cdot tet$, $Ca(hfa)_2 \cdot tet$, and $Cu(hfa)_2$ (hfa = hexafluoroacetylacetonate, tet = tetraglyme) were used as volatile metal sources due to their superior volatility and stability. Tl was introduced into the film via a high temperature post anneal in the presence of a $Tl_2O_3:BaO:CaO:CuO$ pellet (1:2:2:3 ratio). Low O_2 partial pressures were used to reduce the temperature in which the Tl-2223 phase forms and to improve the surface morphology associated with a liquid phase intermediate. Films are highly oriented with the c-axis perpendicular to the substrate and a-b axis epitaxy is seen from x-ray φ-scans. The best films have a resistively measured T_c of 115K and a magnetically derived J_c of 6×10^5 A/cm^2 (77K, 0 T). Preliminary surface resistance measurements, using parallel plate techniques, give $R_s = 0.35$ mΩ at 5K (ω = 10 GHz).

INTRODUCTION

The $Tl_2Ba_2Ca_2Cu_3O_{10}$ (Tl-2223) superconductor, with one of the highest known critical temperatures (125K),[1] has been the focus of considerable of scientific and technological interest. In thin film form, members of this family exhibit excellent current-carrying properties,[2] while the low attainable surface resistance is compatible with a range of microwave device technologies (e.g., passive filters)[3]. To date, high-quality TBCCO films have largely been grown by physical vapor deposition (PVD) techniques such as laser ablation,[4] sputtering,[5] spray pyrolysis,[6] and electron beam evaporation,[7] while metal-organic chemical vapor deposition (MOCVD) techniques have received less attention.[8-11] MOCVD processes offer the attraction of simplicity of apparatus, adaptability to coating large areas and complex shapes, high throughput, and film growth under highly oxidizing conditions. Due to the high Tl_2O vapor pressure of the Tl-2223, an in-situ growth process remains elusive. A two-step process is generally required in which a $BaCaCuO_x$ (BCCO) precursor film is first grown by PVD or MOCVD methods. Then, the superconducting phase is formed by a high temperature anneal of the BCCO film in the presence of a Tl_2O source, often in a sealed quartz tube. We report here an MOCVD process for phase-pure, epitaxial Tl-2223 films which employs fluorinated β-diketonate metal-organic precursors, an improved reactor design, and a flowing gas annealing process. These films are characterized by a variety of microstructural and charge transport techniques.

EXPERIMENTAL

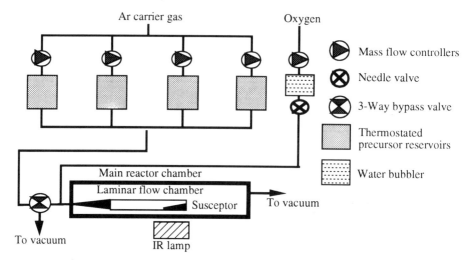

Figure 1. Schematic of MOCVD reactor design.

In the present procedure, BCCO films were grown by MOCVD using the improved reactor design shown in Figure 1. The volatile metal-organic precursors are sublimed from Pyrex reservoirs immersed in constant temperature baths. Argon is used as a carrier gas with flow maintained by Unit UFC-1400A mass flow controllers. The precursor gasses are mixed at a common manifold, and the oxidizing gas is introduced immediately prior to the reaction chamber entrance. To produce a laminar flow, an expansion adapter as well as a quartz laminar flow chamber are placed inside the main quartz reaction chamber. The high aspect ratio (3:1) of the rectangular cross-section in the quartz laminar flow chamber reduces the turbulent flow induced by the thermal buoyancy effects of a heated substrate.[12] A silicon carbide-coated graphite substrate susceptor is angled at 9° to help maintain a uniform diffusion boundary. The substrate susceptor is heated by a water-cooled IR lamp, and all precursor delivery lines are resistively heated to 150° C; the background pressure during film growth is 0.3 Torr. This reactor design affords both a uniform deposition rate (± 10%) and uniform elemental composition (± 1%) over a 20 cm² area.

Ba(hfa)$_2$•tet, Ca(hfa)$_2$•tet, and Cu(hfa)$_2$ (hfa = hexafluoroacetylacetonate, tet = tetraglyme) were used as volatile metal sources due to their superior volatility and thermal stability. Cu(hfa)$_2$ exhibits excellent thermal stability and low sublimation temperatures compared to the more commonly used Cu(acetylacetonate)$_2$. For BCCO precursor film growth, cleaned, single-crystal (110) LaAlO$_3$ substrates were employed. Typical MOCVD growth conditions for BCCO films are shown in Table 1.

Table 1. MOCVD growth conditions for BCCO precursor film

Precursor bath temperatures	115°, 86°, 34° C for Ba,Ca,Cu respectively
Precursor carrier flow	50-100 sccm each
O$_2$/H$_2$O flow	100/250sccm
Total pressure	5 torr
Substrate temperature	730°C
Growth rate	0.6 µm/hr
Substrate	(110) single crystal LaAlO$_3$

An AtomScan 25 ICP atomic emission spectrometer was used to determine the metal content of precursor films dissolved in 1M HNO_3. X-ray diffraction data were collected on a Rigaku DMAX-A diffractometer using Ni-filtered Cu K_α radiation. Film thickness was measured by a Tencor Instruments Alpha Step 200 profilometer

To form the superconducting phase, volatile Tl_2O is introduced into the BCCO precursor film from a Tl_2O_3:BaO:CaO:CuO (1:2:2:3) pellet during a high temperature anneal. Approximately 0.8g of the TBCCO powder is pressed into a 1.0 cm diameter pellet and placed in a small gold crucible. The BCCO film is placed over the pellet, face down, with gold foil used as a spacer to prevent direct film-pellet contact. The crucible is then tightly wrapped with gold foil and a 10% O_2 in Ar gas mixture (flowing at 100 sccm) is used to purge the crucible and the furnace before annealing is initiated. A 10% O_2 gas mixture is used since reduced O_2 pressures, in sealed quartz tube annealing schemes, are known to reduce the temperature for the formation of Tl-2223 and to improve surface morphology for PVD-derived films.[13] A rapid temperature ramp is used to bring the furnace to the setpoint temperature of 820°C within 15 min. This procedure has distinct advantages over sealed-tube thallination methodologies in terms of convenience and amenability to scale-up.

RESULTS AND DISCUSSION

Figure 2. SEM micrograph showing surface morphology of an MOCVD derived Tl-2223 thin film patterned to an 80 μm wide bridge.

To first characterize the MOCVD reactor performance, BaO films were deposited using Ba(hfa)2•tet as a precursor. The effect of precursor temperature, precursor carrier flow rate, total pressure, and substrate temperature on deposition rate was monitored by ICP analysis of the Ba content of various films. There was a strong dependence of precursor temperature on deposition rate with a 10° C increase in temperature resulting in 3x's the deposition rate. The effect of carrier flow was nearly linear, with a slight dilution at high flow rates (>200 sccm). Increasing the total pressure from 5 to 10 torr resulted in 0.5x's the deposition rate and is attributed to slower diffusion. Most important was that the substrate temperature, from 600-730° C, did not affect the deposition rate indicating a mass transfer limited growth. With the MOCVD reactor performance characterized, BCCO films of 2:2:3 stoichiometry were then grown in a reproducible manner. As-deposited films show trace amounts of crystalline BaF_2, Ba_2CuO, and CaO. The addition of water in the oxidizing gas stream increases the oxidation rate and assists in the removal of fluorides derived from the thermolysis of the fluorinated ligands. Previous work with fluorinated MOCVD precursors showed a post-anneal of the BCCO precursor film in the presence of water-saturated O_2 at 720° C for 8 h. was necessary to remove fluoride and allow the formation of the superconducting phase.[14] The present deposition temperature of 730° C in combination with H_2O eliminates the need of an H_2O/O_2 annealing step since the trace amounts of residual fluorides are readily removed from the film during the subsequent thallination anneal (*vide infra*), presumably as volatile TlF.

For the highest T_c films, 2.0:2.2:2.7 Ba:Ca:Cu precursor films were annealed as described for 12 h at 820° C. Figure 2 shows an SEM image of a Tl-2223 film patterned to a 80μm microbridge. The surface features are dominated by platelets oriented parallel to the surface and are indicative of a liquid phase intermediate. A θ-2θ x-ray diffraction scan (Figure 3a) shows essentially phase pure Tl-2223. Only the (00l) peaks are observed, indicating that the film is highly c-axis oriented. The θ-rocking curve (a measure of c-axis orientation) exhibits a full width at half maximum of 0.8° (Figure 3b) which is comparable to values for typical PVD-derived films. Such line widths can be attributed to randomly dispersed intergrowths of Tl-2212 phase layers into Tl-2223 domains, thus reducing the c-axis coherence length.[15] An in-plane φ-scan (Figure 3c) shows the relative orientation of the a-b axes within the c-plane. Reflections appear every 90° as expected for an epitaxial tetragonal TBCCO film.

Transport measurements provide further indication film properties. The resistivity *vs.* temperature behavior of an MOCVD-derived Tl-2223 film (Figure 4a) indicates T_c = 115 K. Several films were patterned using standard photolithography and EDTA etching[16] to define a microbridge structure. Critical current density results from dc and pulsed transport measurements with 80 μm wide microbridges appear in Figure 4b. At 77 K, the highest J_c is

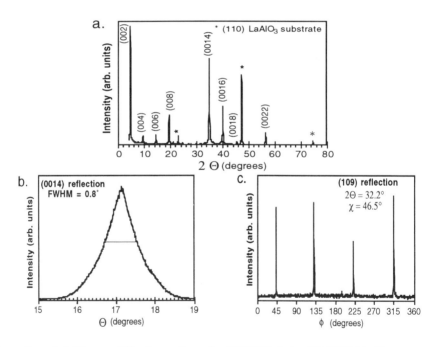

Figure 3. a) θ-2θ x-ray diffraction pattern of an MOCVD-derived Tl-2223 thin film grown on (110) LaAlO3. b) θ-rocking curve of the (0014) reflection. c) φ-scan of the (109) reflection.

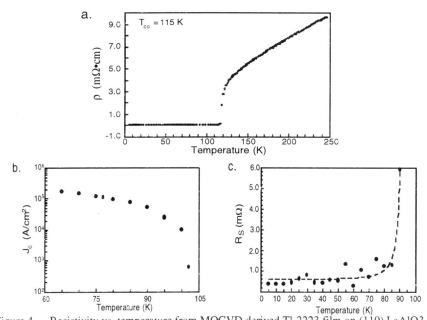

Figure 4. Resistivity vs. temperature from MOCVD derived Tl-2223 film on (110) LaAlO3 showing a Tc of 115K. b) Temperature vs. critical current. c) Temperature Vs surface resistance of a Tl-2223/2212 film, dashed line is London Model with offset.

1.5×10^5 A/cm^2, as defined by a 10 µV/cm offset criterion. For comparison, the J_c of a comparable film was estimated from magnetic hysteresis measurements taken with a SQUID susceptometer. Bean model[17] calculations using these data yield values for J_c of 4×10^6 A/cm^2 at 5 K (0 T) and 6×10^5 A/cm^2 at 77 K (0 T). The 77 K J_c values obtained from the transport and magnetic measurements agree reasonably well, suggesting that intragrain coupling does not adversely affect the transport capacity of these films. The transport properties are comparable to those of the highest quality PVD-derived films reported to date.[18] Figure 4c shows preliminary surface resistance measurements on a film containing some admixed Tl-2212 phase, performed in a parallel plate resonator against a YBCO standard. These measurements yield an R_s = 0.35 mΩ at 5 K, 10 GHz, which approaches the lowest values reported for PVD-derived films.[19]

CONCLUSIONS

In summary high-quality Tl-2223 thin films have been grown in an improved two-step process which utilizes MOCVD. Ba-Ca-Cu-O$_x$ precursor films are grown via MOCVD using fluorinated β-diketonate polyether metal-organic precursors and an improved reactor design. The superconducting phase is formed during a subsequent flowing gas post-anneal in the presence of a Tl-Ba-Ca-Cu oxide mixture. These films show epitaxial growth, good transport properties, and low surface resistance at 10 GHz. Further research is in progress to understand processing/microstructure/charge transport characteristics of Tl-2223 and other TBCCO phases.

ACKNOWLEDGEMENTS

This research was supported by the National Science Foundation through the Science and Technology Center for Superconductivity (Grant No. DMR 9120000) and the Northwestern Materials Research Center (Grant No. DMR 9120521). The authors wuld like to thank Dr. D. L. Schulz for discussions from early work on the TBCCO-MOCVD project and T. P. Hogan for surface resistance measurements. The authors would like to also thank Dr. D. A. Rudman of NIST for supplying a low R_s YBCO thin film.

REFERENCES

1. Z.Z. Sheng, A.M. Hermann, Nature **332**, 55 (1988).
2. M. Okada, T. Nabatame, T. Yuasa, K. Aihara, M. Seido, S. Matsuda, Jap. J. Appl. Phys. **30**, 2747 (1991).
3. R.B. Hammond, G.L. Hey-Shipton, G.L. Matthaei, IEEE Spectrum **30**, 34 (1993).
4. T. Nabatame, Y. Saito, T Kamo, S. Matsuda, Jpn. J. Appl. Phys. **29**, L1813 (1990).
5. W.Y. Lee, S.M. Garrison, M. Kawasaki, E.L. Venturini, B.T. Ahn, R. Boyers, Appl. Phys. Lett. **60**, 772 (1992).
6. J.A. DeLuca, M.F. Garbauskas, R.B. Bolon, J.G. McMullen, W.E. Balz, L. Karas, J. Mater. Res. **6**, 1415 (1991).
7. D.S Ginley, J.F. Kwak, R.P. Hellmer, R.J. Baughman, E.L. Venturini, B. Morosin, Appl. Phys. Lett. **53**, 406 (1988).
8. D.S. Richeson, B.W. Wessels, J. Zhao, J.M. Zhang, H.O. Marcy, T.J. Marks, L.M. Tonge, C.R. Kannewurf, Appl. Phys. Lett. **54**, 2154 (1989).
9. G. Malandrino, D.S.,Richeson, T.J. Marks, D.C. Degroot, J.L. Schindler, C.R. Kannewurf, Appl. Phys. Lett. **58**, 182 (1991).
10. K. Zhang, E.P. Boyd, B.S. Kwak, A.C. Wright, A. Erbil, Appl. Phys. Lett. **55**, 1258 (1989).
11. N. Hamaguchi, R. Gardiner, and P.S. Kirlin, Appl. Surf. Sci. **48/49**, 441 (1991).
12. H. Moffat, K.F. Jensen, J. Cryst. Growth, **77**, 108 (1986).
13. B.T. Ahn, W.Y. Lee, R. Beyers, Appl. Phys. Lett. **60**, 2150 (1992).
14. D.L Schulz, D.S. Richeson, G. Malandrino, D. Neumayer, T.J. Marks, D.C. Degroot, J.L. Schindler, T.P. Hogan, C.R. Kannewurf, Thin Solid Films **216**, 45 (1992).
15. B. Morosin, M.G. Norton, C.B. Carter, E.L. Venturini, D.S. Ginley, J. Mater. Res. **8**, 720 (1993).
16. C.I.H. Ashby, J. Martens, T.A. Plut, D.S. Ginley, Appl. Phys. Lett. **60**, 2147 (1992).
17. C.P. Bean, Phys. Rev. Lett. **8**, 250 (1962).
18. The highest reported transport values for PVD-derived Tl-2223 films are T_c = 121K, J_c = 1x10^6 A/cm^2 (77K). W.Y. Lee, S.M Garrison, M. Kawasaki, E. L. Venturini, B.T. Ahn, R. Boyers, J. Salem, R. Savoy, and J. Vazquez, Appl. Phys. Lett. **60**, 772 (1992).
19. W.L. Holstein, L.A. Parisi, C. Wilker, R.B. Flippen, IEEE Trans. on Appl. Superconductivity **3**, 1197 (1993), R_s = 0.024 mΩ (4.2K, 10GHz).

OPTIMIZATION OF Jc FOR PHOTO-ASSISTED MOCVD PREPARED YBCO THIN FILMS BY ROBUST DESIGN

P. C. Chou, Q. Zhong, Q. L. Li, and A. Ignatiev
Texas Center for Superconductivity and Space Vacuum Epitaxy Center, University of Houston, Houston, Texas 77204-5507

C. Y. Wang, E. E. Deal, and J. G. Chen
Department of Industrial Engineering, University of Houston, Houston, Texas 77204-4812

ABSTRACT

Metalorganic chemical vapor deposition (MOCVD) is emerging as a practical high Tc superconducting thin film preparation technique for industrial application. Intrinsically this technique involves a large number of variable parameters. This is especially critical for the quarternary or higher high Tc materials. Thus, effective methods are required to optimize the parameters for the preparation of high Tc films. A matrix experimental design named Robust Design has been employed for this purpose. The first-phase design was based on a starting knowledge of growth temperature and pressure, and annealing temperature for MOCVD preparation of YBCO thin films. A minimum lab effort of only nine deposition experiments was then used to optimize the process control parameters of precursor oven temperature, carrier gas (Ar) flow rate, O_2 flow rate and N_2O flow rate. The results were then followed by three confirmation depositions. The Robust Design resulted in the growth of YBCO film with Tc consistently in the range of 87.0 K to 90.2 K and Jc improved from about 1.0×10^6 A/cm² to $3\text{-}5 \times 10^6$ A/cm².

INTRODUCTION

Metalorganic chemical vapor deposition (MOCVD) has recently been extended to high-Tc cuprate-based superconducting thin film growth, as seen from a current review[1]. One prominent feature of MOCVD is that it has a large number of parameters available for process control. Certainly, this is an advantage, for example, for variable composition control by adjusting the flow rates of individual precursors. However, it is also a challenge --- how to handle a large amount of processing control parameters simultaneously to obtain high or best quality product films for various specific applications. Thus, an efficient and robust optimization design for the large number of process control parameters is obviously needed.

Optimization of thin film process control parameters of high-Tc superconductors had been reported from time to time[2-5]. However, all of these were performed by the way of guess-and-test (trial-and-error) or under one-parameter-a-time approaches. These methods are not only time consuming, but are also not robust in principle, as they can hardly handle a large number of variables simultaneously and in a balanced manner. For the deposition of thin $YBa_2Cu_3O_{7-x}$(YBCO) films by photo-assisted MOCVD, we have applied a very efficient and robust statistical experimental design, widely known as Robust Design originally developed by G. Taguchi[6,7], to the optimization of YBCO Jc values. Eight selected processing control parameters were reduced to two single parameters and two compound parameters, and then only 9 experiments were needed to optimize Jc.

To our knowledge this is the first systematic statistical optimization study of J_c for cuprate-based high T_c superconducting thin films prepared by any technique, not just by MOCVD. This robust optimization method can now be used to accelerate the prospect of large scale or industrial application of cuprate-based high-T_c thin films prepared by MOCVD.

OPTIMIZATION METHODOLOGY

A schematic diagram of the photo-assisted MOCVD system used for YBCO thin film preparation is shown in Fig. 1. The details of the system design and operation have been reported previously[8,9]. Solid metalorganic precursors, 2,2,6,6-tetramethyl-3,5-heptanedionates (TMHD) of Y, Ba, and Cu, were placed in three different respective precursor ovens. The temperature of the precursor ovens were individually controlled. Four 2 kW tungsten-halogen quartz lamps were used for not only thermal but also photo-stimulation of the reaction[10,11]. Ultrahigh purity argon was used as a carrier gas, and its flow rate was controlled by individual mass flow controllers. During the film growth a mixture of N_2O and O_2 was used as the source of oxidizing agents. For cooling/annealing the specimen after film deposition, only pure O_2 ambient was used. Both the critical temperature T_c, and the critical current density J_c, were measured by the four-point method. The films were lithographically patterned with line dimensions of 40 μm x 150 μm. A 1 μV/cm criterion was used in the 77 K, zero field J_c measurements.

Of the large collection of control parameters in MOCVD growth of YBCO, a number were held as fixed, based on past experience, and four specific parameters were varied as determined by the Robust Design. The fixed parameters are related to the temperatures and times of the growth and annealing process, while the variables parameters listed in Table 1 were O_2 flow rate, N_2O flow rate, Ar flow rate and precursor oven temperature. The last two are compound parameters.

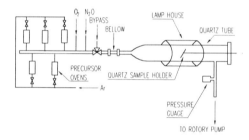

Fig. 1. A schematic of the photo-assisted MOCVD system

Table 1. Control parameters and their levels, and specified level settings for the optimization of J_c of YBCO thin films deposited on $LaAlO_3(100)$ substrates.

Control Parameters			Parameter Levels*		
			I	II	III
A.	O_2 flow rate (cc/min)		200	350	500
B.	N_2O flow rate (cc/min)		200	350	500
C.	Ar flow rate (cc/min) (A compound parameter)	Y	100	200	360
		Ba	150	300	570
		Cu	100	200	360
D.	Precursor Oven Temperature (°C) (A compound parameter)	Y	160	168	176
		Ba	273	287	300
		Cu	160	168	176

* The starting level for each factor is identified by underlining.

In terms of fixed parameters, the YBCO films were deposited on $LaAlO_3$(100) substrates of about 3 mm x 10 mm size. The substrate temperature was first held at 600°C for 1 min and then raised to 760°C for YBCO film deposition for 3 min, with reactor pressure at 3 to 5 Torr. Cooling rate of the sample after deposition was about 6°C/min. The films were then annealed at 460°C in oxygen flow at atmospheric pressure for 2 hr. Finally, the films were cooled to room temperature in 30 min.

Fig. 2 shows a block diagram representation of the general relationship between various input parameters and output response of a product/process. For this study, the product/process box shown in Fig. 2 represents the MOCVD system and the Jc values of the YBCO thin films prepared from the system are considered as the response of the system. The response is also called a quality characteristic being optimized by using the Robust Design. Signal parameters are the parameters that specify the intended value of the product/process's response. Since the intended value of

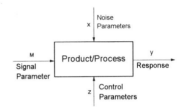

Fig. 2. An input/output block diagram for a product/process system

Jc of the films is of the larger-the-better type, signal parameter can be excluded in this design. Noise parameters are the parameters that cannot be controlled, and/or whose settings are difficult or expensive to be controlled by the designer. Process control parameters are the parameters that can be specified and selected freely by the designer.

The Robust Design is a kind of process parameter design from which an optimum combination of selected process control parameter levels can be identified for a specified quality characteristic of the process, e.g., Jc of this study, and the effects of these control parameter levels expected being least sensitive to processing system variations or noise parameters (robustness)[6,7]. The Robust Design uses a mathematical tool called *orthogonal arrays* to study a large number of control parameters simultaneously with a small number of experiments. It also uses a measure of quality, called *signal-to-noise(S/N) ratio*, or *objective function* η_i to estimate the robustness of a designed experiment.

Estimation of Main Effects

For a product/process optimization study, a number of process control parameters have to be selected, along with two or more level settings for each control parameter. *Main effects* are the effects to the quality characteristics intended to be optimized that are caused by each control parameter level of a Robust Design project.

An orthogonal array is the experiment matrix used in Robust Design that allows the effects of several parameters to be determined simultaneously and efficiently. Table 2 shows the control parameter standard orthogonal array, L_9 (3^4) for Jc optimization, which is a 9 (rows) by 4 (columns) matrix array[6,7]. This matrix experiment design therefore, involves 4 control parameters, each with 3 levels of settings, and needs a minimum of 9 individual experiments to collect all of the data for the Jc optimization analysis. The last column of Table 2, the objective function η_i, represents the observed signal-to-noise (S/N) ratio of a ith designed experiment. For a quality characteristic such as Jc, where the larger-the-better is the optimization goal, and Jc is continuous and nonnegative, the η_i of experiment number i could take the form which is shown in Table 2.

The overall mean value of η_i, m, for the experiment region defined by the array of parameter levels shown in Table 2, is just the overall average of the η_i value. Note all three levels of every control parameter are equally represented in the nine experiments. This means m is a balanced overall mean over the entire experiment region. The effect of a parameter level is defined as the deviation it causes from the overall mean. Now let us consider a parameter level, e.g., A_{II}, which involves experiments 4, 5, and 6, as shown in Table 2. The average η for these three experiments associated with A_{II}, if denoted by $m_{A_{II}}$, is given by:

$$m_{A_{II}} = \frac{1}{3}(\eta_4 + \eta_5 + \eta_6) \quad (1)$$

Table 2. A data list of the 9 samples used for the matrix optimization experiment design. The settings of parameter levels were specified in Table 2. Starting levels are underlined.

Expt No. i	Lab Sample No.	Control Parameters[①]/ Levels				Tc (K)	Jc* (77K)	Log (Jc)	Flow Rate Ratio[②]	Film Thick. (Å)	Obj. Func. η_i[③] (dB)
		A	B	C	D						
1	Y114	I	I	I	I	89.5	0.90	5.95	0.500	5000	$\eta_1 = 119.08$
2	Y119	I	II	II	II	89.0	0.72	5.86	0.364	4500	$\eta_2 = 117.15$
3	Y115	I	III	III	III	90.0	0.42	5.60	0.286	5000	$\eta_3 = 112.04$
4	Y118	II	I	II	III	88.5	1.70	6.23	0.636	5000	$\eta_4 = 124.61$
5	Y113	<u>II</u>	<u>II</u>	<u>III</u>	<u>I</u>	89.1	1.10	6.04	0.500	4000	$\eta_5 = 120.83$
6	Y121	II	III	I	II	89.8	0.98	5.99	0.412	5500	$\eta_6 = 119.82$
7	Y116	III	I	III	II	88.3	0.87	5.94	0.714	4500	$\eta_7 = 118.79$
8	Y117	III	II	I	III	88.8	0.95	5.98	0.588	6000	$\eta_8 = 119.55$
9	Y120	III	III	II	I	90.2	1.30	6.11	0.500	4000	$\eta_9 = 122.88$
Grand Average						89.2	0.99	5.99	0.500	4800	m = 119.35

*: In units of 10^6 A/cm². The next column is $\log_{10}$ Jc.
①: A: O_2 flow rate; B: N_2O flow rate; C: Ar flow rate; D: precursor oven temperature from Table 1.
②: The flow rate ratio of $O_2/(N_2O + O_2)$.
③: For this case study, $\eta_i = -10\log_{10}(1/J_{c_i}^2) = 20\log_{10}(J_{c_i})$.
m: The overall mean of η_i.

Thus, the effect of parameter level A_{II} is given by $(m_{A_{II}} - m)$. $m_{A_{II}}$ is commonly called the *main effect* of parameter level A_{II} to the matrix experiment design. Note that the main effect $m_{A_{II}}$, represents an average η when the parameter level is at A_{II} where the averaging is done over all levels of each of the other three parameters in a balanced way. Main effects of other parameter levels are estimated similarly. After the estimation of all of these main effects, the optimum combination of best parameter level settings was determined by examining the effect of each parameter level separately. Justification for these simple averaging and optimum level selection procedures comes from the use of an orthogonal array to plan the matrix experiment and the use of an additive model as an approximation[7].

Constructing an Orthogonal Array for Jc Optimization

Table 1 indicates that only a minimum number of 9 deposition experiments are needed to optimize Jc. If the study were to be done by a way of guess-and-test(trial-or-error), efficiency would be very low. It could also be performed by the way of one-parameter-at-a-time, while keeping other control parameters at fixed levels. To obtain the same accuracy of the parameter level effect as the Robust Design, a total of 9 + 3 (3 x 2) = 27 experiments are needed. If the identification of the optimum combination of parameter levels from this four, 3-level control parameter case would be conducted by conventional full parametric experiment (that is, conduct experiments under all combinations of control parameter levels), then a total of $3^4 = 81$ experiments are needed for this present case. The efficiency of Robust Design is then clearly seen.

RESULTS AND DISCUSSION

A data list of Tc, Jc, etc., and the calculated values of the objective function η_i of the 9 YBCO thin film samples are presented in Table 2. It is seen from Table 2 that Tc data are limited to a small range of 2 K wide (88.3 - 90.2 K), whereas the Jc values of the YBCO thin film samples varied from 4.20×10^5 to 1.70×10^6 A/cm^2. Thus Tc is not an appropriate indicator of Jc for YBCO thin film samples, as is well-known. However, for the total of 13 YBCO thin film samples (9 + 4 additional for confirmation) used for this Jc optimization study, the minimum Tc was limited to 87 K, for the purpose of minimizing any appreciable Tc effect to Jc measured at 77 K.

Table 3. Main effects (m_{A_I}, $m_{A_{II}}$, ..., $m_{D_{III}}$) of various control parameter levels. For each parameter level, its effect is shown by averaging 3 related η_i values --- with data listed in Table. 2.**

	Control Parameter	Parameter Levels (dB)		
		I	II	III
A.	O$_2$ flow rate	116.09	121.75*	120.21
B.	N$_2$O flow rate	120.83*	119.18	118.05
C.	Ar flow rate	120.39	121.35*	117.22
D.	Oven Temperature	120.73*	118.59	118.73

** Overall mean m = 119.35 dB. Starting levels are underlined, whereas the optimum levels are identified by *. Note C and D are compound parameters.

Four best control parameter levels A$_{II}$, B$_I$, C$_{II}$, and D$_I$, as shown in Table 3, were identified in the experiment. These four levels are selected from the largest main effects among each of those four 3-level process control parameters. Settings of these four optimal parameter levels are listed in Table 1. By utilizing this optimum combination of four best control parameter level settings, Jc values of the YBCO thin film were increased from 1.10×10^6 A/cm^2 initially, as incidentally represented by Sample Y113 shown in Table 1, to about 3×10^6 A/cm^2, as shown for the 3 normal confirmation experiments listed in Table 4, where Jc of the "extra" confirmation test reached 5×10^6 A/cm^2.

An average high film growth rate of 1600 Å/min was found from this study[12]. Variation of flow rates of O$_2$ and N$_2$O in the specified ranges caused no obvious effect to film growth rate, whereas film growth rate increases linearly as precursor oven temperatures increased. No obvious relation between Jc and film growth rate has been seen in this work, however, optimal Jc is found at a oxidizing gas flow rate ratio (O$_2$/(O$_2$+N$_2$O)) of about 0.6. By analyses of variances it can be shown the O_2 flow rate is the most significant control parameter for optimization Jc of this study, whereas the precursor oven temperature is the least important one[13].

Table 4. A data list for the samples of 3 normal and one "extra" confirmation experiments. All of these 4 samples were prepared by the same optimum combination of best control parameter levels, i.e., A_{II}^*, B_I^*, C_{II}^*, and D_I^*. Settings of these level are listed in Table 1.

Expt No. i	Lab Sample No.	Tc (K)	Jc (77K) (in 10^6 A/cm^2)	Log_{10} Jc	Film Thick. (Å)	Obj. Func. η_i (dB)
1	Y135	87.0	2.20	6.34	4000	η_1=126.85
2	Y136	89.1	3.30	6.52	4000	η_2=130.37
3	Y137	88.2	3.00	6.48	4000	η_3=129.54
ex	Y138	90.0	5.00	6.70	1500	η_{ex}=133.98

ex: Indicate sample of the "extra" confirmation experiment --- with film deposition time of 1.5 min and resultant thickness of 1500 Å only.

CONCLUSION

Jc optimization of photo-assisted MOCVD-prepared YBCO thin films has been achieved by using G. Taguchi's Robust Design method. With a minimum effort of only 13 samples prepared, an optimum combination of four best process control parameter levels was identified, and Jc values of the films were improved from the initial value of about 1 x 10^6 A/cm^2 to about 3-5 x 10^6 A/cm^2. From the large increase of Jc, the number of important physical phenomena identified, and the minimum experimental effort involved, the power and efficiency of this optimization method have been clearly demonstrated for the growth of cuprate-based high-Tc films by the industrially highly potentialized technique of MOCVD. The authors gratefully acknowledge the support of the Texas Center for Superconductivity, NASA, and the R. A. Welch foundation.

REFERENCES

1. M. Leskelä, H. Mölsä and L. Nünistö, Supercond. Sci. Technol. **6**, 627(1993).
2. J. Chrzanowski, X. M. Burany, A. E. Curzon, J. C. Irwin, B. Heinrich, N. Fortier, and A. Cragg, Physica C **207**, 25(1993).
3. G. K. Muralidhar, G. M. Rao, J. Raghunathan, and S. Mohan, Physica C **192**, 447(1992).
4. B. Ravean, C. Michel, M. Hervieu, D. Groult, A. Maignan, and J. Provost, J. Superconductivity **5**, 203(1992).
5. B. D. Weaver and G. P. Summers, Appl. Phys. Lett. **61**, 1237(1992).
6. G. Taguchi, Introduction to *Quality Engineering --- Designing Quality into Products and Processes*, Asian Productivity Organization, Minato-Ku, Japan(1986).
7. M. S. Phadke, *Quality Engineering Using Robust Design*, Prentice Hall, Englewood Cliffs, New Jersey(1989).
8. R. Singh, S. Sinha, N. J. Hsu, P. Chou, P. K. Singh, and J. Narayan, J. Appl. Phys. **67**, 1561(1990).
9. R. Singh, S. Sinha, N. J. Hsu, P. Chou, J. Appl. Phys. **67**, 3464(1990).
10. R. Singh, F. Radpour, and P. Chou, J. Vac. Sci. Technol. A7, 1456(1989).
11. R. Singh, P. Chou, and F. Radpour, J. Appl. Phys. **66**, 2381(1989).
12. P. C. Chou, Q. Zhong, Q. L. Li, and A. Ignatiev, In preparation.
13. P. C. Chou, Q. Zhong, Q. L. LI, and A. Ignatiev, To be submitted to J. Appl. Phys.

IN SITU HETEROEPITAXIAL $Bi_2Sr_2CaCu_2O_8$ THIN FILMS PREPARED BY METALORGANIC CHEMICAL VAPOR DEPOSITION

FRANK DiMEO JR.,* BRUCE W. WESSELS,* DEBORAH A. NEUMAYER,** TOBIN J. MARKS,** JON L. SCHINDLER,*** AND CARL R. KANNEWURF***
Science and Technology Center for Superconductivity,
Northwestern University, Evanston, IL 60208-3113
* Northwestern University, Department of Materials Science and Engineering
** Northwestern University, Department of Chemistry
*** Northwestern University, Department of Electrical Engineering and Computer Science

ABSTRACT

$Bi_2Sr_2CaCu_2O_8$ thin films have been prepared *in situ* by low pressure metalorganic chemical vapor deposition using fluorinated β–diketonate precursors. The influence of the growth conditions on the oxide phase stability and impurity phase formation was examined as well as the superconducting properties of the films. Thin films deposited on $LaAlO_3$ substrates were epitaxial as confirmed by x-ray diffraction measurements, including θ-2θ and φ scans. Four probe resistivity measurements showed the films to be superconducting with a maximum T_{c0} of 90 K without post annealing. This T_{c0} is among the highest reported for thin films of the BSCCO (2212) phase, and approaches reported bulk values.

INTRODUCTION

Since the discovery of high T_c superconductors, considerable effort has been directed toward the preparation of thin films because of their potential microelectronic applications. For these applications, films will need to be epitaxial, and have sufficiently high transition temperatures and critical current densities. The BiSrCaCuO system is attractive because of its potential for films with T_c's greater than 90 K. Only recently, however, have (2212) phase films with T_{c0}'s in the 90 K range been prepared [1,2], although post deposition treatments are still needed in some cases to optimize T_{c0}[3]. The *in situ* deposition of BiSrCaCuO (2212) and (2223) thin films via metalorganic chemical vapor deposition (MOCVD) using the β-diketonates $Sr(dpm)_2$ and $Ca(dpm)_2$, (dpm = dipivalomethanate) has been developed [4-7]. The highest Tc reported among these is 97 K for an *in situ* (2223) phase film[4], and is 80 K for *in situ* (2212) films[5,7]. While high quality films have been deposited with the dpm precursors, these precursors suffer from low volatility and poor thermal stability leading to problems with process reproducibility. We have shown previously that fluorinated β–diketonate precursors exhibit improved volatility and stability, and have used them to grow BiSrCaCuO thin films.[8] However, an *ex situ* anneal was required for the formation of the superconducting (2212) phase. These (2212) phase films were epitaxial with T_c's as high as 73 K. In this communication, we report for the first time the *in situ* deposition of superconducting $Bi_2Sr_2CaCu_2O$ (2212) epitaxial thin films via MOCVD using the fluorinated precursors $Sr(hfa)_2$•tetraglyme and $Ca(hfa)_2$•tetraglyme.

EXPERIMENTAL

BiSrCaCuO films were prepared by low pressure MOCVD using the volatile precursors $Sr(hfa)_2$•tetraglyme (hfa = hexafluoroacetylacetonate), $Ca(hfa)_2$•tetraglyme[9, 10], $Cu(acac)_2$ (acac = acetylacetonate) and $Bi(C_6H_5)_3$ as sources. Argon was used as the carrier gas and water saturated O_2 as the reactant gas. $LaAlO_3$ (100) was used as the substrate material, because of its close lattice match, (<1%), to the BiSrCaCuO phases of interest. Typical growth conditions are listed in Table I. The highest source temperature used was 110 °C. This is in contrast to the 180-220°C used previously for the Ca and Sr dpm precursors. Typical film thickness were 5000-8000 Å, and growth rates were 130Å/min.

Table I. Typical growth conditions for the deposition of BiSrCaCuO thin films.

Substrate	(100) $LaAlO_3$
Growth Temperature	800-850 °C
Temperature of triphenyl Bi	100-105 °C
Temperature of $Sr(hfa)_2$•tet	95-100 °C
Temperature of $Ca(hfa)_2$•tet	75- 80 °C
Temperature of $Cu(acac)_2$	105-110 °C
Temperature of H_2O	1.0-5.0 °C
Total pressure	15-20 Torr
Oxygen partial pressure	9-12 Torr

RESULTS AND DISCUSSION

Influence of Growth Conditions

Figure 1. shows $\theta/2\theta$ x-ray diffraction patterns of BiSrCaCuO films grown under similar conditions, but at different substrate temperatures. In Figure 1a, at a growth temperature of 803°C, the 2201 phase is present along with significant amounts of the impurity phases CaF_2 and CuF_2. As the temperature is increased to 832 °C, (Figure 1b) the (2212) phase is now present along with the 2201 phase and the amount of impurity fluoride phases has decreased. At 847 °C, the (2212) phase is the dominant phase, as seen in Figure 1c, with only small amounts of the (2201) phase and there is no detectable amount of the fluoride phases. At temperatures of 850 °C or greater, the ratio of (2201) to (2212) phase is greatly increased. This return to a (2201) rich region with increasing temperature is consistent with the schematic temperature-partial pressure O_2 phase diagram suggested by Strobel, et al[11]. In that case the temperature for the decomposition of (2212) into (2201) was shown to be an increasing function of partial pressure of O_2. In the present study, it was found that the phases stabilized depended similarly upon O_2 partial pressure and temperature.

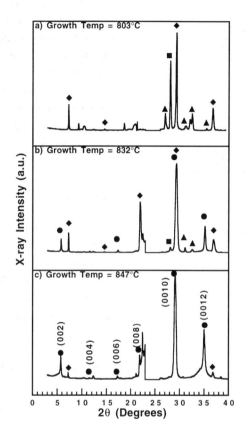

Figure 1. θ-2θ X-ray diffraction patterns of as deposited BSCCO films. In b) and c), the (008) peak of the 2201(c=24.4Å) phase and the (0010) peak of the (2212) (c=30.7Å) are overlapping. ■=CaF_2, ▲=CuF_2, ◆=2201, ●=(2212)

For partial O_2 pressures of ~3 torr, only the (2201) phase was stabilized at a growth temperature of 830 °C, whereas both the (2201) and (2212) phases were stabilized at 10 torr and 832 °C, as shown in Figure 1b. The predominant impurity phases were CuF_2 and CaF_2.

Energy dispersive x-ray analysis (EDAX) using a light element detector by Noran, Inc. was used to qualitatively analyze elemental fluorine content. Spectra were taken at 5 KeV, and minimum detection levels are estimated to be 3 wt%. Although there are more sensitive techniques for fluorine detection, i.e. x-ray photoelectron spectroscopy, and auger electron spectroscopy, these techniques only sample ~30-60 Å into the film. Whereas EDAX, at the energies used, can sample as deep as 2500-5000Å. Figure 2a is a spectrum obtained from a film grown at 840 °C without water vapor, and the fluorine K_α peak is prominent. Figure 2b is a spectrum of the film grown under similar conditions but with water vapor present. Here, the fluorine peak is noticeably reduced, but still present. In Figure 2c, which corresponds to a film grown at 847°C, there is no detectable fluorine. Thus, precise control of the substrate temperature is crucial for the elimination of fluorine impurities as well as for the stabilization of the superconducting phases. This is in agreement with the thermodynamic calculations for the chemical vapor deposition of BiSrCaCuO of Harsta et al[12]. For the precursors Sr(hfa)$_2$, Ca(hfa)$_2$, Cu(hfa)$_2$, and Bi(C_6H_5)$_3$, they predict that there will be a small region of stability for

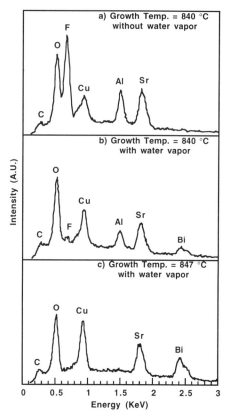

Figure 2. Low energy EDAX spectra of BSCCO thin films showing the dependence of fluorine contamination on growth conditions.

the oxide phase stability only in the presence of water and at high temperatures and high oxygen partial pressures. The agreement with these calculations as well as the agreement with the previous bulk study suggests that growth is occurring near equilibrium.

Film Properties

In Figure 1c, only the (00l) reflections for the (2212) phase are present, indicating a high degree of preferential orientation, with the c-axis normal to the substrate. The small amount of (2201) phase is also preferentially oriented. Figure 3a is a φ scan of the <117> planes of a similar film. For comparison, Figure 3b is a φ scan of the pseudo-cubic <111> planes of the LaAlO$_3$ substrate. The alignment of the film and substrate peaks indicates that the a-b plane of the film is rotated 45° with respect to the a-b plane of the LaAlO$_3$, as has been reported for *ex situ* (2212) films on LaAlO$_3$ [13]. This is expected due to the lattice matching of the film, a=b=5.4Å, with the substrate a = b = 3.79 ≅ 5.4÷√2. The FWHM of the film's φ peak is 0.6°, as compared with the substrate peak which has a FWHM of 0.3°. This indicates a small mosaic spread for the film. A scanning electron micrograph of an *in situ* MOCVD film is shown in Figure 4 and reveals a dense, faceted, well connected, platelet morphology. Small spiral growth islands are also observed. The surface roughness of the platelets was measured to be 19Å with Tencor P-10 profilometer.

Resistivity was measured using the conventional four-probe methods described previously [14]. Figure 5 displays the temperature dependence of resistivity for a superconducting

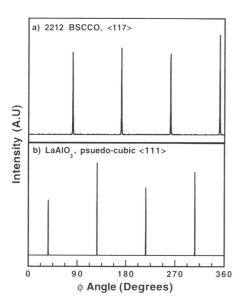

Figure 3. Φ scan x-ray diffraction patterns showing the epitaxial relationship between the (2212) film and the (001) LaAlO$_3$ substrate.

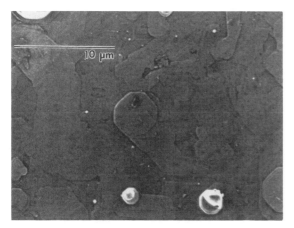

Figure 4. Scanning electron micrograph of an *in situ* superconducting Bi$_2$Sr$_2$CaCu$_2$O$_8$ thin film showing a dense platelet morphology.

(2212) film. The film exhibits linear behavior in the normal state with a resistivity of 12 mΩ•cm at room temperature. The transition is broad, however, starting above 110 K, and having a T$_{c0}$ of 90 K. This T$_{c0}$ is the highest reported for (2212) films deposited by MOCVD. The onset of superconductivity at 110 K is indicative of the presence of the (2223) phase, but no evidence was observed in the x-ray diffraction pattern. A TEM investigation is currently under way to explore the possibility of (2223) intergrowths. Critical current densities were calculated via magnetometer measurements using Bean's model. The applied field was perpendicular to the a-b plane. The critical current density is 2.3x10^5 A/cm^2 at 5 K and zero field.

CONCLUSIONS

In summary, high quality BiSrCaCuO (2212) epitaxial thin films were deposited *in situ* using fluorinated β-diketonate precursors. The deposition temperature, oxygen and water partial pressures, are critical for the *in situ* deposition of the (2212) phase. Deposition of single phase epitaxial (2212) is favored at temperatures in the range of 840-850 °C and oxygen partial pressures in excess of 10 torr on LaAlO$_3$. The epitaxial layers have a T_{c0} of 90 K, which approaches the highest reported values for bulk (2212). With the achievement of superconductivity above 90 K, the BSCCO system may prove to be viable for a variety of microelectronic applications.

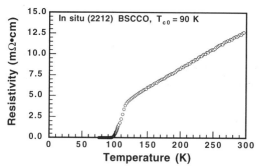

Figure 5. Resistivity as a function of temperature for an *in situ* epitaxial (2212) film on $LaAlO_3$.

ACKNOWLEDGMENTS

The authors would like to thank William Roth of Nissei Sangyo America Ltd., and Roger Teppert of Noran Instruments for their technical assistance in obtaining the SEM and EDAX results. This work was supported by the National Science Foundation (DMR 91-20000) through the Science and Technology Center for Superconductivity.

REFERENCES

1. I. Bozovic, J. N. Eckstein, M. E. Klausmeier-Brown, G. F. Virshup and K. S. Ralls, Mat. Res. Soc. Symp. Proc. **275**, 67 (1992).
2. P. Wagner, F. Hillmer, U. Frey, H. Adrian, T. Steinborn, L. Ranno, A. Elschner, I. Heyvaert and Y. Bruynseraede, Physica C **215**, 123 (1993).
3. G. Balestrino, M. Marinelli, E. Milani, M. Montouri, A. Paoletti and P. Paroli, J. Appl. Phys. **72**, 191 (1992).
4. K. Endo, H. Yamasaki, S. Misawa, S. Yoshidaand K. Kajimura, Physica C **1856**, 1949 (1991).
5. T. Sugimoto, M. Yoshida, K. Yamaguchi, K. Sugawara and S. Tanaka, Appl. Phys. Lett. **57**, 928 (1990).
6. K. Natori, S. Yoshizawa, J. Yoshino and H. Kukimoto, Jpn. J. Appl. Phys **28**, L1578 (1989).
7. M. Schieber, J. Cryst. Growth **109**, 401 (1991).
8. J. M. Zhang, B. W. Wessels, D. S. Richeson, T. J. Marks, D. C. Degroot and C. R. Kannewurf, J. Appl. Phys. **69**, 2743 (1991).
9. K. Timmer, K. I. M. A. Spee, A. mackorand H. A. Meinema, Inorganica Chimica Acta **190**, 109 (1991).
10. D. L. Schulz, D. A. Neumayer and T. J. Marks, submitted to Inorg. Synthesis
11. P. Strobel, W. Korczak and T. Fournier, Physica C **161**, 167 (1989).
12. A. Harsta and J. O. Carlsson, J. Cryst. Growth **114**, 507 (1991).
13. J. Chen, H. A. Lu, F. DiMeo Jr, B. W. Wessels, D. L. Schulz, T. J. Marks, J. L. Schindler and C. R. Kannewurf, J. Appl. Phys. **73**, 4080 (1993).
14. J. W. Lyding, H. O. Marcy, T. J. Marks and C. R. Kannewurf, IEEE Trans. Instrum. Meas. **37**, 76 (1988).

HIGH QUALITY YBaCuO THIN FILM GROWTH BY LOW-TEMPERATURE METALORGANIC CHEMICAL VAPOR DEPOSITION USING NITROUS OXIDE

HIDEAKI ZAMA[*], JUN SAGA[**], TAKEO HATTORI[**] AND SHUNRI ODA[*]
[*]Department of Physical Electronics, Tokyo Institute of Technology, O-okayama, Meguro-ku, Tokyo 152, Japan
[**]Department of Electrical and Electronic Engineering, Musashi Institute of Technology, Tamazutsumi, Setagaya-ku, Tokyo 158, Japan

ABSTRACT

Low-temperature growth of $YBa_2Cu_3O_x$ films by metalorganic chemical vapor deposition using N_2O as an oxidizing agent has been investigated. We have deposited superconducting $YBa_2Cu_3O_x$ on (100)MgO substrates at 500°C for the first time. Films of 15nm-thick show zero-resistivity critical temperature of 80K. Films of as thin as three unit-cell-thick reveal the superconducting onset characteristics. This result suggests that superconductivity is arisen even from effectively monomolecular layer of $YBa_2Cu_3O_x$ when we take into account monomolecular buffer layer and monomolecular cap layer. $YBa_2Cu_3O_x$ films of 9nm-thick grown on (100)$SrTiO_3$ at 600°C with T_c(zero) of 79K and with peak to valley roughness fluctuation of two unit-cell have been obtained.

INTRODUCTION

Metalorganic chemical vapor deposition (MOCVD) has the possibility of the low-temperature growth due to enhanced surface migration of molecular precursors. However, in most reports related to MOCVD of $YBa_2Cu_3O_x$ films [1-7], the growth temperatures were between 650°C and 800°C, the same temperature range as sputtering, laser ablation, and reactive evaporation methods.

Low-temperature growth is desirable to prepare the high quality films with smooth surface and good electrical characteristics, and the hetero epitaxial structure of multilayers, e.g., $YBa_2Cu_3O_x/PrBa_2Cu_3O_x/YBa_2Cu_3O_x$ system for the application to superconducting tunnel devices, avoiding the interdiffusion of atoms between layers and from substrates. It is very important for qualitative evaluation to observe the superconducting characteristics of ultra thin (<10nm, i.e., several unit-cell thickness) films of $YBa_2Cu_3O_x$.

This paper deals with the preparation of superconducting $YBa_2Cu_3O_x$ films by MOCVD using N_2O at 500°C for the first time and the observation of superconducting characteristics of ultra thin (<10nm) films of $YBa_2Cu_3O_x$.

EXPERIMENTAL

We used $Ba(HFA)_2$ (HFA; hexafluoroacetylacetone, $C_5HO_2F_6$) and $Ba(DPM)_2$ (DPM; dipivaloylmetane, $C_{11}H_{19}O_2$) for metalorganic precursors of Ba in the previous reports [6-8]. The problem of $Ba(HFA)_2$ was fluorine impurity incorporated in $YBa_2Cu_3O_x$ films. The problem of $Ba(DPM)_2$ was poor stability. In the present work, for the exclusion of fluorine impurity and the improvement of the stability, Ba(HFA)$_2$ and Ba(DPM)$_2$ were replaced by Ba(DPM)$_2$(phen)$_2$ (phen; 1,10-phenanthroline, $C_{12}H_8N_2$). As a result, single charge of 5g of Ba(DPM)$_2$(phen)$_2$ in the vaporizer were able to deposit the films for cumulative 500 runs while it kept heated at 180-185°C for about one year.

We have also made several modifications to our MOCVD apparatus described in the previous reports [6-8]. We used the mantle heaters covering whole the vaporizers of metal precursors and the gas transfer tubes so as to control temperature precisely. As a result, reproducibility of MOCVD growth improved significantly.

We have used Y(DPM)$_3$ and Cu(DPM)$_2$ for the sources of Y and Cu, respectively. The vaporizing temperatures of Y(DPM)$_3$, Ba(DPM)$_2$(phen)$_2$, and Cu(DPM)$_2$ kept at 87-97°C, 175-185°C, and 80°C, respectively. The reactor pressure was 2Torr, and the partial pressure of N_2O was 1Torr. After deposition, in-situ oxidization carried out at 400°C under the O_2 pressure of 2Torr for 5min.

RESULTS AND DISCUSSION

1. Low-Temperature MOCVD

Figure 1 shows the resistance versus temperature curves of $YBa_2Cu_3O_x$ films prepared for various deposition temperatures, i.e., (a) 640°C, (b) 600°C, (c) 550°C, and (d) 500°C. Measurement current of direct current four probe method was 5µA. Film thickness of the film (a) and (d), evaluated by the cross sectional observation of scanning electron microscopy (SEM), were approximately 20nm and 15nm, respectively. No significant signal of carbon impurity, could be incorporated from metalorganic precursors, was detected in X-ray photoelectron spectroscopy in the film grown at 500°C as well as that grown at 640°C. X-ray diffraction patterns of all films show single phase of c-axis $YBa_2Cu_3O_x$, and the intensity of X-ray diffraction was in the order of (a), (d), (b), and (c). The value of normal resistance was also in the same order as the intensity of X-ray diffraction. All films show almost same superconductivity onset temperature T_c(onset). In films (a) and (d), zero resistivity critical temperature T_c(zero) of higher than 77K were obtained. Normal and superconducting characteristics show the percolation pass of superconducting 90K-transition phase in the samples of 640°C and 500°C was connected and those of 600-550°C was not connected completely in the range of the film thickness of 15-20nm.

It is for the first time to prepare the superconducting $YBa_2Cu_3O_x$ films at a temperature of as low as 500°C. We have already reported the crystal growth of $YBa_2Cu_3O_x$ by MOCVD on sapphire substrates at 500°C [8]. At that time, superconducting characteristics were not obtained presumably because of the poor control technique of metal composition due to lack of stability of raw materials and to the F impurity incorporated from raw materials of Ba(HFA)$_2$. Now, superconducting characteristics have also been obtained from $YBa_2Cu_3O_x$ films prepared on sapphire substrates at 500°C [9].

The reaction mechanism is different between high-temperature (>600°C) MOCVD and low-temperature (around 500°C) MOCVD. The precursors on the growing surface of high-temperature MOCVD are decomposed to reactive atoms, similar to the case of physical vapor deposition such as molecular beam epitaxy, sputtering, laser ablation, and reactive evaporation methods, in which high temperature is required to prepare high quality films by enhancing migration of atomic species on the growing surface. On the other hand, at low temperatures around 500°C, the precursors is partially

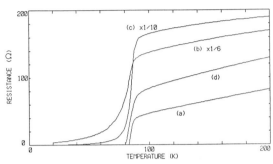

Fig. 1. Resistance vs temperature curves of $YBa_2Cu_3O_x$ films grown by MOCVD on (100)MgO substrates. Deposition temperature was (a) 640°C, (b) 600°C, (c) 550°C, and (d) 500°C.

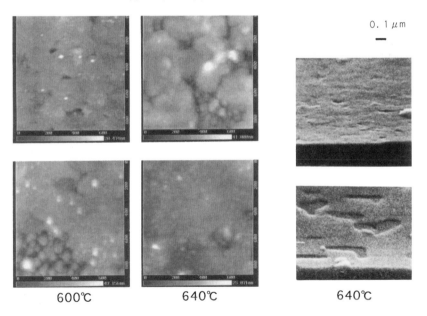

Fig. 2. Atomic force microscope images of $YBa_2Cu_3O_x$ films grown by MOCVD on (100)MgO substrates in the area of approximately 800×900nm^2. (T_{sub}: deposition temperature.)

Fig. 3. Scanning electron microscope images of $YBa_2Cu_3O_x$ films grown by MOCVD on (100)MgO substrates (T_{sub}: deposition temperature.)

decomposed, remaining inert molecular structure, in which migration is enhanced by chemical energy originated from chemical reaction of molecules. We can take full advantages of MOCVD in this temperature range.

The surface morphology of $YBa_2Cu_3O_x$ films prepared at various temperatures were observed by atomic force microscope (AFM) as shown in Fig. 2. The films grown at 600°C and 550°C have very rough surface morphology due to the microscopic separation of superconducting and non-superconducting phase, while those grown at 640°C and 500°C has smooth surface morphology. Moreover, the differences between atom migration and molecule migration were clearly observed by SEM as shown in Fig. 3. The film grown at 640°C has submicron-scale roughness presumably because of interdiffusion, while that grown at 500°C has very smooth surface.

2. Ultra Thin FIlms
2.1 Growth on (100)MgO substrates

The root mean square (rms) surface roughness of (100)MgO substrates measured by AFM was 0.35-0.40nm in the area of approximately $700 \times 600 nm^2$.

Figure 4 shows the resistivity versus temperature curves of $YBa_2Cu_3O_x$ films prepared at 500°C for various deposition time, i.e., (a) 10min, (b) 15min, (c) 20min, and (d) 60min. Film thickness of films (a), (b), (c), and (d) were estimated to approximately 2, 3, 4, and 12 unit-cell-thick, respectively, from the deposition period. The film (a) was characterized by the semiconductor-like behavior. T_c(onset) of 85K was obtained in the film (b), but T_c(zero) was not obtained. As is pointed out by many authors, the topmost one unit-cell-thick layer and the bottom interface layer for ultra thin films of $YBa_2Cu_3O_x$ may be non-superconducting. The bottom interface layer serves as a buffer layer to compensate the lattice mismatch between $YBa_2Cu_3O_x$ films and (100)MgO substrates expanding a- and b-axis lattice length of the $YBa_2Cu_3O_x$ structure, as shown in in-situ RHEED observation during the growth of the $YBa_2Cu_3O_x$ film on (100)MgO substrate reported by Terashima et al. [10], so that it is degraded to a non-superconducting layer. The topmost layer serves as a cap layer covering over the CuO_2 planes of the underlying a $YBa_2Cu_3O_x$ layer to complete the structure of hole doping within the CuO_2 planes [11] and a protection layer protecting underlying layers from the decomposition by H_2O and CO_2 in atmosphere. Hence, the topmost layer may be non-superconducting. Only the intermediate layer serves as a superconducting layer. The superconducting percolation path in the intermediate layer of the film (b), in which the average thickness of superconducting region is effectively one unit-cell-thick, is so deteriorated that only superconducting onset characteristics in the resistivity versus temperature curve is obtained. The film (c) with two intermediate superconducting layer besides the buffer layer and the cap layer shows clear superconducting characteristics with zero-resistivity critical temperature of 30K. These superconducting characteristics of ultra thin films of $YBa_2Cu_3O_x$ were similar to that of the $PrBa_2Cu_3O_x$/$YBa_2Cu_3O_x$/$PrBa_2Cu_3O_x$/$SrTiO_3$ system reported by Terashima et al. [11] and Kwon et al. [12] We report, for the first time, the preparation of ultra thin films, with approximately 3 unit-cell-thick, of superconducting $YBa_2Cu_3O_x$ on (100)MgO substrates by MOCVD.

2.2 Growth on (100)SrTiO$_3$ substrates

The rms surface roughness of (100)SrTiO$_3$ substrates measured by AFM was 0.25nm in the area of approximately 700x600nm^2.

The optimal deposition temperature of YBa$_2$Cu$_3$O$_x$ was different between (100)SrTiO$_3$ and (100)MgO substrates. Figure 5 shows the resistivity versus temperature curve of a YBa$_2$Cu$_3$O$_x$ film prepared at 600°C. Film thickness, evaluated by AFM, was approximately 9nm. T$_c$(onset) of 89K and T$_c$(zero) of 79K were obtained. YBa$_2$Cu$_3$O$_x$ films grown on (100)SrTiO$_3$ have many precipitates on the surface in contrast with (100)MgO. On the other hand, the surface morphology on the matrix between precipitates was very smooth with rms surface roughness of 0.77nm and with peak to valley heights of approximately two unit-cell. Surface morphology of YBa$_2$Cu$_3$O$_x$ films prepared by

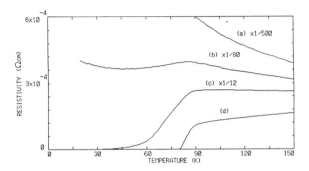

Fig. 4. Resistivity vs temperature curves of YBa$_2$Cu$_3$O$_x$ films grown by MOCVD at 500°C on (100)MgO substrates. Deposition time (and estimated total thickness in unit cell length) was (a) 10min(2uct), (b) 15min(3uct), (c) 20min(4uct), and (d) 60min(12uct).

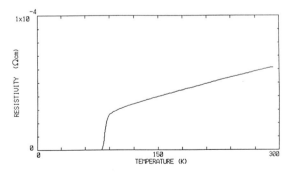

Fig. 5. Resistivity vs temperature curve of a YBa$_2$Cu$_3$O$_x$ film grown by MOCVD at 600°C on a (100)SrTiO$_3$ substrate. Film thickness was 9nm.

our MOCVD system was comparable to the best quality films prepared by reactive evaporation or laser ablation methods.

CONCLUSIONS

We have successfully prepared $YBa_2Cu_3O_x$ superconducting films at as low as $500^\circ C$ by MOCVD using N_2O for the first time. The low-temperature growth of $YBa_2Cu_3O_x$ takes advantage of surface migration of chemical precursors which is the essential merit of MOCVD.

We have also successfully prepared ultra thin (<10nm) films of $YBa_2Cu_3O_x$. On (100)MgO substrates, superconducting $YBa_2Cu_3O_x$ films of as thin as three unit-cell-thick (effectively one unit-cell-thickness besides a buffer layer and a cap layer) grown at $500^\circ C$ were obtained. On (100)$SrTiO_3$ substrates, 9nm-thick $YBa_2Cu_3O_x$ films with T_c(zero) of 79K and with peak to valley heights of approximately two unit-cell were obtained.

MOCVD is a promising method for the deposition of oxide superconductor films in view of comparable film quality to reactive evaporation or laser ablation and lower deposition temperatures.

ACKNOWLEDGEMENTS

This work was supported in part by a Grant-in-Aid for Scientific Research on Priority Areas, "Crystal Growth Mechanism in Atomic Scale" and "Science of High Tc Superconductivity" provided by the Ministry of Education, Science and Culture, by the TEPCO Research Foundation, and by the Yazaki Memorial Foundation.

REFERENCES

1. H. Yamane, M. Hasei, H. Kurosawa and T. Hirai, Jpn. J. Appl. Phys. 30, L1003(1991).
2. T. Tsuruoka, R. Kawasaki and H. Abe, Jpn. J. Appl. Phys. 28, L1800 (1989).
3. Y.Q. Li, J. Zhao, C.S. Chern, W. Huang, G.A. Kulesha, P. Lu, B. Gallois, P. Norris, B. Kear and F. Cosandey, Appl. Phys. Lett. 58, 648(1991).
4. K. Higashiyama, T. Ushida, H. Higa, I. Hirabayashi and S. Tanaka, Jpn. J. Appl. Phys. 30, 1209(1991).
5. H. Ohnishi, H. Harima, Y. Kusakabe, M. Kobayashi, S. Hoshinouchi and K. Tachibana, Jpn. J. Appl. Phys. 29, L2041(1990).
6. H. Zama, S. Oda, T. Miyake and T. Hattori, Physica C 190, 656(1991).
7. H. Zama, T. Miyake, T. Hattori and S. Oda: Jpn. J. Appl. Phys. 31, 3839(1992).
8. S. Oda, H. Zama, T. Ohtsuka, K. Sugiyama and T. Hattori, Jpn. J. Appl. Phys. 28, L427(1989).
9. H. Zama, S. Oda, J. Saga and T. Hattori, Advances in Superconductivity VI (submitted).
10. T. Terashima, K. Iijima, K. Yamamoto, K. Hirata, Y. Bando and T. Takada, Jpn. J. Appl. Phys. 28, L987(1989).
11. T. Terashima, K. Shimura, Y. Bando, Y. Matsuda, A. Fujiwara and S. Komiyama, Phys. Rev. Lett. 67, 1362(1991).
12. C. Kwon, Qi Li, X.X. Xi, S. Bhattacharya, C. Doughty, T. Venkatesan, H. Zhang, J.W. Lynn, J.L. Peng, Z.Y. Li, N.D. Spencer and K. Feldman, Appl. Phys. Lett. 62, 1289(1993).

PART VII

Optoelectronic Materials; Oxide Ceramics

ELECTRO-OPTIC MATERIALS BY SOLID SOURCE MOCVD

R. HISKES[1], S.A. DICAROLIS[1], J. FOUQUET[1], Z. LU[2], R.S. FEIGELSON[2], R.K. ROUTE[3], F. LEPLINGARD[4] and C. M. FOSTER[5]
[1]Hewlett-Packard Laboratories, 3500 Deer Creek Road, Palo Alto, CA 94303
[2]Stanford University, Department of Materials Science and Engineering, Stanford, CA 94305-2205
[3]Center for Materials Research, Stanford University, Stanford, CA 94305-4045
[4]Xerox Palo Alto Research Center, 3333 Coyote Hill Road, Palo Alto, CA 94304
[5]Argonne National Laboratory, 9700 Cass Avenue, Argonne, IL 60439

ABSTRACT

The solid source MOCVD technique[1,2], employing a single powder vaporization source composed of mixed beta-diketonate metalorganic compounds, has been used to grow thin films of a variety of electro-optic materials, including lithium niobate, strontium barium niobate, and potassium niobate. Results for potassium niobate films are quite preliminary, but indicate that a volatile potassium organometallic source can be synthesized which is useful for growing potassium niobate by MOCVD. High quality single phase (001) oriented strontium barium niobate films have been deposited which exhibit waveguiding behavior. The most extensive work has been done on lithium niobate, which has been deposited epitaxially on a variety of substrates. Oriented z-axis (001) films have been grown on c-axis sapphire with and without a (111) oriented platinum base electrode and on a bulk grown lithium niobate substrate. Films grown directly on c-axis sapphire at 700 C exhibit x-ray rocking curve linewidths as low as .044 degrees, nearly perfect in-plane orientation as determined by x-ray phi scans, and peak-to-peak surface roughness less than 40 Å. Optical waveguiding has been demonstrated by a single prism coupling technique on similar films 1175 - 2000 Å thick grown at 500 C, with optical losses of approximately 2 db/cm at 632.8 nm measured over 3.5 cm long films. Polarization vs. electric field measurements on 1100 Å thick films grown on platinum show a hysteresis loop indicating ferroelectric behavior.

INTRODUCTION

Although the beta-diketonate metalorganic complexes were used to successfully grow epitaxial oxide thin films as long ago as 1973[3], interest in this class of metalorganic precursors remained at minimal levels for more than a decade. Intense interest has only recently been rekindled (1987) due to the emergence of the worldwide effort to develop reliable and reproducible thin film deposition techniques for high temperature superconductors, such as yttrium barium copper oxide (YBCO). Many of the heavy metal beta-diketonate complexes, including barium tetramethylheptanedionate (Ba(thd)$_2$), strontium thd and others readily decompose at elevated temperatures. The barium compound, for example, begins to decompose and polymerize almost as rapidly as it volatalizes, even at reduced pressures[4]. Although new classes of more stable compounds have been developed recently[5,6], and techniques implemented which help to stabilize the heavy metal metalorganic complexes in the vapor phase[7], reproducible growth of many of the technologically interesting oxides has often proved difficult by conventional MOCVD, which relies on bubblers containing the metalorganic precursors held at elevated temperatures for extended periods of time to provide the source vapor. Therefore, we developed the Solid Source Metalorganic Deposition (SSMD) reactor which obviates the precursor stability problem and allows reproducible growth of oxide thin films (on substrates up to 4" in diameter) from these marginally volatile heavy metal diketonate sources..

The SSMD technique has been described previously[1,2]. Initially developed for large area growth of the high temperature superconductor yttrium barium copper oxide, the technique has been extended to include various buffer layers such as CeO_2 and MgO needed for reaction barriers on sapphire substrates, magnetic films including haematite (α-Fe_2O_3) and nickel cobalt oxide ($NiCoO_x$), non-oxides such as the low voltage phosphor CaS doped with terbium (a hybrid process using a liquid sulfur precursor), the superionic conductor gadolinium doped barium cerate (Gd:$BaCeO_3$) for solid-state fuel cell development, low resistivity Au/Cr bilayers on quartz and highly resistive MoO_x films on quartz for mass spectrometer applications, $LaSrCoO_3$ base electrodes for ferroelectric applications, SiO_2 films on silicon for integrated optics, piezoelectric films such as lead zirconate titanate and lead magnesium niobate, and recently the electro-optic materials $LiNbO_3$, $KNbO_3$ and $Sr_xBa_{1-x}Nb_2O_6$. The SSMD reactor is beginning to be used by researchers in other laboratories[8,9,10].

Much of the development work for these materials has been published[1,2,8] or is in the process of being prepared for publication. In this paper, in order to illustrate the power of SSMD, we will describe the reactor, briefly mention the ferroelectrics $Sr_xBa_{1-x}Nb_2O_6$ and $KNbO_3$, and then focus on the growth and properties of the technologically important electro-optic thin film $LiNbO_3$, useful for a variety of applications including waveguides, modulators, and ferroelectric memories.

MOCVD REACTOR DESIGN

The reactor configuration is shown in Figure 1. Metalorganic precursors, most of which are readily available as dry inert powders, are ground together (if necessary) in a mortar and pestle. Although a variety of different ligands are available, we have found it advantageous to use the same ligand for all the sources which must be mixed together prior to use to prevent ligand exchange and subsequent decomposition during volatilization. Most of the oxide thin films to be described were deposited using tetramethylheptanedionate (thd) compounds either prepared in-house or obtained from Strem[11].

The powders are packed into a 4 mm diameter precision bore quartz tube. This tube is closed at the bottom and has a 0.4 mm longitudinal slot along one side to permit easy exit of the vapor as the powder is volatilized. The powder density in the precursor tube can be fixed by packing a known weight of the precursors to a fixed depth in the tube, or it can be readily calculated by measuring both the weight of the tube before and after packing and the height of the powder. This procedure permits accurate control of the growth rate from run to run.

The precursor tube is placed in the U shaped quartz reactor and driven down into a very sharp temperature gradient by a precision screw drive on the outside of the reactor. The drive is magnetically coupled to the precursor tube. The sharp temperature gradient is maintained by a quartz halogen lamp placed directly below a water jacket in the reactor. The reactor is maintained at 1 - 760 torr depending on the materials deposited, and the precursor tube is driven down at rates of 0.1 - 5 mm/min. Each of the constituent powders

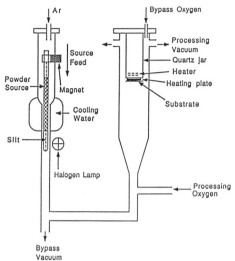

Figure 1. The SSMD reactor.

in a multicomponent mixture vaporizes at its own particular vaporization temperature in the temperature gradient, and as long as the tube velocity is constant, steady state vaporization is obtained and constant vapor composition ensues[10]. The temperature gradient can be as high as 100 °C/mm, and the precursors are vaporized in the hot zone before they can polymerize or decompose.

It takes about 1 second for the vapor to be swept through the reactor to the substrate surface by 300 sccm of argon carrier gas. 300 - 1000 sccm of oxygen for oxide growth or forming gas (4% H_2 in argon) for non-oxides are introduced and mixed with the vapor stream, and laminar flow is maintained by gradually increasing the reactor inside diameter from 14 mm to 200 mm at a 15 degree cone angle. The warm wall reactor is maintained at 150-300 °C by keeping the U shaped portion of the quartz tube in a heated and insulated metal box. Several different modifications of the substrate holder and heating assembly are in use, two of which are described in references 1 and 2. Up to 4" diameter substrates can be accommodated in these reactors. The configuration used for most of the films described here employs a 6" diameter molybdenum disilicide flat (pancake) resistance heater. This heater is quite stable in oxygen atmospheres at temperatures over 1800 C and radiates heat to a 4" nickel susceptor suspended below it. The substrates are mounted in a quartz holder just below the susceptor, but not touching it. Temperature is monitored with a thermocouple embedded in the nickel susceptor, and careful surface measurements of the oxide substrates indicate a 200 C temperature drop between susceptor and the growing film. All temperatures reported in this paper are substrate surface temperatures.

The reactor is held at 1 Torr up to near atmospheric pressure with a roughing pump by adjusting a gate valve in the exhaust line as well as manipulating flows through mass flow controllers. During startup, the bypass exhaust valve is opened while the process vacuum remains closed. After several mm of reactants have vaporized, a steady state vapor composition is obtained. The bypass valve is closed, the process vacuum valve is opened, and film growth begins. Although much of the warm wall reactor is kept at elevated temperatures, all the flanges and valves are at room temperature, eliminating the need for exotic seals.

This type of upflow stagnation point reactor inherently produces very uniform film growth since buoyancy and convective effects in the gas stream are minimized[12,13]. The laminar flow (Reynolds number < 1000) results in thin and uniform momentum, thermal and solute boundary layers across the entire area of the substrate, and film growth occurs by diffusion of the reactants and reaction products across this boundary layer from the well mixed vapor in the reaction chamber to the substrate surface. In this way uniform films can be grown across large area substrates. In principle, the area can be extended indefinitely by making the quartz tube larger in diameter and employing a larger diameter heater while adjusting the total gas flows to keep the Reynolds number constant.

The quartz tube permits visual examination of any portion of the reactor during the process. For example, the vapor/solid interface in the precursor tube is visually monitored and its vertical position can be adjusted by changing the power to the vaporizing lamp. Any condensation or pyrolysis at unwanted areas in the reactor is instantly observable, and wall temperatures can be adjusted accordingly.

STRONTIUM BARIUM NIOBATE

We have grown 2000 - 3000 Å thick single phase (001) oriented $Sr_xBa_{1-x}Nb_2O_6$ (SBN) on (100) MgO substrates. Optical waveguiding was demonstrated in these films. The solubility range of the SBN films grown by SSMD is shifted slightly from the solubility range reported for bulk SBN[14] in the direction away from the Nb apex of an isothermal slice through the ternary oxide phase diagram. Further details of this work are reported in the paper by Lu et al[15].

POTASSIUM NIOBATE

Based on an early literature report of a volatile potassium THD compound[16], we synthesized the metalorganic according to the reported procedure, prepared a 50:50 mixture of the potassium and niobium thd powders, and grew one 2000 Å film on c-plane sapphire at 500 C. Theta-two theta x-ray diffraction analysis indicated the film was not entirely phase pure but contained a significant amount of oriented $KNbO_3$, indicating that SSMD can be used for the deposition of epitaxial thin films of $KNbO_3$ with a suitable adjustment of the source composition to a larger K/Nb ratio. This is consistent with a presumed lower sticking coefficient (higher vaporization kinetics) of potassium versus niobium on the film surface. Further work on this system is in progress.

LITHIUM NIOBATE

Although $LiNbO_3$ thin films have been grown by a variety of standard techniques, including sputtering[17], laser ablation[18], LPE[19], MBE[20], and sol-gel[21], there are to our knowledge only two other reports of $LiNbO_3$ grown by MOCVD[22,23]. In the first case the reported optical loss was very high (40 dB/cm), and in the second the loss was not measured. We have grown epitaxial c-axis (+ z direction) $LiNbO_3$ thin films on bare c-plane sapphire substrates, on epitaxial (111) oriented 500-1000 Angstrom thick Pt buffer layers sputtered on the sapphire substrates, and on bulk c-axis $LiNbO_3$ substrates. The growth rates for these films are in the range 10 to 25 Å/min. The metalorganic precursors $Nb(thd)_4$ and Li(thd) were either prepared according to the synthesis procedures given by Hammond[16], or obtained commercially from Strem[11]. The $Nb(thd)_4$ was found to be quite stable and reproducible from batch to batch whether synthesized at Hewlett-Packard or purchased. However, the quality of the Li(thd) is sensitive to details of the preparation process and both the color (white with yellow overtones) and degree of dryness of the powder varies. This has been found to have a profound effect on the deposition process and quality of the films, and therefore we have used differential scanning calorimetry to look for differences between in-house and commercial powders. The results indicate the commercial Li(thd) is purer, exhibiting only one major peak and one minor peak. The Li(thd) synthesized in-house exhibits an additional peak, is less dry and has less of a yellow cast than the commercial Li(thd). We surmise that the extra peak is due to retained solvent from one of the synthesis steps, and that its effect seems to be to stabilize the Li(thd) against oxidation at higher temperatures in the reactor.

We find it difficult to avoid the formation of a gas-phase-nucleated white powder in the reactor when using commercial Li(thd). X-ray diffraction analysis of this powder indicates it is Li_2CO_3. Substrate surface temperatures higher than 550 C generate so much Li_2CO_3 as the reactants approach the substrate, that the films are often opaque from incorporated particulates. The in-house Li(thd) on the other hand, permits growth higher than 750 C without the formation of this white soot. Adding the neutral thd ligand as a vapor during the growth process with commercial Li(thd) also seems to retard the vapor phase reaction that leads to soot formation. This indicates the source impurity may be retained thd ligand, which inhibits premature oxidation by pushing the chemical equilibrium in the opposite direction.

Higher growth temperatures (≥ 700 C) lead to improved crystallinity in the films, but also can result in thermal stress induced cracking in films thicker than 1500 Å grown on c-plane sapphire. The higher growth temperatures appear advantageous for homoepitaxial growth on $LiNbO_3$ substrates, and the lower temperatures (~ 500 C) are best for growing thick films on c-plane sapphire for waveguiding applications.

Phase Equilibria and Chemical Characterization

At 950 C, the solid solubility field of the $LiNbO_3$ phase in the pseudo-binary Li_2O-Nb_2O_5 phase diagram[24] spans the compositional range from 47.7 to 50.2 mole percent Li_2O as shown in Figure 2. It decreases to approximately 49.2 to 50.2 mole percent Li_2O at 800 C and appears to narrow rapidly at temperatures lower than 800 C. In SSMD, deposition is carried out at much lower temperatures, particularly on sapphire substrates, and as a result the stoichiometry must be carefully controlled. Outside the single phase existance regime, a $LiNbO_3$ + Li_3NbO_4 two phase region exists on the Li-rich side, and a $LiNbO_3$ + $LiNb_3O_8$ two phase region exists on the Li-poor side. At a growth temperature of 640 C, depending on the exact stoichiometry of the Li(thd) we find that phase pure $LiNbO_3$ can be grown for powder compositions between 65 - 67 mole percent Li as shown in Figure 3, whereas (222) Li_3NbO_4 begins to appear at 70% Li and (220) $LiNb_3O_8$ is present at 60% Li. Homoepitaxial growth of thick films (>1 μm) on $LiNbO_3$ bulk substrates must take place at much higher temperatures, since the composition of the bulk $LiNbO_3$ is fixed by

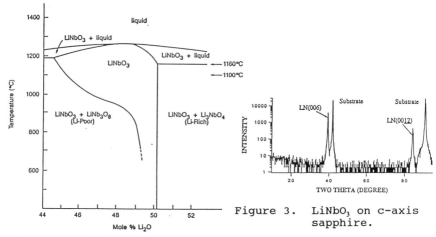

Figure 2. Li_2O-Nb_2O_5 Equilibria.

Figure 3. $LiNbO_3$ on c-axis sapphire.

the congruently melting composition shown in Figure 2, and the substrate may start to decompose when held at lower temperatures in the two phase region during the time needed for film growth. However we have grown 2000 Angstrom films on both +z and -z $LiNbO_3$ substrates at 500 - 700 C, and find that substrate decomposition is not yet apparent during the 1 - 2 hour growth runs. The films grow in the +z direction even if the -z face of the substrate is used for film growth. This was readily determined by etching the films in HF, since the +z face etches very slowly while the -z face is etched quite rapidly to a rough surface.

One interesting application for thick films on bulk $LiNbO_3$ substrates is the reduction of cation impurities commonly found in bulk waveguides, which may cause photorefractive damage in the waveguides upon exposure to intense laser illumination. A SIMS scan through a 5000 Angstrom SSMD film into the $LiNbO_3$ substrate revealed that iron, one of the chief culprits implicated in photorefractive damage, is a factor of 30 lower in the film than in the substrate.

Structural Characterization

An XRD off-axis phi scan of the $LiNbO_3$ (012) peak grown at 700 C shows clear three-fold symmetry about the film plane normal as shown in Figure 4, and the peaks coincide exactly with the sapphire substrate (012) peaks, indicating excellent in-plane orientation with $[10\bar{1}0]LiNbO_3$ || $[10\bar{1}0]$ sapphire. As the growth temperature decreases, more and more 60 degree rotated domains appear as shown in Figure 5. The influence of substrate temperature on other film properties was determined for films grown with a fixed source composition of 65% Li. The results are summarized in Table I. Thicknesses were measured by both Rutherford backscattering (RBS) and by measuring transmission as a function of wavelength with a uv-visible spectrometer. Although RBS cannot detect lithium, the thickness could be determined from the niobium peak by assuming the

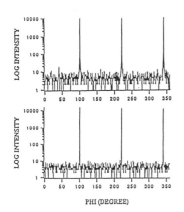

Figure 4. Phi scan for film(top) and Al_2O_3 (bottom).

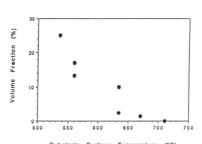

Figure 5. Volume fraction of 60° rotated grains.

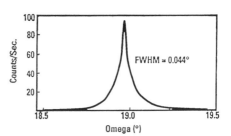

Figure 6. Rocking curve for (006) peak.

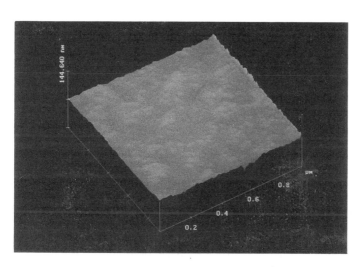

Figure 7. Atomic force microscope image of $LiNbO_3$ film.

stoichiometry. Calculated c-axis lattice parameters are less than the stoichiometric single crystal value of 13.867 Å. The d-spacing of the (012) plane was also measured and found to be slightly larger than the reported bulk value of 3.754 Å. Using the d spacing of the (001) and (012) planes listed in Table I, the a-axis lattice parameters could be calculated and were found to be 5.24, 5.22 and 5.20 Å for films grown at 710, 675 and 640 C respectively. (The accuracy of this calculation is limited by the uncertainty in $d_{(012)}$, and is estimated to be about +0.06 Å.) The a-axis lattice parameters are larger than the single crystal value (5.1478 Å), which suggests the films are under quite a bit of tension at room temperature.

The $LiNbO_3$ (006) rocking curve peak width provides a good indication of film quality since it assesses the spatially averaged out-of-plane misalignment. At a growth temperature of 710 C, a rocking curve FWHM of 0.044 degrees (Figure 6) was obtained on a film grown from a 65% Li source composition, indicating near-perfect out-of-plane alignment. (The sapphire substrate (006) rocking curve FWHM value was 0.007 degrees.). To our knowledge this is the best crystalline quality reported in the literature for a lattice mismatched lithium niobate film. The rocking curve FWHM increases to 0.1 degree as the deposition temperature decreases to 500 C.

The surface morphology of the films as measured by atomic force microscopy was quite smooth as shown in Figure 7 for a 1200 Angstrom film on c-plane sapphire grown at 700 C. The peak-to-peak surface roughness for this film is approximately 40 Å.

Table I. $LiNbO_3$ Film Properties

Ts (°C)	RBS Thickness	no	ne	Spectrometry Thickness	no	ne	d(001) (Å)	d(012) (Å)	Rocking Curve FWHM Value (°)
710	1022 Å	2.32	2.27	1220 Å	2.24	2.12	13.801	3.79	0.044
675	1162 Å	2.31	2.27	1285 Å	2.27	2.18	13.801	3.78	0.053
640	1190 Å	2.33	2.26	1464 Å	2.25	2.11	13.810	3.77	0.098

Note: error bar for RBS thickness is ±55 Å, for no calculated from RBS thickness is ±0.02, for ne calculated from RBS thickness is ±0.04, for d(001) is ±0.006 Å, for d(012) is ±0.025 Å

Optical Characterization

Optical properties were measured on films deposited on 10 mil thick rectangular c-axis sapphire substrates measuring 1 cm by 4 cm long using a single prism (rutile) coupling technique with a 632.8 nm He-Ne laser. All the films measured were less than 2000 Å thick and therefore supported only one TE and one TM mode. The optical loss was measured here and at Xerox PARC using a silicon photodiode coupled to a traveling optical fiber to measure scatter from the guided mode as a function of distance from the coupling spot, and was also measured at Argonne National Laboratory by imaging the scatter with a CCD camera and extracting loss by image analysis techniques. An early 2000 Å thick film was measured at HP to be 25-30 dB/cm. Upon annealing in oxygen at 400 C for 24 hours the loss dropped to 3-5 dB/cm. Further annealing at 500 C for 18 hours increased the loss to 8-9 dB/cm. This film was then measured at Argonne to be ~8dB/cm (a curve fit of the data yielded 7.75 dB/cm for the TE_0 mode and 7.73 dB/cm for the TM_0 mode.) This particular film was grown in two steps of 1000 Å each and the film was cooled to room temperature between the two growth runs, which may account for the high loss. A number of subsequent films with thicknesses of 1100 to 1400 Å have been measured both at HP and at Xerox. In these films the guided mode travels to the edge of the 3.75 cm film and losses of approximately 2 db/cm for the TM_0 modes have been measured. The loss for the TM_0 mode is shown in Figure 8. The reproducibility of the measurement between the three laboratories appears to be about 1 dB/cm. Peaks and valleys in the data in Figure 8 are caused by visually perceptible inhomogeneities in scattered light either in the film or on the film or substrate surface. Annealing one of these low loss films in oxygen at 400 C for three hours significantly increased the scattering instead of decreasing it as in the earlier film. The mechanism for change in optical loss with annealing, also seen by Schwyn[17] for sputtered $LiNbO_3$ films, is not well understood at this time.

From the prism coupling angles, refractive indices were calculated using both RBS and transmission spectra (spectrometer) thickness data and the results are listed in Table I. The transmission spectra thickness values depend on short wavelength data, which are complicated by the fact that accurate refractive indices for wavelengths less than 400 nm are not well known. It was necessary to use extrapolated values calculated from the formula given in[25]. The calculated transmission thicknesses were found to be 10-20% larger than the measured thicknesses. This discrepancy suggests that the refractive indices of the films may slightly different from those measured for bulk single crystals.

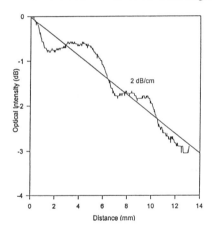

Figure 8. Energy loss for TM_0 mode.

Electrical Characterization

We have grown $LiNbO_3$ films 1000 - 2000 Å thick on sputtered (111) oriented Pt base electrode layers on c-axis sapphire for electrical measurements. An x-ray scan for a 1200 Å thick film (Figure 9) shows nearly 100% c-axis orientation with two unknown impurity peaks. A number of electrical and electro-optic measurements on these films are in progress, including polarization - field hysteresis loops using a Radiant Technologies, Inc. RT66A analyzer[26], and measurements of the electro-optic coefficients. At this time we have preliminary hysteresis loop data. The films clearly exhibit polarization behavior as seen in Figure 9 for an 1100 Å thick film grown at 500 C. The data shown here are the result of the Analysis Software from Radiant Technologies, Inc., which filters out resistive, and capacitive components from the hysteresis loop. Although the remanent polarization P_r is rather low, it may be consistent for such a thin film which can only be driven at very low voltages before breakdown, but it is more likely that the film properties on Pt are not yet optimized.

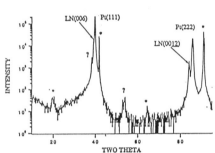

Figure 9. Two theta scan for LiNbO$_3$ on sputtered Pt on c-axis sapphire (* represents sapphire).

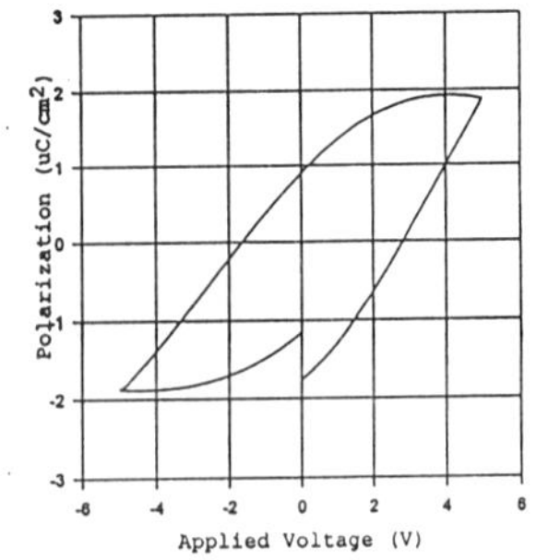

Figure 10. Polarization-voltage loop for LiNbO$_3$ film.

CONCLUSIONS

The solid source MOCVD technique (SSMD) has proven to be useful for the growth of technologically important electro-optic films. Impurity content of the films, particularly for iron which introduces photorefractive damage, is a factor of 30 lower than bulk LiNbO$_3$. The optical loss of LiNbO$_3$ on large area substrates is to date about 2 dB/cm at 633.2 nm for films as thin as 1175 Å. Initial measurements of polarization vs field confirm domain switching in these (001) oriented thin films. The ability of SSMD to grow large area uniform films at practical growth rates make this technique promising for further development in this field.

ACKNOWLEDGMENTS

We are indebted to Bruce Borsberry of Hewlett-Packard for design and construction of the single prism optical loss measurement, to Brady Sih of Stanford University for the differential scanning calorimetry results, and to Professor Angus Kingon of North Carolina State University, Michael Kautzky of Stanford University, and Paul Merchant of Hewlett-Packard for the sputtered (111) oriented platinum base electrode layers. The SIMS analysis was performed by John Turner and Dale Lefforge of Hewlett-Packard.

REFERENCES

1. R. Hiskes, S.A. DiCarolis, J.L. Young, S.S. Laderman, R.D. Jacowitz, and R.C. Taber, Appl. Phys. Lett.,59 (5), 606 (1991).

2. R. Hiskes, S.A. Dicarolis, R.D. Jacowitz, Z. Lu, R.S. Feigelson, R.K. Route, and J.L. Young, J. Cryst. Growth **128**, 781 (1992).

3. M.E. Cowher, T.O. Sedgwick, J. Landerman, J. Elect. Materials **3**, 621 (1974).

4. K. Zhang, E.P. Boyd, B.S. Kwak, A.C. Wright, and A. Erbil, Appl. Phys. Lett. **55**, 1258 (1989).

5. G. Malandrino, D.S. Richeson, T.J. Marks, D.C. DeGroot, J.L. Schindler, and C.R. Kanewurf, Appl. Phys. Lett. **58**, 182 (1991).

6. S.J. Duray, D.B. Buchholz, S.N. Song, D.S. Richeson, J.B. Ketterson, T.J. Marks, and R.P.H. Chang, Appl. Phys. Lett. **59**, 1503 (1991).

7. J. Zhao, C.S. Chern, Y.Q. Li, D.W. Noh, P.E. Norris, P. Zawadski, B. Kear, and B. Gallois, J. Cryst. Growth **107**, 699 (1991).

8. Z. Lu, R.S. Feigelson, R.K. Route, S.A. DiCarolis, R. Hiskes, and R.D. Jacowitz, J. Cryst. Growth **128**, 788 (1992).

9. K. Truman, Conductus, Inc., private communication.

10. G. Meng, G. Zhou, R.L. Schneider, B.K. Sarma, and M. Levy, Appl. Phys. Lett. **63**, 1981 (1993).

11. Strem Chemicals Inc., 7 Mulliken Way, Dexter Industrial Park, P.O. Box 108, Newburyport, MA 01950.

12. C. Houtman, D.B. Graves, and K.J. Jensen, J. Electrochem. Soc. **133**, 961 (1986).

13. J.D. Parsons, J. Cryst. Growth **116**, 387 (1992).

14. J.R. Carruthers and M. Grasso, J. Electrochem. Soc. **117**, 1427 (1970).

15. Z. Lu, R.S. Feigelson, R.K. Routè, R. Hiskes and S.A. DiCarolis, these proceedings.

16. G.S. Hammond, D.C. Nonhebel, and C-W. S. Wu, Inorg. Chem. **2**, 73 (1963).

17. S. Schwyn, H.W. Lehmann, and R. Widmer, J. Appl. Phys. **72**, 1154 (1992).

18. D. Fork and G.B. Anderson, Appl. Phys. Lett. **63**, 1029 (1993).

19. A. Yamada, H. Tamada, and M. Saitoh, Appl. Phys. Lett. **61**, 2848 (1992).

20. R.A. Betts and C.W. Pitt, Electron. Lett. **21**, 960 (1985).

21. K. Nashimoto and M.J. Cima, Mat. Lett. **10** 348 (1991).

22. B.J. Curtis and H.R. Brunner, Mat. Res. Bull. **10**, 515 (1975).

23. A.A. Wernberg, H.J. Gysling, A.J. Filo, and T.N. Blanton, Appl. Phys. Lett. **62**, 946 (1993).

24. L.O. Svaasand, M. Eriksrud, G. Nakhen, and A.P. Gramde, J. Cryst. Growth **22**, 230 (1974).

25. A. Rauber in <u>Current Topics in Materials Science</u>, volume 1, edited by E. Kaldis (1976) p. 481.

26. Radiant Technologies, Inc. 1009 Bradbury Drive SE, Albuquerque, NM 87106.

MOCVD OF MAGNETO-OPTICAL CERAMICS

Masaru OKADA, Shigehisa KATAYAMA and Tadashi KAMIYA
Department of Industrial Chemistry, Faculty of Engineering,
Chubu University 1200, Matsumoto-cho, Kasugai, Aichi 487, Japan

ABSTRACT

Bismuth-substituted yttrium iron garnet, $Bi_xY_{1-x}Fe_5O_{12}$ thin films have been epitaxially grown on a $Gd_3Ga_5O_{12}$(111) substrate at 750°C under reduced pressure (6 Torr), with a deposition rate of 20~50nm·min^{-1} by MOCVD. The maximum value of Faraday rotation was -7.5×10^4 deg·cm^{-1} at a Bi-content of 2.5 atoms per formula unit. The Figure of unit was 14.30 deg., which is significantly larger than that obtained by sputtering. On the other hand, cerium-substituted films could not be obtained even at elevated temperatures, 900°C.

INTRODUCTION

Bismuth-substituted yttrium iron garnet (Bi-YIG) thin films are very attractive for magneto-optical applications because of their high Faraday rotation and weak light absorption. Liquid phase epitaxy[1], sputtering[2~5] and pyrolysis[6] have been used for the fabrication of these films. In addition, it has been reported that completely Bi-substituted garnet films have been successfully prepared by sputtering[3~5]. However, these sputtering methods generally have a low deposition rate (2~10nm·min^{-1}). Metal-organic chemical vapor deposition (MOCVD) has significant advantages over physical deposition, that is, high deposition rate, relatively easy composition control and few surface defects owing to its growth under thermal equilibrium conditions.

In this paper, the microstructure and magneto-optical properties of Bi-YIG films prepared by MOCVD are described in relation to the bismuth content. Cerium-substituted YIG films[6,7] by MOCVD are also described.

EXPERIMENTAL

The metal-organic precursors were triphenyl bismuth, $Bi(Ph)_3$, and both trisdipivaloylmethanato yttrium and iron, $Y(DPM)_3$ and $Fe(DPM)_3$. Table 1 shows the deposition conditions employed for the Bi-YIG film. The CVD apparatus and experimental procedure in the present study are the same as those described in the previous paper [8]. The preparation of Bi-YIG film was carried out at a substrate temperature of 750°C by adjusting the deposition rate of Y_2O_3, Fe_2O_3 and Bi_2O_3 films to achieve a $(Bi_2O_3+Y_2O_3):Fe_2O_3$ composition ratio of 3:5 with various Bi-substituted amount (x). The garnet films were mainly deposited on $10 \times 10 \times 0.5$ mm polished $Gd_3Ga_5O_{12}$, GGG(111) single crystal substrate.

TABLE 1. Deposition conditions of Bi-YIG films.

Source materials	Source temp. (℃)	Carrier gas flow rate (ml/min)	Substrate temp. (℃)
$Bi(Ph)_3$	140		
$Y(DPM)_3$	140	50 ~ 100	750
$Fe(DPM)_3$	130		

Total gas flow rate: 400 ml/min, O_2 concentration: 50 %, Pressure: 6.0 Torr.

Growth rate: 20 ~ 50 nm/min, Film thickness: 1.5 ~ 3.0 μm
Substrate: GGG(111), NGG(111), Quartz

RESULTS AND DISCUSSION

A. Crystallinity and Microstructure

Figure 1 shows the X-ray diffraction (XRD) patterns of garnet film having a Bi-content of 2.5 atoms per formula unit on the GGG (111) substrate. The films were a single phase of the garnet structure, and highly oriented on the GGG(111) substrate. In addition, the films are close to stoichiometric as measured by EDX analysis. Figure 2 shows the dependence of a half width of the

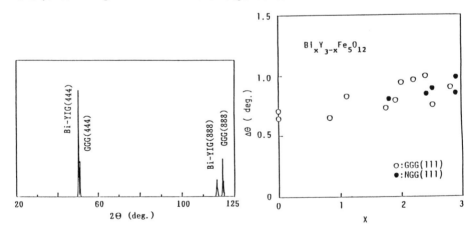

FIG. 1. XRD patterns of $Bi_{2.5}Y_{0.5}Fe_5O_{12}$ garnet films on the GGG(111) substrate at 750℃.

FIG. 2. Dependence of half width of the rocking curves of (444) peaks on Bi content in the garnet films.
Substrate, ○: GGG(111), ●: NGG(111)

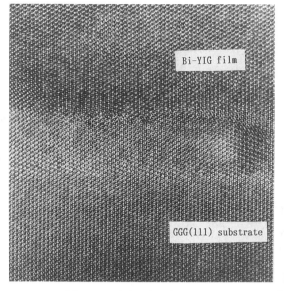

(a): crystal lattice image   (b): electron diffraction pattern

Fig. 3. TEM image of the interface between the garnet film and GGG(111) substrate.

rocking curves of (444) diffraction peaks on the Bi-content in the garnet films. The values of the half width of the rocking curves were very small, 0.6 - 0.8° for the range of the Bi content of 0 to 3.0. The surface morphology of the garnet films deposited on GGG (111) substrate was very smooth and crystal grains or cracks could not be observed, with a Bi content up to 2.7. However, on the film with a Bi content of 3.0 corresponding to the complate substitution, the surface became rough due to the appearance of large crystalline grains.

Figure 3(a) shows the crystal lattice image of the interface between the garnet film (x = 2.5) and GGG(111) substrate using transmission electron microscopy(TEM). Figure 3(b) shows electron diffraction patterns of the same region in a diameter of 100 nm. The TEM image shows that a single crystalline film of garnet could be epitaxially grown along the crystal lattice of GGG(111) substrate.

In the 5nm-thick interface layer between film and substrate, edge dislocations are observed. As is seen in Fig.3(b), the electron diffraction pattern of the garnet film significantly coincides with that of GGG substrate because the mismatch of lattice constants between the film and substrate is small. The films as-deposited at 750℃ were light pink and transparent. Figure 4 shows the optical

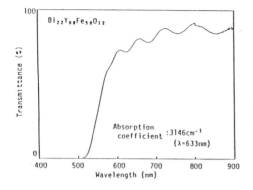

FIG. 4. Optical transmission spectra of the garnet films. Substrate: GGG(111), film thickness: 990 nm

transmission spectra of the garnet film (x = 2.2) deposited on a GGG (111) substrate. The film thickness was 0.99 μm. The interference oscillation corresponding to the flat surface of the film can be observed as shown in Fig.4. The absorption coefficient at a wave length of 633nm was estimated as 3146 cm^{-1}.

B. Magneto-optical properties.

Figure 5 shows typical Faraday hysteresis loops for the YIG and Bi-YIG (x = 2.5) films as-grown at 750℃ on the GGG(111) substrate.

Figure 6 shows the Faraday rotation as a function of the Bi-content in the film. The Faraday rotation linearly increases as the Bi content increases. The maximum value of Faraday rotation at 633nm was -7.5×10^4 deg·cm^{-1}, with a Bi content of 2.5. This value is significantly larger than that obtained by another method. However, when Bi content exceeds 2.5, the values of Faraday rotation decrease because of injection of crystalline defects into the film.

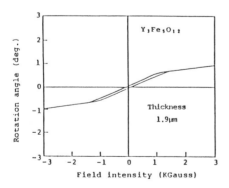
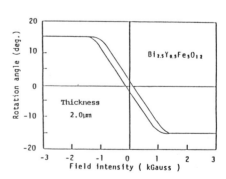

FIG. 5. Faraday hysteresis loops of YIG and Bi-YIG films. Substrate: GGG(111), λ =633 nm

Table 2 shows the figure of merit of the garnet films calculated from the Faraday rotation and absorption coefficient. These values are somewhat larger than those obtained by another method.

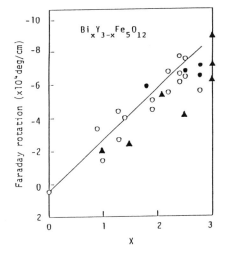

FIG. 6. Dependence of Faraday rotation on the Bi content of Bi-YIG films.
λ =633 nm, ▲ : sputtering method

Substrate, ○ : GGG(111) ● : NGG(111)

TABLE 2. Figure of merit (F/α) of Bi-YIG garnet films.

Substituted Bi content(at./F.U)	Faraday rotation F(deg./cm)	Absorption coeff. α (cm^{-1})	Figure of merit F/α (deg.)
1.3	-2.7×10^4	4438	6.08
2.2	-4.5×10^4	3146	14.30

C. Ce-substituted garnet film.

Also, we tried a preparation of the cerium-substituted garnet film(Ce-YIG), which was expected to have a large magneto-optical effect in the near infrared ragion. Ce(DPM)$_3$ was used as the precursor of cerium oxide instead of Bi(Ph)$_3$. Figure 7 shows XRD patterns of the film deposited at 800℃ with a Ce content of 1.0 at. per f.u.. In Fig.7(c), the peaks originating from the cerium oxide can be observed. This indicates that cerium ion was oxidized to four valences beyond the three valences during the film deposition, and could not be solidly dissolved into the garnet film. At even an elevated temperature of 900 ℃, the results were the same. Accordingly, at present, we have not been able to successfully obtain Ce-substituted garnet film having excellent magneto-optical properties by the MOCVD method.

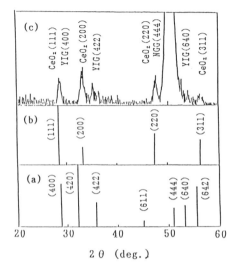

FIG. 7. XRD patterns of $Ce_{1.0}Y_{2.0}Fe_5O_{12}$ garnet films on the NOG(111) substrate at 800 ℃.
(a): YIG powder, (b): CeO_2 powder, (c): Ce-YIG film

CONCLUSIONS

Bismuth-substituted yttrium iron garnet thin films were grown on a GGG(111) substrate at 750℃ under the reduced pressure of 6 Torr, with a large deposition rate of $20\sim50$ nm·min^{-1} by MOCVD. This method permits epitaxial growth during deposition even if the Bi content exceeds 2.5 atoms per formula unit. Large values of Faraday rotation and figure of merit were obtained. This is attributed to the large Bi substitution and the crystalline perfection in the garnet film. The films prepared by MOCVD have significant advantages regarding magneto-optical memory applications.

REFERENCE

1. P. Hansen, C. P. Klages, and K. Witter, J. Appl. Phys. 60, 721 (1986)
2. M. Gomi, T. Tanida, and M. Abe, J. Appl. Phys. 57, 3888 (1985)
3. T. Okuda, N. Koshizuka, K. Hayashi, K. Sato, H. Taniguchi, and H. Yamamoto, J. Appl. Phys. 67, 4944 (1988)
4. K. Shono, H. Kano, N. Koshino, and S. Ogawa, J. Appl. Phys. 63, 3639 (1988)
5. S. Mino, M. Matsuoka, A. Tate, A. Shibukawa, and K. Ono, Jpn. J. Appl. Phys. 31, 1786 (1992)
6. J. Cho, K. Gomi, and M. Abe, Jpn. J. Appl. Phys. 27, 2069 (1988)
7. M. Alex, K. Shono, S. Kuroda, and S. Ogawa, J. Appl. Phys. 67, 4432 (1990)
8. M. Gomi, H. Furuyama, and M. Abe, Jpn. J. Appl. Phys. 29, L99 (1990)
9. M. Okada, S. Katayama, and K. Tominaga, J. Appl. Phys. 69, 3566 (1991)

MOCVD OF GROUP III CHALCOGENIDE COMPOUND SEMICONDUCTORS

ANDREW R. BARRON
Department of Chemistry, Harvard University, 12 Oxford Street, Cambridge, MA 02138

ABSTRACT

A review is presented of recent advances in the metal-organic chemical vapor deposition (MOCVD) of thin films of group III-chalcogenides. The deposition of thermodynamic phases of composition ME and M_2E_3 (M = Ga, In; E = S, Se, Te) will be presented. Also included is a discussion of the development of molecular control over the structure of deposited films and the atmospheric pressure MOCVD growth of the high pressure phase of InS and a meta-stable cubic phase of GaS.

INTRODUCTION

Although not as comprehensively studied as the III-V compound semiconductors, there has been increasing interest in group III chalcogenides, in particular ME and M_2E_3 (M = Ga, In; E = S, Se) because of their possible application in switching devices [1], photovoltaics [2,3], and nonlinear optics [4]. Additional work on gallium sulfide has been prompted by the sulfide passivation of GaAs [5]. While there is an extensive literature dealing with the synthesis of bulk III-VI compounds, their electronic properties [6], photoluminescence properties [7], and structural and phase transformation studies [8], the fabrication of thin films has until recently been the subject of relatively few reports in the literature. Thin films of $In_{2-x}Ga_xS_3$ (x = 0 - 2) have been prepared by the spray pyrolysis of aqueous solutions of $InCl_3$, $GaCl_3$ and thiourea in methanol/water [9], while amorphous films of InSe and In_2Se_3 have been prepared by thermal evaporation [10,11]; the latter has also been produced by electrodeposition [12]. Kanatzidis et al. have reported the synthesis of anionic In/Se clusters, i.e., $[In_2Se_{21}]^{4-}$ and $[In_3Se_{15}]^{3-}$, and suggested their use as precursors for InSe and In_2Se_3 films [13].

Recent studies concerning the growth of III-VI thin films have primarily involved metal-organic chemical vapor deposition (MOCVD) from single-source precursors, and it is this topic that will be discussed herein. This review is not intended to be comprehensive, but rather to serve as a guide to the literature and present results in the area of the molecular control over the structure of deposited films.

MOCVD OF GROUP III CHALCOGENIDES

Table I summarizes the gallium and indium chalcogenide precursor compounds reported for the MOCVD growth of thin films of III-VI compound semiconductors, along with the phase of the resulting films, and the relevant literature citation.

Gallium Sulfide

Gallium sulfide (GaS) thin films have been grown at 380 - 420 °C by atmospheric pressure MOCVD using the single source precursors $[(^tBu)_2Ga(S^tBu)]_2$, $[(^tBu)GaS]_4$ and $[(^tBu)GaS]_7$ [14,15].

Films grown at 400 °C using the dimeric gallium thiolate $[(^tBu)_2Ga(S^tBu)]_2$ (I) as the precursor consisted of stoichiometric GaS, and showed a fine structure consisting of near spherical particles of about 300 Å in diameter [15]. The associated electron diffraction displayed broad diffuse rings indicative of a very poorly crystallized film. The estimated centers of these rings gave d-spacings consistent with the most intense reflections for the thermodynamically

Table I. Summary of the dependence of the group III metal to chalcogen ratio and the crystalline phase with precursor and deposition temperature.

Precursor	Deposition Temp. (°C)	M:E ratio	Phase Observed	Ref.
[(^{t}Bu)$_2$Ga(S^{t}Bu)]$_2$	400	1:1	hexagonal, GaS (poorly crystalline)	15
[(^{t}Bu)$_2$Ga(S^{t}Bu)]$_2$	475	2:3	amorphous (Ga$_2$S$_3$)	15
[(^{t}Bu)GaS]$_4$	< 380	1:1	amorphous (GaS)	14,15
[(^{t}Bu)GaS]$_4$	380 - 400	1:1	cubic GaS	14,15
[(^{t}Bu)GaS]$_4$	> 400	< 20% + 2:3	amorphous + crystalline needles	14,15
[(^{t}Bu)GaS]$_7$	380 - 400	1:1	amorphous (GaS)	15
[(^{t}Bu)GaSe]$_4$	380	1:1	crystalline (unknown)	16
[(^{t}Bu)GaTe]$_4$	380	1:1	crystalline (unknown)	16
[(^{t}Bu)$_2$In(S^{t}Bu)]$_2$	300 - 350	2:1	orthorhombic InS + In rich amorphous phase	17
[(^{t}Bu)$_2$In(S^{t}Bu)]$_2$	400	1:1	tetragonal InS (highly oriented)	17,19
[(^{n}Bu)$_2$In(S^{t}Bu)]$_2$	300 - 350	2:1	orthorhombic InS + In rich amorphous phase	17
[(^{n}Bu)$_2$In(S^{t}Bu)]$_2$	400	1:1	tetragonal InS (highly oriented)	17
[(Me)$_2$In(S^{t}Bu)]$_2$	300 - 350	2:1	amorphous phase	17
[(Me)$_2$In(S^{t}Bu)]$_2$	400	1:1	In$_2$S$_3$ and amorphous phase	17
(Me)In(S^{t}Bu)$_2$	300	1:1	amorphous, crystallizes to β-In$_2$S$_3$ upon annealing	17
(^{n}Bu)In(S^{i}Pr)$_2$	400	2:3	In$_2$S$_3$ (highly oriented)	20
[(Me)$_2$In(SePh)]$_2$	300 - 365	1:1	cubic In$_x$Se$_y$ phase	21
[(Me)$_2$In(SePh)]$_2$	400 - 480	1:1	hexagonal InSe phase cubic In$_x$Se$_y$ (low level)	21
In(SePh)$_3$	400	2:3	hexagonal In$_2$Se$_3$ + cubic In$_x$Se$_y$	21
In(SePh)$_3$	460 - 550	2:3	hexagonal In$_2$Se$_3$ (highly oriented, 001)	21

stable hexagonal phase of GaS (JCPDS No. 30-576) [6]. Thermal annealing of the as deposited film resulted in crystallization consistent with the formation of hexagonal GaS, as indicated by the sharpening of the rings in the electron diffraction pattern. Deposition at temperatures higher than 475 °C yielded amorphous films with a Ga:S ratio of 2:3 (i.e., consistent with Ga$_2$S$_3$). This result suggests that the precursor molecule undergoes premature decomposition at 475 °C, resulting in the rupture of the Ga$_2$S$_2$ core.

Polycrystalline films grown from [(^{t}Bu)GaS]$_4$ (**II**) give an electron diffraction consistent with a face centered cubic lattice with a lattice parameter a = 5.4 Å [14,15]. Upon thermal annealing (15 min. @ 450 °C) these deposited films show significant expansion giving a measured lattice parameter for a free-standing film of 5.5 Å [15]. In contrast to the free standing films, those grown on GaAs (100) are not polycrystalline but near-epitaxial. This is as may be expected since the lattice parameter of the annealed GaS is within 3% of GaAs (5.6532 Å) [6]. The lattice parameter of the near-epitaxial GaS coating on GaAs corresponds to a calculated cell constant of 5.632 Å.

(I)   **(II)**   **(III)**

If film deposition using [(^{t}Bu)GaS]$_4$ is carried out above 450 °C two growth regions are observed in the chamber [14,15]. The upstream deposition zone consists of an amorphous featureless gallium rich matrix containing between 0 and 20% sulfur. Down flow of this amorphous region a polycrystalline film is deposited. The elemental composition of the film was determined by EDX to be between GaS and Ga$_2$S$_3$. However, while the films are crystalline, the electron diffraction did not match any known gallium sulfide phase.

Films deposited from the heptamer [(^{t}Bu)GaS]$_7$ (**III**) show differences to those grown using either [(^{t}Bu)GaS]$_4$ or [(^{t}Bu)$_2$Ga(S^{t}Bu)]$_2$. Although films deposited over the temperatures of 390 ±10 °C consist of a Ga:S ratio of 50:50 (±2), electron diffraction yielded amorphous patterns and the films are essentially featureless [15]. Thermal annealing results in crystallization to an unidentified phase.

Gallium Selenide and Telluride

Thin films of GaSe and GaTe have been deposited using the cubane precursors [(^{t}Bu)GaSe]$_4$ and [(^{t}Bu)GaTe]$_4$ (*c.f.*, **II**), respectively [16]. While the films are stoichiometric GaE (E = Se, Te) and crystalline, their structures are unknown; they do not correspond to any of the reported Ga$_x$E$_y$ phases. It is interesting to note, however, that unlike the sulfide or selenide, which grow as uniform contiguous films, the telluride appears to grow as clusters, presumably due to precursor aggregation in the gas phase.

Indium Sulfide

Films grown using either [(^{t}Bu)$_2$In(S^{t}Bu)]$_2$ or [(^{n}Bu)$_2$In(S^{t}Bu)]$_2$ (**IV**) as the precursor at each of the temperatures 300, 350 and 400 °C were essentially independent of the precursor in terms of morphology and phases present [17]. The microstructural features and chemical composition of the deposited films were, however, observed to have significant dependence on the deposition temperature. Films obtained at 300 °C consisted of an amorphous matrix phase (predominantly indium in composition with little or no sulphur) and a crystalline island phase. The latter was found to be InS (1:1) and exhibited an electron diffraction pattern consistent with orthorhombic InS (JCPDS No. 19-588) [6]. The overall composition of these films was found to be in the region of *ca*. In 66%, S 33%, consistent with a mixture of InS and In [17]. Films grown at 350 °C were found to be similar in nature to those grown at 300 °C. In contrast, films deposited at 400 °C were found to be a single phase, the tetragonal high pressure phase of InS (JCPDS No. 36-643) [18], with a distinct oriented growth morphology [19].

$$\text{(IV) } R = {}^tBu, {}^nBu, Me \qquad \text{(V)}$$

Films deposited from the methyl substituted precursor, [(Me)$_2$In(S^{t}Bu)]$_2$, show marked differences to those grown using [(^{t}Bu)$_2$In(S^{t}Bu)]$_2$ or [(^{n}Bu)$_2$In(S^{t}Bu)]$_2$. At 300 °C the deposited film are amorphous, with a composition of In 65%, S 35%, consistent with the presence of InS and an In rich material, since In$_2$S is unstable except in the gas phase. Films deposited from the methyl precursor at 400 °C yielded a crystalline diffraction pattern, In$_2$S$_3$ (JCPDS No. 25-390) [6], superimposed on a background of broad amorphous rings. EDX analysis of the overall film indicated a composition of 50% indium and 50% sulfur, which in light of the identification of In$_2$S$_3$ by electron diffraction suggests that an indium rich phase should also be present.

Films deposited from the sulfur rich precursor, [(^{t}BuS)(Me)In(S^{t}Bu)]$_2$ (**V**), at a deposition temperature of 300 °C, yielded amorphous electron diffraction patterns and gave a composition of 50% In: 50% S by EDX analysis. Annealing the sample in-situ by focusing the electron-beam on the sample for a short period of time gave dark precipitates can clearly be observed which yield a diffraction pattern consistent with β-In$_2$S$_3$ (JCPDS No. 32-456) [6]. Similarily, Nomura et al. have reported that highly oriented In$_2$S$_3$ may be grown using (^{n}Bu)In(S^{i}Pr)$_2$ [20].

Indium Selenide

Thin films of InSe have been grown using a modified MOCVD technique; Spray-MOCVD [21]. During spray-MOCVD a toluene solution of the precursor, [(Me)$_2$In(SePh)]$_2$ (**VI**) is transported as a mist, in argon carrier gas, into a heated reactor zone containing the substrate. With GaAs as a substrate films grown at 365 °C were found to be near stoichiometric InSe. However, the X-ray diffraction pattern revealed that the films consist of a new cubic phase of InSe (In:Se *ca.* 1.2), whose lattice parameter (a = 5.72 Å), when compared to that of cubic GaS (*vida supra*) is consistent with the relative size of gallium versus indium, and sulfur versus selenium. Furthermore, based on SEM the cubic InSe films grow with a preferred orientation, essentially perpendicular to the substrate surface. Deposition using the same precursor system, at growth temperatures of 400 - 476 °C resulted in films containing a low level of the cubic phase

(**VI**)

and major levels of the thermodynamically stable hexagonal phase (JCPDS No. 34-1431) [6]. Elemental analysis of these films indicated the stoichiometry to be InSe (1:1) [21].

Use of the *tris*-selenolate, In(SePh)$_3$ as a precursor for spray-MOCVD on GaAs (100) substrates resulted in the deposition of highly oriented In$_2$Se$_3$ films (JCPDS No. 23-294) over the temperature range 460 - 550 °C. Deposition on Si (100) substrates at 400 °C gave in addition to the hexagonal In$_2$Se$_3$ small quantities of the cubic phase [21].

MOLECULAR CONTROL OVER THE PHASES OF MOCVD GALLIUM AND INDIUM SULFIDES.

In principle, MOCVD offers significant advantages over physical vapor deposition (PVD) methods: these examples include simple apparatus, mild process conditions, control over composition, high deposition rates, and possible large scale processing. Perhaps the most significant advantage that CVD has over other methods is the production of metastable materials. Most CVD processes operate far from equilibrium conditions so that kinetically rather than thermodynamically favored products are produced [22]. However, an important question arises: Can the molecular structure of the precursor control the phase structure of a solid product? While several groups have shown that thermodynamically stable phases can be prepared by MOCVD from molecules designed to represent the smallest fragment of the solid state structure, it is only resently that workers have demonstrated that new or high energy phases may be prepared by MOCVD from single molecular precursor.

The results discussed above prompt a number of interesting questions concerning the molecular control exhibited over the phase and composition of the group III-VI thin films deposited from single source precursors. For example: Why are the films deposited from [(^{t}Bu)$_2$Ga(S^{t}Bu)]$_2$ poorly crystalline while those from [(R)$_2$In(S^{t}Bu)]$_2$ (R = ^{t}Bu, ^{n}Bu) highly crystalline? Why do the gallium sulfide films deposited, under otherwise identical conditions, using [(^{t}Bu)$_2$Ga(S^{t}Bu)]$_2$ and [(^{t}Bu)GaS]$_4$ as single source precursors have the same gallium:sulfur ratio but completely different structures? Why are the gallium sulfide films deposited from [(^{t}Bu)GaS]$_7$ amorphous?

Highly Crystalline versus Poorly Crystalline.

While the films deposited from [(^{t}Bu)$_2$Ga(S^{t}Bu)]$_2$ are of the thermodynamic phase of GaS they are poorly crystalline [15]. In contrast, those grown from [(R)$_2$In(S^{t}Bu)]$_2$ (R = ^{t}Bu, ^{n}Bu) are not only highly crystalline but also of the kinetically stable high pressure phase [17,19].

From a consideration of the structures of the orthorhombic (thermodynamic) and tetragonal (high pressure) phases of indium sulfide (InS) it appears as though the dimeric structure of [(R)$_2$In(S^{t}Bu)]$_2$ (R = ^{t}Bu, ^{n}Bu) acts to define the identity of the deposited phase. While the tetragonal phase of InS may be considered to consist of fused four membered In$_2$S$_2$ rings (see Figure 1a) (as is present in the precursor in the gas phase), the orthorhombic phase [6] consists of fused six-membered cycles, in a chair conformation (see Figure 1b). Thus, while the tetragonal form may be built up from the precursor molecules In$_2$S$_2$ core, ring rupture, or rearrangement to In$_3$S$_3$ must occur for the orthorhombic form to be deposited.

In the case of films deposited from [(^{t}Bu)$_2$Ga(S^{t}Bu)]$_2$, if the precursor's core remains essentially intact during deposition then the resulting phase would be expected to consist of four-membered rings, which neither of the previously known phases of GaS (hexagonal and rhombohedral) consist of, both consisting of a layered S-Ga-Ga-S structure built up of fused six-membered rings (e.g., Figure 2). However, if one considers the mode of packing of the four-fold layers shown in Figure 2, it may be seen that hexagonal GaS does consist of *highly* distorted Ga$_2$S$_2$ rings, see Figure 3. While, the intra-layer Ga-S distance in hexagonal GaS (2.337 Å) is similar to that expected for [(^{t}Bu)$_2$Ga(S^{t}Bu)]$_2$ (2.35 - 2.45 Å), the inter-layer Ga···S distance is much larger (3.727 Å).

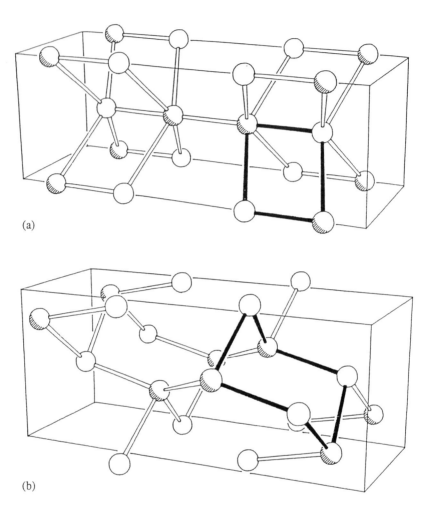

Figure 1. Unit cell of the tetragonal InS (a) and orthorhombic InS (b). The solid bonds represent the smallest cyclic structural fragment; which in (a) is related to the In_2S_2 cyclic core of $[(^tBu)_2In(S^tBu)]_2$.

Thus, while the core of $[(^tBu)_2In(S^tBu)]_2$ is structurally matched to tetragonal InS, severe distortion of the $[(^tBu)_2Ga(S^tBu)]_2$ core is required to match hexagonal GaS. As a consequence, the GaS films deposited from $[(^tBu)_2Ga(S^tBu)]_2$ are poorly crystallized, whereas the InS film grown under identical conditions are highly crystalline with strong preferred orientation. Upon thermal annealing the Ga_2S_2 dimers presumably undergo the requisite rearrangement to allow for crystallization of the hexagonal phase.

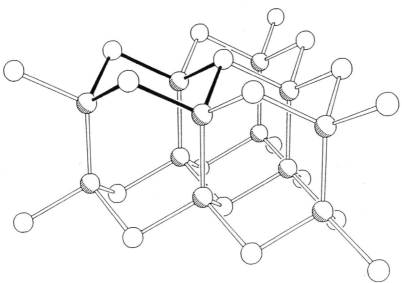

Figure 2. The unit cell structure of hexagonal GaS. The solid bonds represent the smallest cyclic structural fragment.

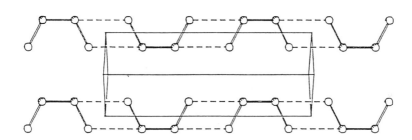

Figure 3. Inter layer stacking of hexagonal GaS viewed perpendicular to the {110} plane.

Hexagonal versus Cubic versus Amorphous.

The formation of hexagonal GaS would be expected if film growth was thermodynamically controlled, and we have rationalized the reason for the formation of poorly crystalline hexagonal GaS from [(tBu)$_2$Ga(StBu)]$_2$. However, the formation of a new cubic phase of GaS by the MOCVD of [(tBu)GaS]$_4$ must be explained in terms of the retention of the cubane precursor core in the deposited film.

In a simplistic view, the cubic core presumably relaxes in the deposited film and rearranges to a cubic lattice (Figure 4) [14]. The electron diffraction pattern of the as-deposited films grown from [(tBu)GaS]$_4$ is consistent with a face-centered cubic lattice.

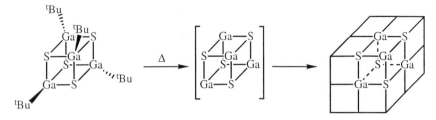

Figure 4. Schematic representation of the formation of a fcc-cubic phase from [(^{t}Bu)GaS]$_4$.

Of the common face-centered cubic structures two have a 1:1 stoichiometry: halite (NaCl) and zinc blende (cubic ZnS), which for a single compound (e.g., GaS) would have near identical diffraction patterns, the only differences being the relative intensities of certain reflections. A survey of halite and zinc blende structures of a similar cell constant to the cubic GaS, formed from [(^{t}Bu)GaS]$_4$, show that the only consistent relative intensity between the structures is that the {111} reflection is more intense than the {200} reflection for the zinc blende structure, while the {200} is more intense than the {111} for halite structures. However, given the difficulties in quantification of electron density patterns, the exact structure could be open to question. Further studies lead to the conclusion that while the films may initially grow as shown in Figure 4, cubic-GaS adopts the zinc blende structure. The following evidence and observations are presented to support this conclusion:

(i) The preference of gallium for tetrahedral over octahedral coordination sites is well documented, and exemplified by the structure of the stable oxide phase, β-Ga$_2$O$_3$, in which the gallium is divided between tetrahedral and octahedral sites [23]. This is in comparison with alumina where all the aluminum atoms are in octahedral vacancies. The tetrahedral preference for gallium should therefore favor the zinc blende structure for cubic GaS.

(ii) The cell parameter for cubic-ZnS is 5.409 Å. Given the adjacent position of Ga and Zn in the Periodic Table the cell parameter for cubic-GaS would be expected to be the same or slightly larger. This is indeed observed; 5.4 and 5.5 Å for the as-deposited and thermally annealed cubic-GaS, respectively.

(iii) If cubic-GaS adopts the zinc blende structure then the Ga-S distance calculated from the observed lattice parameters would be 2.33 and 2.38 Å for the as-deposited and thermally annealed cubic-GaS, respectively, and 2.42 Å for the GaS grown on GaAs (see above). These values compare favorably to those we have previously reported for molecular gallium-sulfide clusters (2.31 - 2.44 Å), and the hexagonal phase of GaS (2.337 Å).

(iv) Using the method of Mooser and Pearson [24], which predicts the crystal structure of normal valence compounds based on the principal quantum number and electronegativities [25] of the component atoms, a zinc blende (tetrahedral coordination) structure is expected for cubic GaS.

If we assume that the cubane core of the [(^{t}Bu)GaS]$_4$ precursor is retained during the deposition, then as described in Figure 4 a halite structure would ensue. However, the relationship between halite and zinc blende structure involves a simple (^{1}/$_4$, ^{1}/$_4$, ^{1}/$_4$) shift of one of the interpenetrating fcc lattices, or the movement of one element from octahedral to a tetrahedral environment. It is thus reasonable to propose that this halite to zinc blende transformation be facile on the surface of the growing film. Furthermore, during deposition on GaAs (100) the substrates surface will promote such a change. Thus, while a cubic phase is determined by the precursor, its growth is enhanced by the cubic-GaAs substrate.

The use of [(^{t}Bu)GaS]$_7$ as a single-source precursor does not allow a crystalline phase to be deposited, irrespective of substrate. If, as assumed for the dimer and tetramer the core of the heptamer is not disrupted during the deposition process, it is possible to propose why no crystalline phase is produced. The Ga$_7$S$_7$ core of [(^{t}Bu)GaS]$_7$ has a C$_3$ axis (see **III**) with no possibility of ordered close packing. Thus, during deposition a random array of gallium and sulfur atoms would be incorporated into the film. This concept is consistent with the amorphous nature of the deposited films.

CAN WE ASSUME MOLECULAR CONTROL TO EXIST?

It is clear from the discussion above that the molecular structure of the precursor will define the structure of a deposited phase. However, the MOCVD of indium sulfide and selenide also teaches us that molecular control may not be sufficient. For example: Why are indium sulfide films deposited from $[(R)_2In(S^tBu)]_2$ (R = ^tBu, ^nBu) deficient in sulfur at low deposition temperatures but not at higher deposition temperatures? Why does the deposition temperature control the phase, but not the composition of InSe grown from $[(Me)_2In(SePh)]_2$?

Sulfur Deficient versus Stoichiometric.

Why is it that lower deposition temperatures produce fragmentation of the In_2S_2 molecular core of the precursors $[(R)_2In(S^tBu)]_2$ (R = ^tBu, ^nBu), while at higher temperatures the core is retained? Despite the counter-intuitive nature of this observation, it may readily be explained in terms of surface-catalyzed versus gas phase decomposition. If we consider the high temperature deposition as the simplest case it is commonly accepted that dimeric group III pnictide compounds $[R_2M(ER_2')]_2$ decompose cleanly to yield the M_2E_2 core and the appropriate hydrocarbon side products. A similar mechanism may be proposed for the decomposition of $[(R)_2In(S^tBu)]_2$ at 400 °C (see Table I), resulting in the retention of the In_2S_2 units [17]. At lower deposition temperatures, however, insufficient thermal energy is imparted to the molecule and it undergoes chemisorption with decomposition. The $^tBuSS^tBu$ is thus formed by the elimination of surface bound ^tBuS groups. As discussed above, the use of the mono-thiolate complexes $[(R)_2In(S^tBu)]_2$ (R = ^tBu, ^nBu) as precursors for deposition below 350 °C results in films having a low sulfur content, as a consequence of loss of $^tBuSS^tBu$. It is reasonable to propose that if this decomposition pathway is a constant for dimeric alkylindium thiolate compounds, then we may assume that the decomposition of a bis-thiolate complex $[(^tBuS)(Me)In(S^tBu)]_2$, should yield stoichiometric InS below 350 °C, which is indeed the case:

$$[(^tBuS)MeIn(\mu-S^tBu)]_2 \xrightarrow[350\ °C]{\Delta} 2\ InS + {}^tBuSS^tBu + 2\ H_2C=CMe_2 + 2\ CH_4 \quad (1)$$

Hexagonal versus Cubic

The phase of InSe grown from $[(Me)_2In(SePh)]_2$ is controlled by the deposition temperature. This is unlike the growth of InS from structurally similar precursors, since the composition would be expected to change due to premature decomposition or fragmentation of the precursor, as is the case in the MOCVD of indium selenide from $In(SePh)_3$.

As with the InS films (above) consideration of the high temperature deposition as the simplest case should result in clean decomposition of $[(Me)_2In(SePh)]_2$ above 400 °C (see Table I) resulting in the retention of the In_2Se_2 unit, and formation of the thermodynamically stable hexagonal phase. At lower deposition temperatures, however, a different decomposition mechanism must be dominant.

It has been observed that gallium selenolates $[(R)_2Ga(SeR)]_2$ decompose upon thermolysis to yield the cubane compound $[(R)GaSe]_4$, and hydrocarbon side products.

$$2\ [(R)_2Ga(SeR)]_2 \xrightarrow{\Delta} [(R)GaSe]_4 + 4\ R\text{-}R \qquad (2)$$

If a similar reaction occurs for $[(Me)_2In(SePh)]_2$ then the actual precursor present in the gas phase (or at the growth surface) would be $[(Me)InSe]_4$, which, given the results observed for the growth of cubic GaS (above), should result in the deposition of a cubic phase of InSe, as is indeed observed from X-ray diffraction.

$$2\ [(Me)_2In(SePh)]_2 \xrightarrow[-4\ MePh]{\Delta} [(Me)InSe]_4 \xrightarrow[-4\ Me\cdot]{} 4\ InSe \qquad (3)$$

CONCLUDING REMARKS

At this time the commercial application of MOCVD of group III-VI compound semiconductors is in its infancy. However, the application of cubic-GaS as a passivation layer for III-V semiconductors [26] and as a new gate material for GaAs metal-insulator-semiconductor (MIS) devices [27] has already shown significant promise, and industrial interest. Even without the prospect of commercial application, the MOCVD and spray-MOCVD of group III-VI compounds has already uniquely demonstrated some important concepts in the MOCVD growth of thin films. Thus, the growth of cubic-GaS and InSe thin films, as well as tetragonal (rather than orthorhombic) InS, has given conclusive proof for the concepts of solid state synthesis being controlled by the structure of a predesigned molecular motif. However, it may also be shown from these studies that the simplistic approach often applied to molecular design of single-source precursors for solid state materials, so often promoted in the literature, is fraught with pitfalls. We propose, however, that increased understanding of the intimate mechanism of precursor decomposition and thin film crystal growth may provide sufficient information for further modifications towards the design of successful single source precursors for group III-VI thin films.

ACKNOWLEDGMENTS

Financial support during the writing of this review was provided by the National Aeronutics and Space Administration (NASA) and the National Science Foundation. The author aknowledges Dr Andrew N. MacInnes and Christopher C. Landry for their assistance with this review, and Dr. Henry J. Gysling for making his work available.

REFERENCES

1   (a) T. Nishino and Y. Hamakawa, Jpn. J. Appl. Phys. **16**, 1291 (1977).
        (b) R.S. Becker, T. Zheng, J. Elton, and M. Saeki, Sol. Energy Mater. **13**, 97 (1986).
2   (a) L.I. Man, R.M. Imanov, and S.A. Semiletov, Sov. Phys. Crystallogr. **21**, 255 (1976).
        (b) W.-T. Kim, and C.-D. Kim, J. Appl. Phys. **60**, 2631 (1986).
        (c) R. Nomura, K. Kanaya, A. Moritake, and H. Matsuda, Thin Solid Films **167**, L27 (1988)
3   (a) L.I. Man, R.M. Imanov and S.A. Semiletov, Sov. Phys. Crystallogr. **21**, 255 (1976).
        (b) T. Nishino and Y. Hamakawa, Jpn., J. Appl. Phys. **16**, 1291 (1977).
        (c) R.S. Becker, T. Zheng, J. Elton, and M. Saike, Sol. Energy Mater. **13**, 97 (1986).
        (d) W.-T. Kim and C.-D. Kim, J. Appl. Phys. **60**, 2631 (1986).

    (e) R. Nomura, K. Kanaya, A. Moritake, and H. Matsuda, Thin Solid Films **167**, L27 (1988).
4   A. M. Mancini, G. Micocci, and A. Rizzo, Mater. Chem. Phys. **9**, 29 (1983)
5   See for example: (a) E. Yablonovitch, C.J. Sandroff, R. Bhat, and T. Gmitter, Appl. Phys. Lett. **51**, 439 (1987).
    (b) M.S. Carptenter, M.R. Melloch, M.S. Lundstron, and S.P. Tobin, Appl. Phys. Lett. **52**, 2157 (1988).
    (c) Y. Nannichi, J.F. Fan, H. Oiwawa, A. Koma, Jpn. J. Appl. Phys. **27**, L2367 (1988).
    (d) J.F. Fan, Y. Kurata, and Y. Nannichi, Jpn. J. Appl. Phys. **28**, L2255 (1989).
6   O Madelung, Ed., *Semiconductors, other than Group IV Elements and III-V Compounds.* (Springer-Verlag, New York, 1992).
7   M. Balanski, C. Julien, A. Chevy, K. Kambas, Solid State Commun. **59**, 423 (1986)
8   (a) C. DeBlasi, A. V. Drigo, G. Micocci, A. Tepore, J. Cryst. Growth **94**, 455 (1989)
    (b) C. DeBlasi, D. Manno, G. Micocci, A. Tepore, J. Cryst. Growth **96**, 947 (1989)
9   W. -T. Kim and C. -D. Kim, J. Appl. Phys. **60**, 2631 (1986).
10   M. A. Kenway, A. F. El-Shazly, M. A. Afifi, H. A. Zayed, H. A. Zahid, Thin Solid Films **200**, 203 (1991).
11   M. Persin, A. Persin, B. Celustka, and B. Etlinger, Thin Solid Films **11**, 153 (1972).
12   J. Herrero and J. Ortega, Solar Energy Mater. **16**, 477 (1987).
13   M. G. Kanatzidis, Chem. Mater. **2**, 353 (1990).
14   A. N. MacInnes, M. B. Power, A. R. Barron, Chem. Mater. **4**, 11 (1992).
15   A. N. MacInnes, M. B. Power, A. R. Barron, Chem. Mater. **5**, 1344 (1993).
16   A. N. MacInnes and A. R. Barron, unpublished results.
17   A. N. MacInnes, M. B. Power, A. F. Hepp, A. R. Barron, J. Organomet. Chem. **449**, 95 (1993).
18`   C. D. Wagner, L. H. Gale, R. H. Raymond. Anal. Chem. **51**, 466 (1979).
19   A. N. MacInnes, W. M. Cleaver, M. B. Power, A. F. Hepp, A. R. Barron, Adv. Mater. Optics Electron. **1**, 229 (1992).
20   R. Nomura, S.-J. Inazawa, K. Kanayou and H. Matsuda, Appl. Organomet. Chem. **3**, 195 (1989).
21   H. J. Gysling, A. A. Wernberg, T. N. Blanton, Chem. Mater. **4**, 900 (1992).
22   B.S. Meyerson, F.K. LeGoues, T.N. Nguyen, Tand D.L. Harame, Appl. Phys. Lett. **50**, 113 (1987).
23   See Greenwood, N. N.; Earnshaw, A. "Chemistry of the Elements", Pergamon Press, Oxford, Great Britain, 250 (1984).
24   Mooser, E.; Pearson, W. B. Acta Cryst. **12**, 1015 (1959).
25   Gordy, W.; Thomas, W. J. O. J. Chem. Phys. **24**, 439 (1956).
26   A. N. MacInnes, M. B. Power, A. R. Barron, P.P. Jenkins, A. F. Hepp, Appl. Phys. Lett. **62**, 711 (1993).
27   M. Tabib-Azar, S. Kang, A. N. MacInnes, M. B. Power, A. R. Barron, P.P. Jenkins, A. F. Hepp, Appl. Phys. Lett. **63**, 625 (1993).

CHEMICAL VAPOR DEPOSITION OF VANADIUM OXIDE THIN FILMS

OGIE STEWART*, JOAN RODRIGUEZ*, KEITH B. WILLIAMS*, GENE P. RECK*, NARAYAN MALANI**, AND JAMES W. PROSCIA**
*Department of Chemistry, Wayne State University, Detroit, MI 48202.
**Ford Motor Company, Glass Division, Dearborn, MI 48120.

ABSTRACT

Vanadium oxide thin films were grown on glass substrates by atmospheric pressure chemical vapor deposition (APCVD) from the reaction of vanadium(IV) chloride with isopropanol and t-butanol. Films were deposited in the temperature range 250 to 450°C. The as-deposited films were a dark greenish color consistent with formation of a lower oxide of vanadium. Annealing a film deposited on Corning 7059 glass in air converted the material to a yellow film. X-ray diffraction of the yellow film revealed the presence of V_2O_5. Optical spectra of the films are presented. Glass substrates previously coated with conductive fluorine doped tin oxide were coated with V_2O_5 and evaluated for electrochromic activity.

INTRODUCTION

Vanadium oxide thin films have thermochromic and electrochromic applications. The thermochromic oxides of vanadium are VO_2[1,2] and V_2O_3 [3]. These materials can be used to enhance the solar control properties of glass windows in buildings and cars by restricting the transmission of solar radiation within a desired temperature range. This phenomenon occurs because vanadium oxide undergoes a phase transition from semiconductor to metal at a certain transition temperature (T_t). V_2O_5 is an electrochromic oxide of vanadium [4]. When lithium ions are electrochemically inserted into this material, the color changes from yellow to colorless. This color change is reversed by reversing the applied voltage.

Vanadium oxide films have previously been prepared by sputtering [5], evaporation, organometallic chemical vapor deposition [6], and dip-coating methods [7]. In the present paper, atmospheric pressure chemical vapor deposition (APCVD) is used to deposit vanadium oxide films. APCVD is a particularly useful deposition process for the coating of large substrates at high deposition rates. Films were prepared by reacting vanadium (IV) chloride with either isopropanol, t-butanol, or water as the primary oxygen source. When the as-deposited green films were annealed in air the color changed to yellow. X-ray diffraction indicated that V_2O_5 was present in the yellow films.

EXPERIMENTAL

Thin films of vanadium oxide were produced by APCVD in the one inch diameter tube reactor illustrated in Figure 1. Both quartz and glass tubes were used as the reaction chamber. Reactant gases were introduced into the reactor tube via three concentric tubes and allowed to flow over samples placed on a semi-cylindrical sample holder with a tapered front to provide for mixing. Vanadium (IV) chloride contained in a bubbler constructed from polyvinylchloride was entrained in a nitrogen carrier gas and flowed through the center tube, a nitrogen separator gas was flowed through the next tube, and various alcohols entrained in a nitrogen or air carrier gas were flowed through the outermost tube. The reactor tube was heated in a Lindberg Mini-Mite Model 55035 tube furnace. A stainless steel tube through which cooling air flowed was positioned along the top of the reactor tube. This cooling of the top surface of the quartz tube creates a temperature gradient which allowed thermophoresis to levitate gas phase nucleation products (powder) away from the substrates keeping the deposited films pinhole free. The exhaust from the reactor was passed through a water trap and then vented into a laboratory hood. Substrates coated included Corning 7059 glass, microscope slide glass, and single crystal silicon wafers. Vanadium oxide films were grown in the temperature range 250- 400 °C.

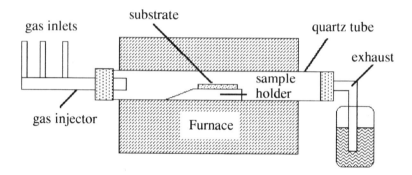

Figure 1. Schematic of APCVD reactor

Optical reflectance and transmission spectra were taken on a Perkin-Elmer lambda 9 spectrophotometer. Scanning electron microscopy (SEM) was performed on a Jeol JEM-100CX electron microscope. Samples for SEM were over-coated with a thin film of platinum-paladium to make them more conductive. The films were further characterized by x-ray diffraction, Rutherford backscattering spectroscopy (RBS) and by x-ray photoelectron spectroscopy (XPS). The samples were sputtered prior to recording the XPS in order to get beyond any surface contamination.

RESULTS

The as-deposited films were dark green in appearance. The growth rate for a film deposited from vanadium chloride and t-butanol at 350 °C was about 1000 Å/min. Annealing the as-deposited films in air converted the green films to yellow. Figure 2 shows the transmission spectrum of a film deposited from vanadium chloride and t-butanol at 300 °C before and after annealing in air. XPS and RBS revealed that the apparent stoichiometry of both the annealed and unannealed films was $VO_{(1.6 \text{ to } 2)}$. The XPS analysis did not reveal any significant differences for films deposited from 300 to 400 °C on Corning 7059 glass. Similarly, the XPS for films using water, isopropanol, or t-butanol as the oxygen source were indistinguishable. Chlorine and carbon were not observed in either the RBS or XPS spectra of films deposited on Corning 7059 glass. However, significant levels of chlorine were observed in the RBS spectrum of a film deposited on crystalline silicon. Low angle x-ray diffraction did not reveal any peaks for the as-deposited film deposited from vanadium chloride and t-butanol at 300 °C. The annealed film made under these conditions showed x-ray diffraction peaks consistent with V_2O_5. SEM of the films deposited on Corning 7059 glass (low sodium containing glass) revealed a smooth featureless surface. In contrast films deposited on microscope slides (high sodium containing glass) had strikingly large features (figure 3).

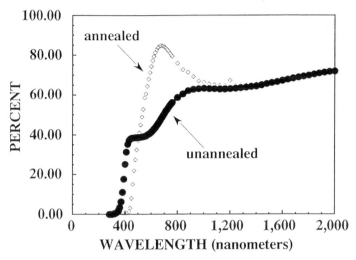

Figure 2. Transmission spectrum of a film deposited from vanadium chloride and t-butanol at 300 °C on Corning 7059 glass before and after annealing in air.

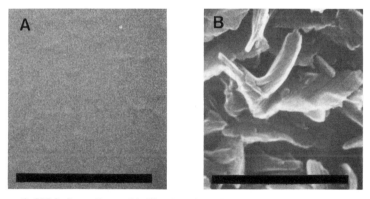

Figure 3. SEM of vanadium oxide film deposited by reacting vanadium (IV) chloride and t-butanol: at 300 °C on Corning 7059 glass (A) and at 350 °C on microscope glass (B). The line equals 1 micron.

An annealed vanadium oxide film was deposited on a piece of Corning 7059 glass coated with a conductive fluorine doped tin oxide film. Lithium was electroinserted into the vanadium oxide by placing the film in a 0.1M LiClO4 in isopropanol solution applying a negative bias of 6 volts. The film was observed to change from a yellow to a colorless state. Figure 4 gives the transmission and reflection spectra of the vanadium oxide-tin oxide stack with and without lithium inserted. The reflectance spectrum was observe to change slightly in the near IR, while the transmission spectrum showed significant change in the visible.

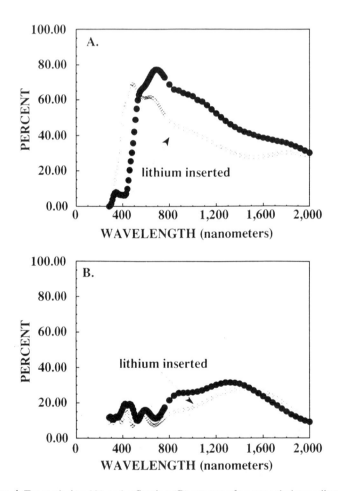

Figure 4. Transmission (A) and reflection (B) spectra of an annealed vanadium oxide film with and without lithium inserted.

DISCUSSION

Reactions of metal halides with water and alcohols to produce oxide films is well known [8]. The XPS and RBS analysis of the as-deposited films seem consistent with the stoichiometry being between V_2O_3 and VO_2. When the films were heated in air the color of the films to changed to yellow. Heating the films in the absence of oxygen did not result in a color change. The x-ray diffraction pattern and optical spectra of the yellow films were consistent with V_2O_5. The stoichiometry of the annealed films indicated by both RBS and XPS was still consistent with a lower vanadium oxide. This discrepancy might be explained if the as-deposited amorphous films underwent an incomplete phase transition to polycrystalline V_2O_5.

The annealed vanadium oxide films readily incorporated lithium ions and became colorless. These films are suitable as counter-electrodes for electrochromic devices that utilize an active layer such as such as tungsten oxide. When the active layer is suitably biased the lithium moves in the active layer and the device darkens. When the bias is reversed the lithium moves out of the active layer and into the vanadium oxide layer. The device then becomes transparent.

CONCLUSION

The reaction of vanadium (IV) chloride with isopropanol and t-butanol produces green vanadium oxide films. These films are most likely mixture of lower oxides of vanadium with a stoichiometry of about $VO_{1.5-2}$. Annealing the films in air converted at least a portion of the films to V_2O_5. These yellow films were found to be electrochromically active.

ACKNOWLEDGEMENTS

This work was funded by the Ford Motor Company, Glass Division. The scanning electron microscopy was performed by Robert Benoit. RBS analysis was performed at the Michigan Ion Beam Laboratory of the University of Michigan. XPS was performed at the Material Research Center of the University of Michigan.

RFERENCES

1. M.S.R. Khan, K. A. Khan, W. Estrada, C. G. Granqvist, J. Appl. Phys. **69**, 3231 (1991).
2. S.-J. Jiang, C.-B. Ye, M.S. Khan, C. G. Granqvist, Appl. Opt. 30, 847 (1991).
3. F, C. Case, J. Vac. Sci. Technol. **A 9(3)**, 461 (1991).
4. S. F. Cogan, N. M. Mguyen, S. J. Perrotti, and R. D. Rauh, J. Appl. Phys. **66**, 1333 (1989).
5. D. Wruck, S. Ramamurth, M. Rubin, Thin Solid Films **182**, 79 (1989).
6. C. B. Greenberg, Thin Solid Films **110**, 73 (1983).
7. Y. Takahashi, M. Kanamori, H. Hashimoto, Y. Moritani, Y. Masuda, J. Mater. Sci. 24, 192 (1989).
8. W. H. Bennett, U.S. Patent #3,075,861 (1963).

ATOMIC LAYER CONTROLLED DEPOSITION OF Al$_2$O$_3$ FILMS EMPLOYING TRIMETHYLALUMINUM (TMA) AND H$_2$O VAPOR

A.C. Dillon, A. W. Ott, and S.M. George, Dept. of Chemistry and Biochemistry, University of Colorado, Boulder, Colorado 80309; and J.D. Way, Dept. of Chemical Engineering and Petroleum Refining, Colorado School of Mines, Golden, CO 80401

ABSTRACT

Sequential surface chemical reactions for the controlled deposition of Al$_2$O$_3$ films were studied using transmission Fourier transform infrared spectroscopy (FTIR). Experiments were performed *in situ* in an ultrahigh vacuum UHV chamber using high surface area alumina membranes. Trimethylaluminum [Al(CH$_3$)$_3$] (TMA) and H$_2$O vapor were employed sequentially in an ABAB... binary fashion to achieve atomic layer controlled growth. An optimal Al$_2$O$_3$ growth procedure was established that employed TMA/H$_2$O exposures at .3 Torr and 500 K. The experiments revealed that each reaction was self-terminating and atomic layer controlled growth was dictated by the surface chemistry. The controlled deposition of Al$_2$O$_3$ may be employed on silicon surfaces for the formation of high dielectric gate and passivation layers.

INTRODUCTION

The development of atomic layer controlled deposition processes has been a focus of recent research. The controlled growth of Al$_2$O$_3$ thin films has several important technological applications. For example, Al$_2$O$_3$ deposition on silicon surfaces is useful for the low temperature formation of high dielectric insulators. Passivating layers deposited at low temperatures are important because of the problem of autodoping that can be observed during high temperature growth (1).

Trimethylaluminum [Al(CH$_3$)$_3$] (TMA) and H$_2$O vapor may be employed for controlled Al$_2$O$_3$ deposition using the binary reactions:
  A) AlOH + Al(CH$_3$)$_3$ --> Al-O-Al(CH$_3$)$_2$ + CH$_4$
  B) Al-O-Al(CH$_3$)$_2$ + 2H$_2$O --> Al-O-Al(OH)$_2$ + 2CH$_4$.
The growth of thin films by the sequential adsorption of water and TMA has been reported for both alumina (2) and silicon surfaces (3). Similarly, a TMA and H$_2$O$_2$ binary sequence was employed for Al$_2$O$_3$ deposition on GaAs and related compounds (4). However, these previous studies did not confirm the surface reaction mechanism. The earlier investigations also did not probe a wide range of reaction conditions in order to establish an optimal Al$_2$O$_3$ growth process.

In the current study, optimal conditions were established for the controlled deposition of Al$_2$O$_3$ on alumina. Alternate H$_2$O and TMA exposures at 3 Torr for 1 min. at 500 K provided the most efficient deposition. In addition, periodic annealing to 1000 K between sets of cycles was necessary to prevent excess hydroxyl accumulation on the surface. Previous studies have indicated that alumina film growth processes are nearly independent of the starting substrate (4). Thus, the optimal Al$_2$O$_3$ growth recipe established here may be employed in a variety of applications.

EXPERIMENTAL

Sensitivity requirements limit transmission FTIR studies to high surface area materials because typical infrared cross sections are low (5). ANOPORE porous alumina membranes were obtained from Anotec Inorganic Membrane Technology. Membranes with .2 μ pore diameters, 60 μ thickness and a porosity of ≈ 50 % were employed in these studies. These samples provided a surface area of 4 m^2/g that was sufficient for facile transmission infrared studies.

A portion of a circular 13mm diameter ANOPORE membrane was mounted on a 10 x 11 mm^2 single crystal Si(100) wafer with a thickness of 400 μ. Aremco 571 high temperature ceramic adhesive was applied at the top and bottom edges of the ANOPORE membrane. The silicon crystal mounting design and techniques for sample heating have been described previously (6).

A schematic of the UHV chamber for *in-situ* transmission FTIR studies has also been given earlier (7). For these studies, the UHV chamber was equipped with a diffusion pump which was employed to achieve pressures of approximately 1×10^{-3} Torr. An isolation valve could then be opened to a 190 l/sec Balzers turbomolecular pump. The turbomolecular pump enabled pressures of 5-6 $\times 10^{-6}$ Torr to be obtained within ~ 10 min.

A Nicolet 740 FTIR spectrometer and an MCT-B infrared detector were employed in these studies. The infrared beam passed through a pair of .5 inch thick CsI windows on the vacuum chamber. The CsI windows could be isolated from the chamber using gate valves in order to prevent species from being deposited inadvertently on the windows.

The surface species produced following TMA and H$_2$O adsorption were studied as a function of exposure and alumina surface temperature. In these experiments, the alumina substrate was held at a constant surface temperature for a specified TMA or H$_2$O exposure. Upon evacuation of the chamber, FTIR spectra were recorded at 300 K. Each new porous alumina membrane was annealed to 1000 K for 10 min. prior to the adsorption studies. This procedure was sufficient to convert the amorphous alumina membrane to crystalline γ-alumina (8). Infrared spectra were referenced to background spectra of the newly annealed substrate.

RESULTS AND DISCUSSION

Changes in the infrared absorption spectra of porous alumina membranes versus H$_2$O or TMA exposure at 300 K and 500 K are displayed in Figs. 1 and 2. The broad infrared feature between 3800-2600 cm^{-1} is characteristic of AlO-H stretching vibrations (9). An analysis of bonds assigned to isolated and hydrogen-bonded hydroxyls will be given later (10). The sharp infrared features between 2942-2838 cm^{-1} are consistent with AlC-H$_3$ stretching vibrations (11).

The initial spectrum in Fig. 1 was recorded following a 2 Torr, 5 min H$_2$O exposure on an alumina surface previously annealed to 1000 K. This exposure was sufficient to achieve a saturation AlOH coverage. The alumina membrane was then exposed to TMA at 2 Torr for 5 min. at 300 K. The infrared spectrum in Fig. 1, following the 300 K TMA exposure, reveals a decrease in the broad absorbance of the AlO-H stretching vibrations and a concurrent increase in the infrared absorbance of the C-H$_3$ stretching vibrations. Following a second 2 Torr, 5 min. TMA exposure at 300 K, no significant change was observed in the infrared spectrum.

As displayed in Fig. 1, the broad AlO-H stretching vibration is observed to virtually disappear following a 2 Torr, 5 min. TMA exposure at 500 K. Concurrently, a very large increase is observed in the AlC-H$_3$ stretching region. These results indicate that a reaction temperature of 500 K is necessary to achieve complete reaction between TMA and surface AlOH species.

Figure 2 displays the changes in the infrared spectra of a porous alumina membrane versus sequential .3 Torr, 1 min. H$_2$O and TMA exposures at 500 K. The infrared spectrum at the top of Fig. 2 was recorded following H$_2$O exposure to a sample previously annealed to 1000 K. The spectrum reveals a broad infrared absorbance in the AlO-H stretching region. Upon TMA exposure, this broad absorption feature disappeared and a concurrent growth was observed in the C-H$_3$ stretching region. Subsequent exposure to H$_2$O resulted in the disappearance of the infrared absorbance of the C-H$_3$ stretching vibrations and an increase in the AlO-H stretching region. These spectral results are consistent with aluminum oxide growth on the alumina surface according to the binary reactions: A) AlOH + Al(CH$_3$)$_3$ --> Al-O-Al(CH$_3$)$_2$ + CH$_4$; and B) Al-O-Al(CH$_3$)$_2$ + 2H$_2$O --> Al-O-Al(OH)$_2$ + 2CH$_4$.

The final infrared spectrum in Fig. 2, corresponding to the second TMA exposure, reveals a slight absorbance in the AlO-H stretching region. An additional 2 Torr, 5 min. TMA exposure at 500 K did not result in the disappearance of this AlO-H absorption feature This behavior is consistent with a slight accumulation of unreactive hydroxyls on the alumina surface. Subsequent H$_2$O/TMA exposures resulted in a gradual increase in AlOH species which were not lost upon TMA exposures.

Annealing the alumina surface to 1000 K for 10 min. resulted in the disappearance of detectable absorbance in the AlO-H stretching region. The unreactive AlOH species may be attributed to the formation of an amorphous Al$_2$O$_3$ surface with sites inaccessible to the TMA molecules. Annealing the alumina to 1000 K results in the decomposition of the AlOH species (10), as well as the crystallization of the deposited alumina film (8). The prevention of excess hydroxyl accumulation may be achieved by annealing the alumina surface to 1000 K following a set of 10 TMA/H$_2$O cycles.

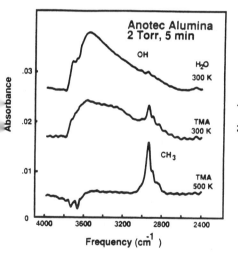

Figure 1 Changes in the infrared absorption spectra of porous alumina versus 2 Torr, 1 min. H$_2$O or TMA exposures at 300 K and 500 K.

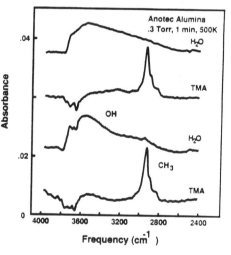

Figure 2 Changes in the infrared absorption spectra of porous alumina versus .3 Torr, 1 min. H$_2$O or TMA exposures at 500 K.

Figures 3 and 4 display the normalized integrated absorbance vs .01 Torr TMA and H_2O exposures, respectively, at 500 K. The initial TMA exposure in Fig. 3 was on an alumina surface with a saturation coverage of AlOH species produced by H_2O exposure at 500 K. The initial H_2O exposure in Fig. 4 was on an alumina surface that had reacted to completion with TMA at 500 K. The integrated absorbances in both Figs. 3 and 4 were normalized to the maximum values.

Figure 3 displays an initial rapid decrease in the AlO-H stretching vibration, followed by a subsequent gradual disappearance of infrared absorbance in the AlO-H stretching region. Concurrently, an initial rapid increase is observed for the infrared absorbance of the C-H_3 stretching vibrations of the surface AlCH$_3$ species. This initial rapid AlCH$_3$ uptake rate is followed by decreasing uptake rates which eventually indicate the completion of the reaction. For TMA exposures > 210 s., no significant changes were observed in the infrared spectra.

Figure 4 displays an initial rapid decrease in the C-H_3 stretching vibration followed by the gradual disappearance of the C-H_3 infrared absorbance. Concurrently, an initial rapid increase is observed for the infrared absorbance of the AlO-H stretching vibrations. This initial rapid AlOH uptake rate is followed by progressively decreasing uptake rates. For H_2O exposures > 180 s., no significant changes were observed in the infrared spectra. Figures 3 and 4 indicate that the reaction of H_2O with AlCH$_3$ alumina surface species is slightly more rapid than the reaction of TMA with a hydroxylated alumina surface.

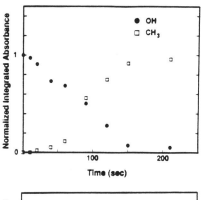

Figure 3 The normalized integrated absorbances of the AlO-H and AlC-H_3 stretching vibrations versus a .01 Torr, TMA exposure at 500 K.

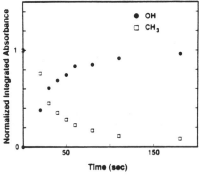

Figure 4 The normalized integrated absorbances of the AlO-H and AlC-H_3 stretching vibrations versus a .01 Torr H_2O exposure at 500 K.

Al$_2$O$_3$ thin film deposition employing the TMA/H$_2$O binary reaction sequence was demonstrated on a single crystal Si(100) surface. The Si(100) surface was annealed to 800 K for 4 min in order to remove the surface hydride species (12,13). The surface was then alternately exposed to TMA or H$_2$O at .1 Torr for 1 min. at 700 K. Atomic layer controlled deposition of Al$_2$O$_3$ employing the TMA/H$_2$O reaction sequence at temperatures > 700 K is not possible because of pyrolosis of the TMA molecule (14). Between exposures, the chamber was evacuated to approximately 1×10^{-4} Torr.

Figure 5 reveals the changes in the infrared absorption spectra as a function of TMA/H$_2$O reaction cycles. The broad infrared absorbance between 1032-540 cm^{-1} is characteristic of amorphous Al$_2$O$_3$ lattice vibrations (15). The integrated absorbance of the infrared region between 1032-540 cm^{-1} is plotted versus number of TMA/H$_2$O reaction cycles in Fig. 6. Figure 6 reveals that Al$_2$O$_3$ film growth employing the TMA/H$_2$O reaction sequence is roughly linear with the number of reaction cycles.

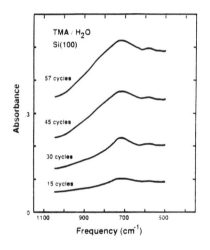
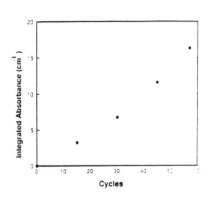

Figure 5 Changes in the infrared absorption spectra of Si(100)2x1 versus .1 Torr, 1 min. TMA/H$_2$O exposures at 700 K.

Figure 6 The normalized integrated absorbance of the Al$_2$O$_3$ lattice vibrations versus .1 Torr, 1 min. TMA/H$_2$O exposures at 700 K.

PROPOSAL FOR Al$_2$O$_3$ CONTROLLED DEPOSITION

Both water and TMA are liquids with high vapor pressures. These characteristics facilitate their use in film deposition processes. The FTIR experiments revealed that each reaction in the binary TMA/H$_2$O sequence was self-terminating. Atomic layer controlled growth was a direct result of the surface chemistry. Based on these results, an optimal low temperature controlled Al$_2$O$_3$ deposition process would be:

1. .3 Torr, 1 min. TMA exposure on a hydroxylated surface at 500 K.
2. .3 Torr, 1 min.. H$_2$O exposure on a methylated surface at 500 K.
3. Repeat steps 1 and 2 for a total of 10 cycles.
4. Anneal to 1000 K, 10 min. to prevent accumulation of AlOH surface species.
5. Repeat steps 1-4.

CONCLUSIONS

The controlled deposition of Al$_2$O$_3$ may be employed on silicon surfaces for the low temperature formation of high dielectric films. In this study, sequential surface chemical reactions for atomic layer controlled Al$_2$O$_3$ deposition were monitored using transmission FTIR spectroscopy. Trimethylaluminum (TMA) and H$_2$O vapor were employed to achieve atomic layer controlled growth using the binary reactions: A) AlOH + Al(CH$_3$)$_3$ --> Al-O-Al(CH$_3$)$_2$ + CH$_4$ and B) Al-O-Al(CH$_3$)$_2$ + 2H$_2$O --> Al-O-Al(OH)$_2$ + 2CH$_4$. An optimal Al$_2$O$_3$ growth procedure was established which employed .3 Torr, 1min. TMA/H$_2$O exposures at 500 K. Periodic annealing to 1000 K between sets of 10 binary reaction cycles was necessary to prevent excess hydroxyl accumulation on the surface. The experiments revealed that each reaction was self-terminating. The atomic layer controlled growth was a direct consequence of the binary surface chemical reactions. In addition, Al$_2$O$_3$ film growth employing the TMA/H$_2$O reaction sequence on Si(100) was roughly linear.

ACKNOWLEDGMENTS

This work was supported by the National Science Foundation under Contract No. CTS-9303762 and the Office of Naval Research under contract No. N00014-92-J-1353. SMG acknowledges the National Science Foundation for a Presidential Young Investigator Award (1988-1993).

REFERENCES

1. C.W. Pearce in VLSI Technology 2nd ed., edited by S.M. Sze. (McGraw-Hill, New York, (1988) Chap. 2.
2. C. Soto and W.T. Tysoe, J. Vac. Sci. Technol. A **9**, 2686 (1991).
3. G.S. Higashi and C.G. Flemming, Appl. Phys. Lett. **55**, 1963 (1989).
4. J.F. Fan, K. Sugioka and K. Toyoda, Jap. J. Appl. Phys., 30, L1139 (1991).
5. L.A. Pugh, and K.N. Rao, in Molecular Spectroscopy: Modern Research edited by K.N. Rao (Academic, New York 1976).
6. P. Gupta, V.L. Colvin and S.M. George, Phys. Rev. B **37**, 8234 (1988).
7. A.C. Dillon, M.B. Robinson, M.Y. Han and S.M. George, J. Electrochem. Soc. **139**, 537 (1992).
8. E. Highwell, Anotec Inorganic Membrane Technology, (*private communication*).
9. J.B. Peri, J. Phys. Chem. **69**, 211 (1965).
10. A. C. Dillon, A.W. Ott, S.M. George and J.D. Way, (*in preparation*).
11. T. Ogawa, Spectrochimica Acta, **24**, 15 (1968).
12. M.L. Wise, B.G. Koehler, P. Gupta, P.A. Coon, S.M. George, Surf. Sci. **258**, 166 (1991).
13. B.G. Koehler, C.H. Mak, D.A. Arthur, P.A. Coon and S.M. George, J. Chem. Phys. **89**, 1709 (1988).
14. T.M. Mayer, J.W. Rogers Jr. and T.A. Michalske, Chem. Mater. **3**, 641 (1991).
15. T. Maruyama and S. Arai, Appl. Phys. Lett. **60**, 322 (1992).

DEPOSITION OF CERIUM DIOXIDE THIN FILMS ON SILICON SUBSTRATES BY ATOMIC LAYER EPITAXY

Heini Mölsä and Lauri Niinistö, Laboratory of Inorganic and Analytical Chemistry, Helsinki University of Technology, FIN-02150 Espoo, Finland

ABSTRACT

CeO_2 overlayers up to 360 nm thick were deposited on Si(100) substrates in a flow-type ALE reactor from $Ce(thd)_4$ (thd = 2,2,6,6-tetramethyl-3,5-heptanedione) precursor and ozone. The growth rate was studied as a function of deposition and source temperatures, reactor pressure and pulse durations. The films were characterized for crystallinity, thickness and composition by using XRD, profilometry, XRF, RBS, XPS and SIMS techniques. Films deposited at 375 °C showed a preferential (110) orientation while at 425 °C they were (111) preferentially oriented. Due to the steric hindrance caused by the bulky precursor the growth rate was only 0.4 Å/cycle.

INTRODUCTION

Cerium dioxide is an attractive buffer layer material for silicon technology because it exhibits great chemical stability even at high temperatures and close lattice parameter matching with silicon. Several authors have demonstrated the use of CeO_2 as a buffer layer for the deposition of high temperature superconductor (HTSC) thin films on silicon [1,2], $LaAlO_3$ [3], MgO [4] and sapphire [5-10]. CeO_2 has a fluorite structure with a=5.411 Å [11]; the lattice mismatch between CeO_2 and cuprate superconductors seems to be large but by rotating 45° in the CeO_2 basal plane the lattice mismatch will be less than 1%. In addition to the buffer layers, CeO_2 can be used as a compatible intermediate material in fabrication of superconductor-insulator-superconductor (SIS) multilayer structures [12]. Furthermore, cerium is one of the components of the Nd-Ce-Cu-O superconductor.

Cerium dioxide is also a suitable material for other than superconducting applications. Epitaxial insulating layers on a silicon substrate have been of great interest since a hetero-epitaxial junction between insulator and semiconductor is desirable for the development of highly integrated devices. CeO_2 as well as other fluorite type oxides are very promising for this purpose since they have been reported to grow epitaxially on silicon without twinning. The large dielectric constant (26) makes CeO_2 especially advantageous for stable capacitor devices for large scale integration. CeO_2 has high refractive index and good transmittance in the visible spectrum and is therefore an interesting material for optical devices. It is also a promising material for resistive type oxygen sensors for combustion gases [13,14].

Both physical and chemical vapor deposition methods have been used to deposit

CeO_2 thin films. Evaporation [15,16] and laser deposition [17-20] methods have commonly been used, but there are also reports on the use of sputtering [14,21] and molecular beam epitaxy (MBE) [22,23]. CeO_2 films can be grown by Chemical Vapor Deposition (CVD) from the chloride [24] but the use of metalorganic precursors offers considerable advantages. The preparation of CeO_2 by Metalorganic CVD (MOCVD) has been reported by Gerfin et al. [25]. The films on (100)Si substrate had a preferential orientation in the [100] direction. Substrate temperature was in the range of 350-600°C and $Ce(thd)_4$ (thd=2,2,6,6-tetramethyl-3,5-heptanedione) was used as the source material, but other cerium ß-diketonate derivatives were studied as possible MOCVD source materials as well [26]. Hiskes et al. [27] reported recently the preparation of CeO_2 thin films by a solid source MOCVD process. (100) oriented films were obtained on sapphire substrates at a temperature less than 700°C. There is also a report on the CeO_2 growth by electrochemical vapor deposition [28] and by spray ICP technique [29].

In this work, we report on the deposition of CeO_2 overlayers on silicon substrates in a flow type Atomic Layer Epitaxy (ALE) reactor using $Ce(thd)_4$ and ozone as reactants. Earlier we have structurally characterized $Ce(thd)_4$ [30] and used it as source chemical in the ALE process to dope electroluminescent thin films [31]. The influence of the deposition parameters such as substrate and source temperatures, pressure and pulse durations on the film growth has now been studied in detail. This work is a part of our investigation into the growth of high T_c superconductors by ALE and CVD [32].

EXPERIMENTAL

Cerium oxide thin films were deposited in a flow type Atomic Layer Epitaxy (ALE) reactor described elsewhere [33]. The source material for cerium was $Ce(thd)_4$, which was synthesized by the method of Eisentraut and Sievers [34] from $Ce_2(CO_3)_3$ (99.9%, Treibacher Chemische Werke AG, A9330 Treibach, Austria). The crucible for the solid source material was glass. Ozone was used as oxygen source. It has been previously used in the MBE and MOCVD growth of oxide superconductors enabling the film growth at lower temperatures [35-37]. Ozone was produced by feeding oxygen gas into the reactor through an ozone generator (Fischer model 502). The concentration of ozone was about 10% (60 g/m^3) and the gas flow rate during the pulse was about 60 cm^3/min (measured for the oxygen gas). The reactant pulses were separated by nitrogen gas purging. Silicon (100) with native oxide layer was used as the substrate. The film growth was studied by varying the total pressure (0.5-3 mbar, measured before the reaction chamber), the temperature of the substrate (300-500°C) and source furnace (150-200°C) as well as the durations of reactant and purge gas pulses (0.2-2.4s). Typical pulse durations were 1.8, 0.6 and 2s for $Ce(thd)_4$, O_3 and N_2, respectively. Typically 5000 cycles were grown; the average thickness was around 250 nm.

Crystal structure and crystallite orientation were determined by X-ray diffraction measurements with a Philips powder diffractometer MPD1880 using CuK$_\alpha$ radiation. Thicknesses of the films were measured by profilometry (Dektak 3030 from Veeco Instruments) and by X-ray fluorescence (Philips PW1480). XRF offers a practical way to determine the film thickness without destroying the sample [38]. RBS measurements were carried out at the Accelerator Laboratory of University of Helsinki, ESCA measurements at the University of Turku and SIMS measurements at the Center for Chemical Analysis of Helsinki University of Technology. TG and DTA curves were recorded simultaneously in a SEIKO TG/DTA 320 instrument of the SSC5200 series. A pressure of 1-2 mbar and nitrogen atmosphere were chosen to simulate the growth conditions in the ALE reactor. Typical heating rate was 10 °C/min and sample weight about 10 mg. The UHV high resolution spectra were recorded in a JEOL MS DX-303/DA5000 instrument. The solid samples were introduced into the mass spectrometer through a sample stage equipped with programmable heating.

RESULTS AND DISCUSSION

The TG/DTA studies indicated a practically complete and sharp volatilization of the Ce-precursor at around 200 °C (Fig. 1). Depending on the heating rate there was only a small whitish residue in the order of few percent. This residue was also found in the source crucible of the ALE reactor after thin film depositions. Unlike Becht et al. [26] who found an endothermic peak at about 200 °C in the DTA curve measured in air or argon, no signs of melting before vaporization were observed in the DTA curve of Ce(thd)$_4$ measured under reduced pressure.

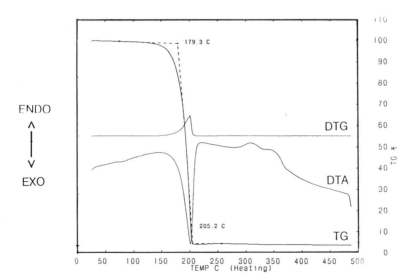

Fig. 1 DTA/TG curves for Ce(thd)$_4$ recorded with a heating rate of 10 °C/min in flowing nitrogen at the pressure of 1-2 mbar. Sample weight was 9.0 mg.

The mass spectrum and mass chromatograms of Ce(thd)$_4$ are presented in Fig. 2. The dominant peaks at m/e 127, 506 and 690 correspond to $C_3H_2O_2C(CH_3)_3^+$, Ce(thd)$_2$ and Ce(thd)$_3$, respectively. Dimeric units were not observed in the spectrum. The main peaks were present at temperatures above 200 °C. However, the volátilization temperatures in various experiments (TG, ALE, MS) are not directly comparable with each other because of pressure differences.

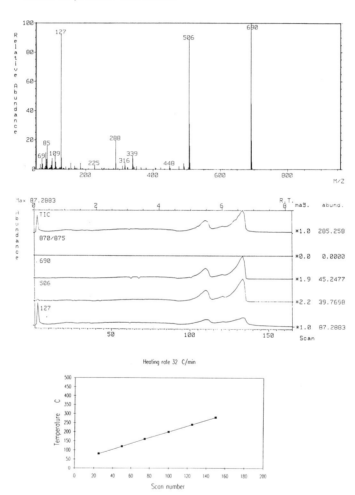

Fig. 2. Mass spectrum and mass chromatograms of Ce(thd)$_4$ showing the relative intensities of the main peaks as function of temperature (see the heating profile).

The sublimation behaviour of Ce(thd)$_4$ in the growth conditions was studied by varying the source temperature. The deposition temperature was 400 °C and the total pressure was 2 mbar. The pulse durations were 600 ms for cerium and 400 ms for ozone. The total consumption of Ce(thd)$_4$ increased linearly as the source temperature increased. Constant growth rate of the film was achieved at the source temperature of 190 °C. However, the amount of the white residue in the crucible, indicating partial decomposition of the precursor, was larger than at lower source temperature. When the cerium pulse duration was increased to 1800 ms the saturation of the growth rate was observed at the source temperature of 160 °C. At the pressure of 0.5 mbar the growth rate increased until 190 °C after which it suddenly decreased.

The behaviour of Ce(thd)$_4$ was also dependent on the deposition pressure. The consumption of Ce(thd)$_4$ decreased as the pressure increased from 0.5 to 3 mbar indicating larger evaporation rate at lower pressure. The growth rate of the film was not affected by the pressure below 2 mbar and a white residue in the crucible was also observed under lower pressure. The pressure of 2 mbar was chosen for further experiments since the thickness variation of the film was larger at lower pressures which could clearly be seen from the interference colours. This might have been due to the low flow rate of the carrier gas. A significant lengthening of the cerium pulse duration might improve the situation but then the deposition time would be much longer.

The growth rate was dependent on the growth pulse durations of the source materials (Fig. 3). The growth rate saturated to a constant value of about 0.4 Å/cycle when the pulse durations were longer than 1800 ms for cerium and 400 ms for ozone (at the source temperature of 160 °C and deposition temperature of 400 °C). The saturation of the growth rate indicates that a certain minimum pulse duration is needed to get a full coverage of the substrate. When the full coverage is reached the situation is stable. However, the growth rate saturates at much less than one monolayer per cycle since the theoretical growth rate for cubic CeO$_2$ should be about 5.4 Å/cycle. The low growth rate obtained here is probably caused by steric hindrance due to the bulky Ce(thd)$_4$ molecule.

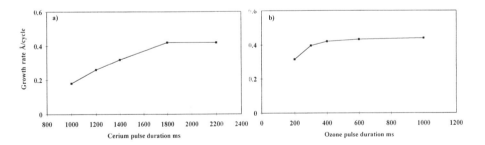

Fig. 3. Dependence of the growth rate on a) the cerium and b) the ozone pulse duration. Deposition and source temperatures were 400 °C and 160 °C, respectively.

The dependence of the growth rate on the growth temperature is shown in Fig 4. The growth rate increases until 375 °C where it reaches the maximum value and then starts to decrease. Below 350 °C the temperature is probably too low for a complete reaction. On the other hand, a possible reason for the decrease of the growth rate at higher temperatures is the decomposition of Ce(thd)$_4$ and/or a change in the growth mechanism. It is difficult to find an "ALE-window" where a self-saturated growth would occur independently on growth conditions, though the differences in the growth rates are not large in the temperature range of 350-400 °C.

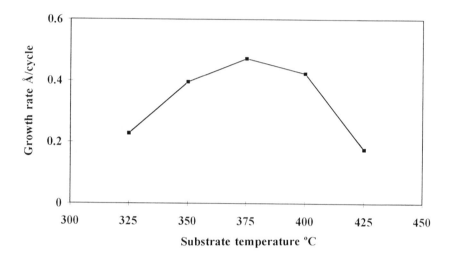

Fig. 4. Dependence of the growth rate on the deposition temperature. Pulse durations were 1800 ms for cerium and 600 ms for ozone. Source temperature was 160 °C.

Fig. 5 shows the growth rate as a function of reaction cycles at 375 °C. The pulse durations in these experiments were 1800 ms for cerium and 600 ms for ozone and the source temperature was 160 °C. The growth rate is larger at the beginning of the growth (with cycles less than 7000) but after that the thickness approaches the linear dependence on the number of cycles. This behaviour might be due to changes in the morphology of the film during the growth process. Preliminary AMF (Atomic Force Microscopy) roughness data appear to support this assumption.

The XRD spectrum of a CeO$_2$ film deposited at 375 °C for 10000 cycles is shown in Fig 6. All peaks can be identified as belonging either to CeO$_2$ or to the substrate. The film shows a preferential orientation in the [110] direction but some reflections due to other orientations can also be seen. This is not surprising since the deposition temperature of 375 °C was expected to be too low for epitaxial growth. (110) oriented growth of CeO_2 on Si(100) has been observed by other groups, too [16]. When the number of cycles were less than 7000 the films were not so clearly oriented. The film grown at 425 °C showed a preferential orientation in the [111] direction but the growth rate was about half of that at 375 °C.

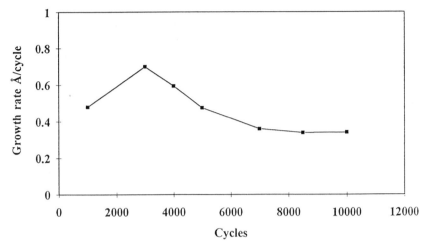

Fig. 5. The growth rate as a function of reaction cycles. Deposition temperature was 375 °C and other parameters the same as in Fig. 4.

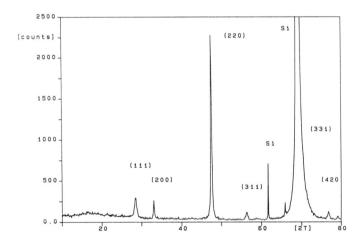

Fig. 6. X-ray powder diffraction pattern (Cu K_α radiation) of a CeO_2 film prepared at 375 °C on Si(100) substrate. The film thickness was about 350 nm.

RBS measurements of the films with various thicknesses (130-360 nm) were carried out with a 2 MeV ^{4}He beam. Preliminary results suggested that the O:Ce ratio in the films is nearly 2. This is consistent with the XRD results showing only the presence of CeO_2.

ESCA spectra of a 330-nm-thick film were collected to analyse the impurities in the films and to determine the oxidation state of cerium. After sputtering for 3 minutes only a small C1s peak was detected. The charge-up shifts depending on the insulating sample were corrected using the C1s signal (284.6 eV) from the unsputtered sample. The F1s peak was found at 684.7 eV corresponding to the binding energies of rare-earth fluorides or oxofluorides. The O1s spectrum consists of two peaks at 531.6 eV and 529.3 eV. The small peak at higher binding energy is attributed to hydroxide or more probably to carbon-oxygen compounds like carbonates, while the dominant peak at 529.3 eV is assigned to the oxide. To determine the oxidation state of the cerium the sample surface was mechanically cleaned since sputtering has been observed to cause the reduction of oxides. However, the ESCA spectra of cerium are rather complicated due to the existence of several satellite peaks and since the initial work of Burroughs et al. [39] many theoretical and experimental studies have been made to find the absolute assignment of the Ce3d photoemission lines. By comparing the intensities of the u''', v' and v'' transitions (labeled following Burroughs et al. [39]) in the Ce3d spectrum of the sample with the CeO_2 reference material and by comparing the spectra with those of the literature [39-41] we came to the conclusion that our samples may also contain trivalent cerium. However, owing to the complexity of the Ce3d spectrum nothing certain of the Ce(III)/Ce(IV) ratio can be said but the amount of Ce(III) appears to be low if any.

SIMS analysis confirmed the results of the ESCA measurements of trace impurities but in addition to the fluorine and carbon contamination also chlorine was detected. All the trace elements (F, Cl, C) were evenly distributed in the films and a well-defined film/substrate interface indicated no significant interdiffusion. Carbon was not unexpected because it is a common impurity in the films prepared from metalorganic precursors. Fluorine contamination of the films was found to be due to the fluorine content in the Ce(thd)$_4$ precursor originating from the $Ce_2(CO_3)_3$ starting material. The precursor and the quartz reactor used are responsible for the chlorine contamination.

Thus the precursor was found mainly to be responsible for the trace impurities of the films. Carbon is due to the decomposition of the Ce(thd)$_4$ ligand and cannot be completely avoided but the contents of chlorine and especially that of fluorine can be reduced and completely excluded by a more careful selection of the chemicals and the synthetic procedure used in the preparation of Ce(thd)$_4$.

CONCLUSIONS

Cerium oxide thin films were deposited on Si(100) substrates in a flow type ALE reactor from Ce(thd)$_4$ and ozone. Among the deposition parameters the source temperature of Ce(thd)$_4$ and the pulse durations of the reactants had the most critical effect on the growth rate. However, above the certain minimum pulse durations the growth rate saturated to a value of about 0.4 Å/cycle which is significantly lower than the theoretical value.

The growth temperature had a marked influence on the growth rate as well as the orientation of the films. The growth rate reached the maximum value at 375 °C and the film was preferentially oriented in the [110] direction whereas at 425 °C the films were (111) preferentially oriented. A preferential orientation and constant growth rate was observed only in the films thicker than 2500 Å (7000 cycles). This might be due to the changes in the morphology of the film during the growth.

Acknowledgements

The authors wish to thank Dr Eero Rauhala for RBS measurements, Dr Ilkka Kartio for ESCA measurements, Ms. Sari Lehto, M.Sc. for SIMS measurements, Prof. Mauri Lounasmaa and Mr Pertti Sarkio, M.Sc. for MS measurements as well as Dr Eero Ristolainen and Prof. Markku Leskelä for useful discussions. This work was supported in part by the Academy of Finland.

REFERENCES

1. Luo, L., Wu, X.D., Dye, R.C., Muenchausen, R.E., Foltyn, S.R., Coulter, Y., Maggiore, C.J. and Inoue, T., Appl. Phys. Lett. **59** (1991) 2043.
2. Chin, T.S., Huang, J.Y., Perng, L.H., Huang, T.W., Yang, S.J. and Hsu, S.E., Physica C **192** (1992) 154.
3. Wu, X.D., Dye, R.C., Muenchausen, R.E., Foltyn, S.R., Maley, M., Rollett A.D., Garcia, A.R. and Nogar, N.S., Appl. Phys. Lett. **58** (1991) 2165.
4. Tanimura, J., Kuroda, K., Kataoka, M., Wada, O., Takami, T., Kojima, K. and Ogama, T., Jpn. J. Appl. Phys. **32** (1993) L254.
5. Wu, X.D., Foltyn, S.R., Muenchausen, R.E., Cooke, D.W., Pique, A., Kalokitis, D., Pendrick, V. and Belohoubek, E., J. Supercond. **5** (1992) 353.
6. Maul, M., Schulte, B., Häussler, P., Frank, G., Steinborn, T., Fuess, H. and Adrian, H., J. Appl. Phys. **74** (1993) 2942.
7. Denhoff, M.W. and McCaffrey, J.P., J. Appl. Phys. **70** (1991) 3986.
8. Merchant, P., Jacowitz, R.D., Tibbs, K., Taber, R.C. and Laderman, S.S., Appl. Phys. Lett. **60** (1992) 763.
9. Holstein, W.L., Parisi, L.A., Face, D.W., Wu, X.D., Foltyn, S.R. and Muenchausen, R.E., Appl. Phys. Lett. **61** (1992) 982.
10. Cole, B.F., Liang, G.C., Newman, N., Char, K., Zaharchuk and Martens, J.S., Appl. Phys. Lett. **61** (1992) 1727.
11. Natl. Bur. Stand. (U.S.) Monogr. 25, **20** (1983) 38.
12. Kusumori, T. and Iguchi, I., Jpn. J. Appl. Phys. **31** (1992) L956.
13. Beie, H.-J. and Gnörich, A., Sensors and Actuators B **4** (1991) 393.

14. Schwab, R.G., Steiner, R.A., Mages, G. and Beie, H.-J., Thin Solid Films **207** (1992) 283, 288.
15. Inoue, T., Yamamoto, Y., Koyama, S., Suzuki, S. and Ueda, Y., Appl. Phys. Lett. **56** (1990) 1332.
16. Inoue, T., Ohsuna, T., Luo, L., Wu, X.D., Maggiore, C.J., Yamamoto, Y., Sakurai, Y. and Chang, J.H., Appl. Phys. Lett. **59** (1991) 3604.
17. Yoshimoto, M., Nagata, H., Tsukahara, T. and Koinuma, H., Jpn. J. Appl. Phys. **29** (1990) L1199.
18. Amirhaghi, S., Beech, F., Vickers, M., Barnes, P., Tarling, S. and Boyd, I.W., Electr. Lett. **27** (1991) 2304.
19. Amirhaghi, S., Archer, A., Taguiang, B., McMinn, R., Barnes, P., Tarling, S. and Boyd, I.W., Appl. Surf. Sci. **54** (1992) 205.
20. Sanchez, F., Varela, M., Ferrater, C., Garcia-Guenca, M.V., Aguiar R. and Morenza, J.L., Appl. Surf. Sci. **70/71** (1993) 94.
21. Sundaram, K.B., Wahid, P.F. and Sisk, P.J., Thin Solid Films **221** (1992) 13.
22. Nagata, H., Tsukahara, T., Gonda, S., Yoshimoto, M. and Koinuma, H., Jpn. J. Appl. Phys. **30** (1991) L1136.
23. Nagata, H., Yoshimoto, M. and Koinuma, H., J. Cryst. Growth **118** (1992) 299.
24. Taylor, H.L. and Trotter, J.D., Proc. Int. Conf. Chem. Vap. Deposition, 3rd, (1972) 475.
25. Gerfin, T., Becht, M. and Dahmen, K.-H., Ber. Bunsenges. Phys. Chem. **95** (1991) 1564.
26. Becht, M., Gerfin, T. and Dahmen, K.-H., Chem. Mater. **5** (1993) 137.
27. Hiskes, R., DiCarolis, S.A., Jacowitz, R.D., Lu, Z., Feigelson, R.S., Route, R.K. and Young, J.L., J. Cryst. Growth **128** (1993) 781; Idem., Ibid. p. 788.
28. Jue, J.-F., Jusko, J. and Virkar, A.V., J. Electrochem. Soc. **139** (1992) 2458.
29. Suzuki, M., Kagawa, M., Syono, Y. and Hirai, T., J. Cryst. Growth **112** (1991) 621.
30. Leskelä, M., Sillanpää, R., Niinistö, L. and Tiitta, M., Acta Chem. Scand. **45** (1991) 1006.
31. Leppänen, M., Leskelä, M., Niinistö, L., Nykänen, E., Soininen, P. and Tiitta, M., SID 91 Digest (1991) 282.
32. Leskelä, M., Mölsä, H. and Niinistö, L., Supercond. Sci. Technol. **6** (1993) 627.
33. Suntola, T., Pakkala, A. and Lindfors, S., U.S. Patent 4,389,973, (1983).
34. Eisentraut, K.J. and Sievers, R.E., J. Am. Chem. Soc. **87** (1965) 5254.
35. Berkley, D.D. et al., Appl. Phys. Lett. **53** (1988) 1973.
36. Nakayama, Y., Ochimizu, H., Maeda, A., Kawazu, A., Uchinokura, K. and Tanaka, S., Jpn. J. Appl. Phys. Lett. **28** (1989) L1217.
37. Ohnishi, H., Harima, H., Kusakabe, Y., Kobayashi, M., hoshinoughi, S. and Tachibana, K., Jpn. J. Appl. Phys. **29** (1990) L2041.
38. Lehto, S., Niinistö, L. and Yliruokanen, I., Fresenius J. Anal. Chem. **346** (1993) 608.
39. Burroughs, P., Hamnett, A., Orchard, A. F. and Thornton, G., J. Chem. Soc., Dalton Trans. **17** (1976) 1686.
40. Strydom, C.A. and Strydom, H.J., Inorg. Chim. Acta **161** (1989) 7.
41. Abi-aad, E., Bechara, R., Grimblot, J. and Aboukais, A., Chem. Mater. **5** (1993) 793.

GROUP 2 ELEMENT CHEMISTRY AND ITS ROLE IN OMVPE OF ELECTRONIC CERAMICS

WILLIAM S. REES, JR.*, Department of Chemistry and Materials Research and Technology Center, The Florida State University, Tallahassee, Florida 32306-3006

I. INTRODUCTION

Coordination compounds of alkaline-earth metals with simple monodentate ligands were mentioned in the literature as early as 1820, when Faraday reported 'metal-ammonias'.[1] The number of heavy alkaline earth element coordination compounds remained limited for many years, however, as it was thought to be unlikely for the large earth alkaline cations to form such complexes. In 1967 Pedersen discovered that cyclic oligoethers (crown ethers) can serve as suitable ligands for divalent alkaline earth cations.[2] His findings were extended a few years later by Lehn who found that macrobicyclic multidentate ligands (cryptands) are efficient ligands for alkaline earth cations.[3] In each of these examples, the primary mode of metal-ligand interaction is electrostatic in origin. These examples demonstrate the lack of well-defined covalent bonding for Ca, Sr and Ba compounds. In the 25 years which have passed since Pedersen's seminal discovery, a large number of coordination compounds containing alkaline earth metals have been synthesized. In recent years, the emphasis has shifted towards the preparation of group 2 element-containing compounds which potentially can be used as precursors in the preparation of metal oxides by chemical vapor deposition [4-8]. In light of the general agreement that no "perfect" barium source presently exists for OMVPE purposes, this article will focus on general themes in group 2 element chemistry and, where relevant, correlate those themes within an integrated approach to design of new compounds of greater potential utilization for the preparation of electronic materials in thin film form.

Suitable CVD precursors have been developed for the early group 2 elements; however, significant challenges still remain for the late group 2 elements. The emergence of electronic materials containing barium has prompted growing interest in the preparation of barium precursors. The main challenge is to develop barium precursors which possess a high vapor pressure. Another requirement which plays an important role in the development of a barium precursor is that the compound should be thermally stable at evaporation and transport temperatures.

---

* Address all correspondence to this author at: School of Chemistry and Biochemistry and School of Materials Science and Engineering, Georgia Institute of Technology, Atlanta, Ga 30332-0400

II. COORDINATION CHEMISTRY

A. Neutral Ligands

1. Oxygen-based Donors

There are many examples of oxygen donor complexes with the group 2 elements. The oxygen donors may be as simple as diethyl ether or as complex as a Schiff-base. One of the most traditional types of these complexes are those of the Grignard reagents. Diethyl ether forms an oxygen donor complex with both PhMgBr [9] and EtMgBr.[10] Dioxane is another ether which can coordinate to group 2 metals. Recent examples show 1,4-dioxane coordinating either through one oxygen atom or both oxygen atoms.[11] In a calcium compound, $Ca[HC(TMS)_2]_2(1,4\text{-dioxane})_2$, each dioxane donates electron density to the calcium center through one oxygen. In a strontium compound, $[Sr(NR_2)_2(\mu\text{-}1,4\text{-dioxane})]_\infty$, both oxygens are involved in electron donation, one to each of the two metal centers.

Oligoethers also can be involved in electron donation to metal centers. Strontium and barium β-diketonate compounds exist with tetraglyme as the oxygen donor.[12-13] One of these compounds, $Ba(hfac)_2(L)$, has been found to be sufficiently volatile for use in MOCVD applications. A second class of oligoethers, the crown ethers, mentioned in the introduction, can be useful as oxygen donors. Metal/crown ether complexes with counter ion examples specific to group 2 elements include halides, carboxylates, and triflates.[14] One of these such compounds, $Ba(O_2CCMe_3)_2(L)$, has been characterized structurally.[15] Despite initial reports of substantial vapor pressure, this compound appears to be unsuited for OMVPE growth of thin films of group 2 element-containing electronic materials, due to its unacceptabily low vapor pressure.

Other non-ether-containing molecules also can coordinate to metal centers. One of these types of molecules is diacetamide, MeC(O)NHC(O)Me, which can coordinate through one or both of the carbonyl oxygens. Compounds of calcium, strontium, and barium exist, and some of the solid state structures have been determined.[16-19]

Triarylphosphine oxides also are known to coordinate to metal centers through the lone electron pairs on the oxygen. Triphenylphosphine oxide forms complexes with magnesium and calcium tetrafluoroborates [20] and perchlorates.[21]

Among the more complex ligands to coordinate to group 2 metals are Schiff-bases. The solid state structure of the $Ba(ClO_4)_2$ salt with a Schiff-base obtained from the condensation of 2,6-diacetylpyridine and 3,6-diazaoctane-1,8-diamine has been determined.[22]

2. *Nitrogen-based Donors*

Since the early 1800's alkaline earth metals have been known to be complexed by simple monodentate nitrogen-containing ligands. The elemental forms of the metals readily dissolve in liquid ammonia [1] to give dark blue/black solutions usually described as "solvated electrons". Upon removal of the ammonia, bronze-colored metal amides can be isolated. More recently, the stoichiometry of the group 2 metal dihalide/ammonia adducts have been determined.[23] The increase in the number of coordinated ammonia ligands coincides with the increase in ionic radii of the elements as one moves down the group. The group 2 metal halides also dissolve in a variety of multidentate amines and ethers. Of these, ethylenediamine (en) is the simplest bidentate amine, and has been shown to dissolve MCl_2 (M = Ca, Sr, Ba) to give solvated compounds, $MCl_2(en)_n$, where n = 6, 6, and 4, respectively.[24] Other bidentate amines, such as *ortho*-phenanthroline (phen), give crystalline group 2 element complexes with strontium and barium, $[M(phen)_2(OH_2)_4](ClO_4)_2 \cdot 2$ phen.[25] Each metal ion is a distorted cubeoctahedron consisting of four oxygen atoms from water molecules and four nitrogens (*trans*) from two phen ligands. The remaining two phen molecules are non-bonding. The group 2 element halides and pseudohalides were found to be soluble in diethylenetriamine (trien).[26] It was postulated, by comparison of IR data with known transition metal-trien complexes, that the group 2 element ions were nine coordinate. Other occurances of higher coordination numbers also can be found. For example, $Sr(NO_3)_2$ complexed with triethanolamine gives an eight coordinate crystalline metal complex, $Sr\{[N(CH_2CH_2OH)_3]_2\}(NO_3)_2$.[27] The strontium atom interacts with eight heteroatoms of two triethanolamine molecules to give an approximately cubic geometry around the metal center. $Ba(OAc)_2$ also forms a complex with triethanolamine, in which the barium atoms have a coordination number of nine in the solid state.[28]

B. *Anionic Ligands*

1. *Oxygen-based Anions*
a. *Non-chelating*

The synthesis of group 2 element alkoxides has been plagued with the formation of large clusters or aggregates, rather than isolation of discreet monomeric species. One example, $Ba[(OC_2H_4)N(C_2H_4OH)_2]_2 \cdot 2$ EtOH,[29] is hydrogen bonded extensively through the lattice ethanol molecules, forming a two-dimensional network. A second example of such a large molecular aggregate is $Ca_9(OCH_2CH_2OMe)_{18}(HOCH_2CH_2OMe)_2$ [30] in which the bidentate ligands adopt a structure to coordinatively saturate the metal center. In attempts to reduce the molecularity of the group 2 element alkoxides, larger and more sterically demanding ligands have been employed. This has resulted in the formation of aggregates such as HBa$_5$(O)(OPh)$_9$(THF)$_8$,[31] Sr$_4$(OPh)$_8$(PhOH)$_2$(THF)$_6$,[32] [Ba$_3$(OSiPh$_3$)$_6$(THF)] · 0.5 THF [33] and Ba$_2$(OSi^{t}Bu$_3$)$_4$(THF) [34], some of which have been structurally characterized.

While these "spatially-controlled" attempts have resulted in clusters with a nuclearity as low as three, syntheses combining bulky aryloxide ligands with solvent molecule incorporation into the coordination sphere of the metal have been successful in creating the mononuclear species Ca(OC$_6$H$_2$-^{t}Bu$_2$-2,6-Me-4)$_2$(THF)$_3$ · THF,[35] and Sr(OC$_6$H$_2$-^{t}Bu$_3$-2,4,6)$_2$(THF)$_3$ · 0.5 THF.[36] These calcium and strontium examples have been characterized by single crystal x-ray diffraction structures. However, to date, no comparable derivative has been reported for barium.

All of the group 2 element alkoxides mentioned so far have been crystalline solids, but not discreet monomers. The group 2 compounds of HO(CH$_2$CH$_2$O)$_n$Me (n = 2, 3) are ambient temperature liquids, were determined to be monomers in benzene solution by cryoscopy, and, to date, are the only examples of monomeric barium *bis*(alkoxide) compounds.[37]

b. *Chelating: Group 2 Element Bis(β-diketonates)*
i. *Introduction*

A great deal of focus has been placed on the use of β-diketonates as ligands in CVD precursors, due to their ability form six-membered MO$_2$C$_3$ rings upon resonance stabilization of the alkoxide, ene-one configuration. The only commercially available barium compound which meets both the volatility and thermal stability requirements is Ba(fod)$_2$ (fod = 1,1,1,2,2,3,3-heptafluoro-7,7-dimethyloctane-3,5-dionate).[38-39] Although fluorinated compounds show the necessary properties to be useful precursors in CVD, they yield films which contain significant amounts of BaF$_2$. The films, therefore, require a post-deposition hydrolysis step to remove the fluorine, making the use of these fluorinated compounds slightly more difficult in an overall process, relative to the potential for F-free examples.

Recent research has been focused heavily on developing thermally stable source compounds with suitable volatility. The bulk of the research presently being conducted is centered on methods to increase the volatility of the existing non-fluorinated metal β-diketonate compounds by the formation of Lewis type acid-base adducts. Due to the large ionic radius of barium (134 pm) [40] combined with its +2 charge, barium β-diketonate complexes, in which the β-diketonate ligand is bidentate, often are coordinatively unsaturated, and, therefore, possess a tendency to oligomerization and/or solvent molecule coordination. In the presence of neutral oxygen- or nitrogen-containing donor molecules, however, the metal ions approach coordinative saturation, inhibiting their ability to oligomerize, and resulting in an increase in their volatility.

ii. *Preparation of Group 2 Element M(β-diketonate)$_2$ Compounds*

Group 2 element *bis*(β-diketonate) complexes typically are prepared by one of two methods, redox or metathesis. The redox route is the most commonly used method of obtaining anhydrous metal β-diketonate complexes. The reaction can be done in a variety of solvents, however, increasingly this type of reaction is being done in ammonia/ethereal solutions. The second method, metathesis, can be carried out either in an aqueous or a non-aqueous solvent. Traditionally, the aqueous metathesis route has been used to form group 2 element *bis*(β-diketonate) compounds, however, this preparation invariably leads to water-solvated products.[41] The non-aqueous metathesis preparation is not as commonly used as the other routes, due to its smaller driving force, *i. e.*, a group 2 element dihalide to a group 1 element monohalide. This route could have possibilities, however, as it might be predicted to produce anhydrous reaction products.

iii. *Properties of Group 2 Element M(β-diketonate)$_2$ Compounds*

As mentioned above, many factors determine the stability and volatility of group 2 element *bis*(β-diketonate) complexes. Various group 2 element β-diketonate compounds and some of their respective properties have been discussed earlier..

iv. *Intermolecular Stabilization of Group 2 Element Bis(β-diketonate) Compounds*

In order to fulfill the coordination number of group 2 element *bis*(β-diketonate) compounds, several research teams have employed intermolecular stabilization, garnered by addition of Lewis bases. This has lead to the isolation of some new and interesting complexes. In the cases of magnesium *bis*(β-diketonate) compounds, the preferred coordination number of six for magnesium has been well-established, independent of the β-diketone and/or the selected Lewis base. For the cases of the later group 2 element - β-diketone derivatives there exists a range of coordination numbers for each metal. For most of the compounds reported, there is no volatility data given. It is interesting to note that for those compounds which have volatility data available, only one compound, Ba(tmhd)$_2$(MeOH)$_2$(H$_2$O)$_2$, reportedly remains intact upon sublimation.[42] The barium coordination number appears nominally to be only eight in this molecule, which is highly unusual.

Enhanced volatility is observed when a neutral oxygen-containing donor ligand such as an oligoether is added to fluorinated Ba-*bis*(β-diketonate) compounds, as discussed above. Although the volatility of the barium source is increased, the wide-spread employment of these complexes is slightly more difficult, due to significant amounts of BaF$_2$ in the produced films.

This material has to be removed in a post-synthesis step, producing large volumes of HF, which may not be of a friendly nature to scale-up. The compound Ba(tmhd)$_2$ · tetraglyme also exists; however, it does not possess the stability required for utilization in CVD. Nitrogen-containing bases, such as amines or ammonia, also have been used to increase the volatility and stability of Ba-*bis*(β-diketonate) compounds.[43] The bases can be added neat, or diluted, and, under such conditions, some Ba-*bis*(β-diketonate) compounds have been claimed as volatile between 130 and 230°C (at 1 atmosphere) (depending on the amine used), without undergoing any visible decomposition. Initially, no isolable complexes of Ba-*bis*(β-diketonate) compounds and nitrogen-containing donor ligands were reported. However, we have isolated a complex between [Ba(tmhd)$_2$]$_n$ and NH$_3$ and have characterized this compound by single crystal X-ray diffraction.[44] The compound [Ba(tmhd)$_2$ · 2 NH$_3$]$_2$ was found to be a highly symmetric dimer. The coordinated ammonia molecules are detached upon volatization, and it is believed that the primary purpose of the added NH$_{3(g)}$ is to break up the oligomerized [Ba(tmhd)$_2$]$_n$ species, yielding smaller and more volatile species.

The compounds Ba(hfac)$_2$ · tetraglyme,[45] and Ba(tmhd)$_2$(diglyme)$_2$,[46] are examples of the recent developments which have been made in the area of intermolecular stabilization. It is particularly noteworthy that Ba(hfac)$_2$ · tetraglyme has been used to produce very high quality films of barium-containing materials by CVD.[47]

v . Intramolecular Stabilization

The search for metal β-diketonate complexes of low molecularity, and thus enhanced volatility, has prompted researchers to design ligands which are able to form intramolecularly-stabilized complexes. This intramolecular stabilization is achieved through the addition of a "tail" onto the β-diketone ligand precursor. The main aim of this intramolecular stabilization is to coordinatively saturate the metal, and, thereby, to limit oligomerization. In the case of barium, which prefers a coordination number of 8 - 12, this tail allows for an approach toward coordination saturation, which is not achieved by employment of bidentate β-diketonate ligands lacking a "tail." One example of such a ligand precursor is 2,2-dimethyl-8-methoxyoctane-3,5-dione (Hdmmod).[48] The copper and barium complexes of this deprotonated ligand have been prepared, with the later being the initial report of an ambient-condition liquid barium compound based on β-diketonate ligands. Two other examples of potentially intramolecularly-stabilized β-diketonates are the barium complexes of 2-methoxy-2,6-trimethyl-heptane-3,5-dione, and 2-methoxy-2,6-dimethylheptane-3,5-dione.[49]

Marks, *et al*. have reported the synthesis of several β-ketoiminate complexes containing appended oligoether "lariats."[50] The β-ketoiminate ligands, designated as diki and triki, have been attached to barium, yielding Ba(diki)$_2$ and Ba(triki)$_2$, both colorless crystalline solids which can be sublimed with some decomposition at 80 - 120°C at 10^{-5} Torr. OMVPE experiments carried out using Ba(diki)$_2$ and Pb(tmhd)$_2$ as a precursor produced crystalline films of BaPbO$_3$, with some BaO contamination.

Although further progress needs to be made, intramolecular stabilization seems, at this juncture, to be the most promising approach to ensure a low degree of association, and, therefore, enhanced volatility of the barium β-diketonate complexes.

2. *Nitrogen-based Anions*
a. *Group 2 Element Bis(bis(trimethylsilyl)amide) Compounds*

The only group 2 element *bis*(amide) compounds known to date are the group 2 element *bis*(*bis*(trimethylsilyl)amide) compounds. They have been prepared by several routes.[51] When prepared in an ethereal solvent, they generally are obtained as adducts with two solvent molecules per barium center. These compounds loose one solvent molecule upon heating in an inert solvent, and a dimer is formed. The last solvent molecule is lost upon sublimation or heating in vacuum. Some of the adducts and the base-free dimers have been structurally characterized.

b. *Group 2 Element Bis(benzamidinate) Compounds*

These compounds are formed by the reaction of group 2 element *bis*(*bis*(trimethylsilyl)amide) compounds with benzonitrile. They also form adducts with solvent molecules and/or additional benzonitrile.[51]

3. *Carbon-based Anions*
a. *Group 2 Element Carbides*

The simplest compounds of this category (containing only the group 2 element and carbon) are the group 2 element carbides. They are high-melting solids which decompose with water to the appropriate metal hydroxide and acetylene.[52]

b. *Group 2 Element Bis(alkynyl) Compounds*

Such compounds may be synthesized by the reaction of a terminal alkyne with the metal in liquid NH$_3$. They are all moisture-sensitive yellow solids with polymeric structures.[53]

c. Group 2 Element Bis(cyclopentadienyl) Compounds

Members of this group of compounds can be obtained by two different routes, either directly from the metal and the appropriate cyclopentadiene compound, or via metathesis reactions of a group 2 element halide (usually iodide) with an alkali metal salt of the cyclopentadienide anion. The tendency to form adducts with Lewis bases decreases with the steric demand of the substituents on the cyclopentadienyl ring system, and the stability of these compounds increases in the same order.[51] Cyclopentadienide derivatives containing more than one bulky substituent (*e. g.*, *i*-Pr, *t*-Bu, TMS), therefore are preferred. The heavily studied C_5Me_5-compounds form intermolecular adducts with donor molecules. The solvent-free molecules form polymer chains in the solid state. Some efforts have been directed to obtain intramolecular stabilization of group 2 element *bis*(cyclopentadienyl) compounds by tethering tails containing one or more donor atoms (O, N) directly to the cyclopentadiene ring system.[54-56]

d. Other Group 2 Element-Carbon Bonded Compounds

This category may be further subdivided in σ-bonded and π-bonded compounds. The application of the label of σ-bonded compounds is somewhat misleading, since the bonds in these compounds have high ionic character. Only group 2 element aryl compounds and *bis*- and *tris*-arylmethane type compounds have been reported so far. Compounds containing π-bonded ligands are related closely to the *bis*(cyclopentadienyl) compounds, described above. One or two phenyl rings are fused to the C_5 ring, and thereby impart enhanced stability to these compounds.[53]

4. Miscellaneous Element-Based Anions

Compounds containing "less traditional" interactions between a group 2 element and additional elements also exist.[57] One of the early works discussed $Ba(SiPh_3)_2$,[58] which also described the calcium and strontium compounds. They were found to be much less stable than the Ba compound, which itself has limited stability. Other examples of non-traditional bonding include a series of calcium-, strontium- and barium carboxolates of the formula $M(O_2CR)_2$ • n RCO_2H • m H_2O.[59] The most recent example of an unusual M-E interaction is the compound $[Ba(P\{SiMe_3\}_2)_2(THF)_2]_n$.[60]

III. SUMMARY

Much research effort in several groups has been directed at the problem of precursors designed for OMVPE of group 2 element-containing species for potential insertion into electronic materials areas.[61-62] Over the past several years a few trends have emerged; however, no single "success story" has come through.[63-66] Thus, the field is fertile and, hopefully, some additional insight into fundamental source compound design will grow out of these investigations.

IV. ACKNOWLEDGEMENTS

It has been the author's privilege to be affiliated with several good research group members over the past four years at Florida State University who have, in various degrees, worked in the area of group 2 element chemistry. Gratitude is expressed to K. Dipple, W. Hesse, C. Caballero, M. Carris, D. Moreno, H. Luten, U. Lay, H. Ly and E. Doskocil. Postdoctoral fellowship awards (K. D. and W. H.) from the Deutsche Forschungsgemeinschaft, seed funding from DARPA, and instrumentation support from the Materials Research and Technology Center are acknowledged. Single crystal X-ray structures by Professor V. L. Goedken (deceased) and Dr. B. Neumüller (Marburg, Germany) have complemented this work.

V. REFERENCES

1. M. Faraday, Prog. Ann. **1**, 20 (1820).
2. C. J. Pedersen, J. Am. Chem. Soc. **89**, 7017 (1967).
3. B. Dietrich, J. M. Lehn, J. P. Sauvage, Tetrahedron Lett. 1969, 2885.
4. J. V. Mantese, A. B. Catalan, A. H. Hamdi, A. L. Micheli, Appl. Phys. Lett. 1969, 2885.
5. L. G. Hubert-Pfaltzgraf, Nouv. J. Chem. **11**, 663 (1987).
6. M. J. Benac, A. H. Cowley, R. A. Jones, A. F. Tasch, Jr., Chem. Mater. **1**, 289 (1989).
7. W. S. Rees, Jr. and A. R Barron, Materials Science Forum, Vol. 137-139, pp. 473-494 (1993).
8. A. R. Barron and W. S. Rees, Jr., Advanced Materials for Optics and Electronics, Vol 2, 271 (1993).

9. G. Stucky, R. E. Rundle, J. Am. Chem. Soc. **86**, 4825 (1964).
10. L. J. Guggenberger, R. E. Rundle, J. Am. Chem.Soc. **90**, 5375 (1968).
11. F. G. N. Cloke, P. B. Hitchcock, M. F. Lappert, G. A. Lawless, B. Royo, J. Chem. Soc., Chem. Commun. 1991, 724.
12. S. R. Drake, S. A. S. Miller, M. B. Hursthouse, K. M. M. Malik, Polyhedron **12**, 1621 (1993).
13. R. Gardiner, D. W. Brown, P. S. Kirlin, A. L. Rheingold, Chem. Mater. **3**, 1053 (1991).
14. K. Timmer, H. A. Meinema, Inorg. Chim. Acta **187**, 99 (1991).
15. A. L. Rheingold, C. B. White, B. S. Haggerty, P. S. Kirlin, R. A. Gardiner, Acta Cryst. **C49**, 808 (1993).
16. J. P. Roux and J. C. Boeyens, Acta Cryst. **B26**, 526 (1970).
17. J. P. Roux and G. J. Kruger, Acta Cryst. **B32**, 1171 (1976).
18. P. S. Gentile, M. P. Dinstein, J. G. White, Inorg. Chim. Acta **19**, 67 (1976).
19. P. S. Gentile, J. White, S. Haddad, Inorg. Chim. Acta **13**, 149 (1975).
20. M. W. G. De Bolster, I. E. Kortram, W. L. Groeneveld, Inorg. Nucl. Chem. **34**, 575 (1972).
21. N. M. Karayannis, C. M. Mikulski, M. J. Strocko, L. L. Pytlweski, Inorg. Nucl. Chem. **32,** 2629 (1970).
22. M. G. B. Drew, C. V. Knox, S. M. Nelson, J. Chem. Soc., Dalton Trans. 1980, 942.
23. S. Westman, P. Werner, T. Schuler, W. Raldow, Acta Chem. Scand. **A35**, 467 (1981).
24. H. S. Isbin, K. A. Kobe, J. Am. Chem. Soc. **67**, 464 (1945).
25. G. Smith, E. J. O'Reilly, C. H. L. Kennard, A. H. White, J. Chem. Soc., Dalton Trans. 1977, 1184.
26. P. S. Gentile, J. Carlotto, T. A. Shankoff, J. Inorg. Nucl. Chem. **29**, 1427 (1967).
27. P. J. C. Voegele, J. Ficher, R. Weiss, Acta Cryst. **B30**, 66 (1974).
28. J. C. Voegele, J. C. Thierry, R. Weiss, Acta Cryst. **B30**, 70 (1974).
29. O. Poncelet, L. G. Hubert-Pfaltzgraf, L. Toupet, J. C. Daran, Polyhedron **10**, 2045 (1991).
30. S. C. Goel, M. A. Machett, M. Y. Chiang, W. E. Buhro, J. Am. Chem. Soc. **113**, 1844 (1991).
31 K. G. Caulton, J. Chem. Soc. 1990, 1349.
32. S. R. Drake, W. E. Streib, M. H. Chisholm, K. G. Caulton, Inorg. Chem. **29**, 2707 (1990).
33. K. G. Caulton, M. H. Chisholm, S. R. Drake, W. E. Streib, Angew. Chem. Int. Engl. Ed. **29**, 1483 (1990).

34. S. R. Drake, W. E. Streib, K. Folting, M. H. Chisholm, K. G. Caulton, Inorg. Chem. **31**, 3205 (1992).
35. K. F. Tesh, T. P. Hanusa, J. C. Huffman, Inorg. Chem. **31**, 5572 (1992).
36. S. R. Drake, D. J. Otway, M. B. Hursthouse, K. M. A. Malik, Polyhedron **11**, 1995 (1992).
37. a. W. S. Rees, Jr. and D. A. Moreno, J. Chem. Soc. Chem. Commun. **1991**, 1759.
    b. W. S. Rees, Jr. and D. A. Moreno, in N. R. Johnson, W. N. Shelton and M. A. El-Sayed, Eds "Spectroscopy and Structure of Molecules and Nuclei"; ., World Scientific; Singapore, 1992, pp. 367 - 374.
38. A. D. Berry, D. K. Gaskill, R. T. Holm, E. J. Cukaukas, R. Kaplan, R. L. Henny, Appl. Phys. Lett. **52**, 1743 (1988).
39. H. Yamane, H. Kurosawa, T. Hirai, Chem. Lett. **1988**, 939.
40. Fluck's VCH "Periodic Table of the Elements", VCH Verlagsgesellschaft, Weinheim, 1985.
41. S. B. Turnispeed, R. M. Barkley, R. E. Sievers, Inorg. Chem. **30**, 1164 (1991).
42. A. Gleizes, D. Medus, S. Sans-Lenain, MRS Spring Meeting, San Francisco, CA,1992,Abstract N4.7/P3.7.
43. A. R. Barron, J. M. Buriak, R. Gordon, US Patent, 5, 139, 999 (1992).
44. W. S. Rees, Jr., M. W. Carris, W. Hesse, Inorg. Chem. **30**, 4479 (1991).
45. A. L. Spek, P. van der Sluis, K. Timmer, H. A. Meinema, Acta Cryst. **46C**, 1741 (1990).
46. S. R. Drake, M. B. Hursthouse, K. M. Abdul Malik, S. A. S. Miller, Inorg. Chem. **32**, 4653 (1993).
47. S. R. Gilbert, B. W. Wessels, D. A. Neumayer, T. J. Marks, MRS Fall Meeting, Boston, MA, 1993, Abstract Y2.5.
48. W. S. Rees, Jr., C. R. Caballero, W. Hesse, Angew. Chem. **104**, 786 (1992).
49. N. Kuzmina, personal communication, 1992.
50. D. L. Schultz, B. J. Hinds, C. L. Stern, T. J. Marks, Inorg. Chem. **32**, 249 (1993).
51. T. P. Hanusa, Chem. Rev. **93** (1993) in press.
52. Gmelin's Handbuch der Anorganischen Chemie, 8th ed., Vol. 30, 1932.
53. Comprehensive Organometallic Chemistry, G. Wilkinson, F. G. A. Stone, A. W. Abel, eds., Pergamon Press, Oxford, Vol 1, 1982.
54. W. S. Rees, Jr. and K. A. Dippel, Org. Prep. Proc. Intl. **24**, 531 (1992).
55. W. S. Rees, Jr., U. W. Lay, K. A. Dippel, J. Organomet. Chem. 1994, submitted for publication.

56. W. S. Rees, Jr. and K. A. Dippel, in L. L. Hench, J. K. West, D. R. Ulrich, Eds., "Ultrastructure Processing of Ceramics, Glasses, Composites, Ordered Polymers, and Advanced Optical Materials V", Wiley, 1992, pp. 327 - 332.
57. W. S. Rees, Jr. "The Inorganic Chemistry of the Group 2 Elements", in, "Encyclopedia of Inorganic Chemistry", R. Wells, Ed. Wiley, 1994, in press.
58. E. Wiberg, O. Stecher, H. J. Andrascheck, L. Kreuzbichler, E. Staude, Angew. Chem. **75**, 516 (1963)
59. A. N. Koreva, N. P. Kuzmina, K. M. Dunaeva, Russ. J. Inorg. Chem. **37**, 366 (1992).
60. S. R. Drake, P. Hall, R. Lincoln, Polyhedron **12**, 2307 (1993).
61. W. S. Rees, Jr., H. A. Luten, M. W. Carris, C. R. Caballero, W. Hesse, V. L. Goedken, MRS Symposium Proceedings, Spring 1993 Meeting, San Francisco, CA., "Ferroelectric Thin Films III", E. R. Myers, S. B. Desu, B. A. Tuttle, P. K. Larson, Eds., in press.
62. W. S. Rees, Jr. K. A. Dippel, M. W. Carris, C. R. Caballero, D. A. Moreno, W. Hesse, MRS Symposium Proceedings, **271**, 1992, 127 - 134.
63. W. S. Rees, Jr., H. A. Luten, M. W. Carris, E. J. Doskocil, V. L. Goedken, MRS Sympopsium Proceedings, 1992, **271**, 141 - 147.
64. W. S. Rees, Jr., Y. S. Hascicek, L. R. Testardi, MRS Symposium Proceedings, 1992, **271**, 925.
65. W. S. Rees, Jr., in R. Rasmussen, Ed. Proceedings of the Fourth Florida Microelectronics and Materials Conference, University of South Florida Press, Tampa, Florida, **1992**, 83.
66. W. S. Rees, Jr. Ceramic Industries International, April 1993, 22 - 26.

Author Index

Andras, Maria T., 227
Athavale, Satish D., 3

Bai, G.R., 35
Barron, Andrew R., 317
Bernard, C., 139
Bihari, Bipin, 87
Boatner, L.A., 75
Bu, Y., 21

Chang, H.L.M., 35
Chen, Gang, 183
Chen, H., 35
Chen, J.G., 279
Chen, Zhi J., 233
Chern, C.S., 29, 107, 113
Chiu, Chien C., 183, 233
Chou, P.C., 279
Chour, Kueir-Weei, 65
Chu, Wei-Kan, 3
Clark, Eric B., 227
Cole-Hamilton, David J., 249
Collins, J., 171
Cook, M.J., 101
Cook, Stephen L., 249

Deal, E.E., 279
Degroot, Donald C., 273
Desu, Seshu B., 53, 93, 183, 215, 233
DiCarolis, S.A., 59, 299
Didier, N., 209
Dillon, A.C., 335
DiMeo Jr., Frank, 285
Dubourdieu, C., 209
Duraj, Stan A., 227

Ebihara, Kenji, 261
Economou, Demetre J., 3
Erbil, A., 75

Fanwick, Phillip E., 227
Feigelson, R.S., 59, 299
Feng, Z.C., 75
Foster, C.M., 299
Fouquet, J., 299
Fujishima, Tomoyuki, 261

Gao, Feng, 177
Gardiner, R.A., 203, 221
George, S.M., 335
Gilbert, S.R., 41
Gillen, Greg, 47
Gilliland, Douglas D., 249
Gokoglu, Suleyman A., 171
Gordon, D.C., 203
Gordon, Roy G., 9
Greenwald, Anton C., 123
Gurary, A., 241

Haase, R., 209
Han, Bin, 273
Hattori, Takeo, 291
Hehemann, David G., 227
Hendricks, Warren C., 93, 215
Hepp, Aloysius F., 227
Hillig, W.B., 101
Hinds, Bruce J., 273
Hiskes, R., 59, 299
Hitchman, Michael L., 249
Ho, Pauline, 131
Hoffman, David M., 3
Holstein, William L., 165
Hudner, J., 209
Hudson, J.B., 101
Hwang, Cheol-Seong, 47

Ignatiev, A., 279

Jia, Quanxi, 261
Jiang, Xin Li, 87

Kaiser, Debra L., 47
Kalkhoran, Nader M., 123
Kaloyeros, Alain E., 123
Kamiya, T., 311
Kannewurf, Carl R., 41, 273, 285
Katayama, S., 311
Kear, B.H., 29, 113
Kirlin, P.S., 221
Kumar, Jayant, 87
Kwak, B.S., 75

Leplingard, F., 299
Li, Q.L., 279
Liang, S., 29, 107, 113
Lin, M.C., 21
Lin, Ray Y., 177
Linden, Hans L., 193
Liu, D., 35
Liu, Jia-Rui, 3
Lu, P., 29, 113
Lu, Y., 107, 113
Lu, Z., 59, 299

Madar, R., 139
Malani, Narayan, 329
Marks, Tobin J., 41, 273, 285
McKee, M., 241
Meinema, Harry A., 193
Melius, Carl F., 131
Mölsä, Heini, 341
Moore, J.F., 81
Mossang, E., 209
Moy, K., 241

Nagahori, Atsushi, 117
Namavar, Fereydoon, 123
Neumayer, D.A., 41, 273, 285

Niinistö, Lauri, 341

Oda, Shunri, 291
Okada, M., 311
Ostrander, R.L., 203
Ott, A.W., 335

Patibandla, N., 101
Peng, Chien H., 93
Pisch, A., 139
Proscia, James W., 329

Raj, Rishi, 117
Rangarajan, Sri Prakash, 3
Reck, Gene P., 329
Rees Jr., William S., 351
Rheingold, A.L., 203
Richards, Barbara C., 249
Robins, Lawrence H., 47
Rodriguez, Joan, 329
Rosner, D.E., 171
Rotter, Lawrence D., 47
Route, R.K., 59, 299
Ruckman, M.W., 81

Safari, A., 29, 107, 113
Saga, Jun, 291
Scheiman, Daniel A., 227
Schindler, Jon L., 41, 273, 285
Schumaker, N.E., 241
Senateur, J.P., 209
Shamlian, Sarkis H., 249
Shi, Z.Q., 29, 107, 113
Shiga, Masanobu, 261
Spee, Carel I.M.A., 193
Stall, R.A., 241
Stathakos, Ioannis, 123

Stauf, Gregory T., 87
Stewart, G.D., 171
Stewart, Ogie, 329
Strongin, D.R., 81
Strongin, M., 81

Thomas, O., 209
Thompson, Simon C., 249
Timmer, Klaas, 193
Tompa, G.S., 241
Tsai, Ching Yi, 215, 233

Vaartstra, Brian A., 203
Van Buskirk, Peter C., 87, 221
Vaudin, Mark D., 47
Vijay, Dilip P., 53

Wang, C.Y., 279
Wang, Guangde, 65
Way, J.D., 335
Weiller, Bruce H., 159
Weiss, F., 139, 209
Wessels, Bruce W., 41, 285
Williams, Keith B., 329
Wolak, E., 241
Wu, P.K., 101

Xu, Ren, 65

Yen, B.M., 35
Yoo, In K., 53
Yoon, S., 29

Zama, Hideaki, 291
Zawadzki, P.A., 241
Zheng, Zongshuang, 3
Zhong, Q., 279

Subject Index

ab initio calculations, 131
alumina, 81, 335
aluminum nitride, 101
atmospheric pressure, 329
atomic layer epitaxy, 341

barium titanate, 53, 87
$Ba_{1-x}Sr_xTiO_3$, 29, 41
β-diketonate complex, 193
β-SiC, 183
bismuth-substituted yttrium iron garnet, 311
$Bi_2Sr_2CaCu_2O_8$ thin films, 285
Bragg reflectors, 241
breakdown voltage, 107

capacitance vs. temperature, 29
carboxylate, 227
cerium dioxide, 341
Ce(thd)4 precursor, 341
chemical vapor deposition, 3, 81, 123, 171, 177, 203, 209, 329
  delivery apparatus, 221
  controlled deposition, 335
cubic zirconia, yttria stabilized, 123

deposition
  activation energy, 249
  mechanism, 249
dielectric constant, 41, 107, 113
double metal alkoxide, 65

electronic-structure calculations, 131
electrooptic, 299
enthalpy of sublimation, 249
epitaxy, 87, 183

Faraday rotation, 311
fluorine doping, 329
FTIR, 335

GaAs, 241
gas-phase nucleation, 171
germanium nitride films, 3
glass, 329
glycol ether, 203

heat of formation, 131
heteroepitaxial growth, 123
high T_c superconductor, 165, 193, 241, 249, 261, 285
HN_3, 21

II-IV compound semiconductors, 241
III-V compound semiconductors, 241
indium oxide, 227
infrared sensors, 29
InN, 21
in situ, 285
isotope effect, 159

J_c optimization, 279

kinetics, 159, 177, 215

lead
  bis-tetramethylheptadionate, 215
  oxide, 215
  titanate thin films, 35, 75, 93
leakage current, 107, 113
lithium
  niobate, 299
  tantalate, 65
low-temperature growth, 291
LPCVD, 21, 101, 233
  hot-wall, 183
  laser-assisted, 21

magneto-optical ceramics, 311
metalorganic, 299
MgO, 87
MOCVD, 35, 41, 47, 59, 75, 87, 93, 123, 139, 159, 193, 215, 241, 273, 285, 291, 311
  electron-cyclotron resonance plasma assisted, 117
  photo-assisted, 279
  plasma-enhanced, 29, 107, 261
    microwave, 261
    RF, 261
  solid-source, 299
modelling, 131, 171, 177
monomolecular layer, 291
multicomponent oxide, 65
multidentate ligand, 193

NH_3, 159
N_2O, 291
numerical analysis, 171

optical waveguide, 299
orientation, 87
  highly oriented lead titanate, 35
  in-plane grain, 59

phase diagram, 233
polarization vs. electric field, 299
polyamine, 203
precursor, 193, 209, 227
  barium, 193, 249
  calcium, 193
  low volatility, 221
  MOCVD, 101, 117
  molecular, 65
  strontium, 193
  yttrium, 193, 249

Raman
  scattering, 75
  spectroscopy, 47

reactor
  hot-wall, 177
  impinging jet, 171
  rotating disk, 241
robust design, 279

second harmonic generation, 87
silicon(-)
  carbide, 177, 233
    β-SiC, 183
  containing compounds, 131
  substrates, 341
SIMS, 47
solar cells, 227
Soret transport, 171
$Sr_xBa_{1-x}Nb_2O_6$, 59
stoichiometric vapor deposition, 65
strontium titanate, 113
  epitaxial, 107
  single crystal thin films, 113
surface morphology, 113
synchrotron radiation, 81
synthesis, 227

TEM, 47
tetramethylheptadionates, 209
thermodynamics, 139, 165
  data, 131
thin films, 59, 81, 341
  polycrystalline, 53

TiN, 159
$Ti(NMe_2)_4$, 159
titania, 117, 171
titanium tetra-iso-propoxide, 171
$Tl_1Ba_2Ca_2Cu_3O_{10-x}$ thin films, 165, 273
TlF, 165
Tl_2O, 165
TlOH, 165
trimethylaluminum, 335
tris-dimethylaluminum amide, 101
tungsten bronze, 59

UPS, 21

vanadium oxide, 329
volatile barium, 203

water vapor, 335

XPS, 21
x-ray crystallography, 227

YBaCuO, 261
$YBa_2Cu_3O_x$, 291
$YBa_2Cu_3O_{7-x}$ films, 139, 209
YBCO thin films, 279
YSZ, 123
yttrium precursor, 193

ZrO_2, 123